Plant Functional Ecology

Nianpeng He • Guirui Yu • Congcong Liu
Ying Li • Ruili Wang

Plant Functional Ecology

From Organ to Ecosystem

Springer

Nianpeng He
School of Ecology
Northeast Forestry University
Harbin, China

Congcong Liu
College of Life and Environmental Science
Minzu University of China
Beijing, Beijing, China

Ruili Wang
College of Forestry
North West Agriculture and Forestry University
Yangling, Shaanxi, China

Guirui Yu
Institute of Geographic Sciences and Natural Resources Research
Chinese Academy of Sciences
Beijing, Beijing, China

Ying Li
School of Grassland Science
Beijing Forestry University
Beijing, Beijing, China

National Natural Science Foundation of China 32430067

ISBN 978-981-96-8263-8 ISBN 978-981-96-8264-5 (eBook)
https://doi.org/10.1007/978-981-96-8264-5

Translation from the Chinese language edition: "Plant Functional Ecology: From Organ to Ecosystem" by Nianpeng He et al., © Science Press 2024. Published by Science Press. All Rights Reserved.

Jointly published with China Science and Technology Press Co., Ltd.

The print edition is not for sale in China (Mainland). Customers from China (Mainland) please order the print book from: China Science and Technology Press Co., Ltd.

This Springer imprint is published by the registered company Springer Nature Singapore Pte Ltd.
The registered company address is: 152 Beach Road, #21-01/04 Gateway East, Singapore 189721, Singapore

Contents

1 Current and Future Development of Plant Functional Traits in Macroecology 1
1.1 Introduction 2
1.2 Origin and Development of Traits 2
1.2.1 Traits Are Closely Related to the History of Human Development 2
1.3 Standardization of Traits 3
1.3.1 Defining "Traits" and "Functional Traits" 3
1.3.2 Standardization of Trait Measurement 5
1.4 Bibliometric Analysis of Advances in Plant Functional Traits 6
1.5 Major Advances in Plant Functional Trait Research 7
1.5.1 Intraspecific Variations of Plant Functional Traits 7
1.5.2 Interspecific Variation of Plant Functional Traits 8
1.5.3 Trait–Trait Relationships 9
1.5.4 Plant Traits and Community Assembly 11
1.5.5 Plant Functional Traits and Ecosystem Functions 12
1.6 Development of Plant Functional Trait Databases 14
1.6.1 Species-Level Plant Trait Databases 14
1.6.2 Plant Functional Trait Databases Combined with Community Structure 15
1.6.3 Plant Functional Trait Databases Based on in Situ Community Surveys and Multi-trait Measurement 15
1.7 Potential Directions for Studying Plant Functional Traits 18
1.7.1 Development and Application New Plant Functional Trait Parameters 19
1.7.2 Cooperative Adaptation and Optimal Allocation of Functional Traits Among Different Organs 20
1.7.3 Multidimensional Adaptation and Response Functional Traits in Plants 22

1.7.4 Plant Trait Frequency and Community Structure 23
1.7.5 Plant Functional Traits and Ecosystem Functions 24
1.7.6 Multidisciplinary Integration of Plant Functional Traits 25
1.8 Summary 28
References 28

2 Variation, Adaptation, and Optimization Mechanism of Plant Functional Traits at the Organ Level 33
2.1 Variations in Leaf Morphological Traits Along the Environmental Gradient 35
2.1.1 Large-Scale Spatial Variations in Leaf Morphological Characteristics 36
2.1.2 Factors Affecting Spatial Variations in Leaf Morphological Traits 37
2.2 Spatial Variations in Leaf Anatomical Traits Along Environmental Gradients 39
2.2.1 Geographical Patterns of Leaf Anatomical Traits 39
2.2.2 Factors Influencing the Spatial Variations in the Anatomical Traits of the Leaves 39
2.3 Spatial Variation in Leaf Vein Traits Along Environmental Gradients 41
2.3.1 Large-Scale Geographical Pattern of Leaf Vein Traits 42
2.3.2 Influencing Factors of Spatial Variation in Leaf Vein Traits 43
2.4 Spatial Variations in Leaf Stomatal Traits Along the Environmental Gradient 44
2.4.1 Geographic Pattern of Stomatal Morphological Traits in Leaves 47
2.4.2 Factors Influencing the Spatial Variations in Stomatal Traits 48
2.5 Variations in Leaf Functional Elements Along Environmental Gradients 49
2.5.1 Geographical Patterns of Leaf Functional Elements 49
2.5.2 Factors Influencing the Spatial Variations in Leaf Functional Elements 52
2.6 Variations in Fine Root Traits Along Environmental Gradients 55
2.6.1 Geographic Pattern of Root Morphological Traits 55
2.6.2 Spatial Variations in Root Morphological Traits and Their Influencing Factors 56
2.7 Changes in the Environment That Affect the Functional Elements of Fine Roots 60
2.7.1 Geographic Patterns of Fine Root Functional Elements 60
2.7.2 Factors Influencing the Spatial Variations in Fine Root Function Elements 61

2.8 Challenges and Prospect 62
References 64

3 Coevolution and Optimization of Functional Traits Among Different Plant Organs 71
3.1 Structure and Functions of Different Plant Organs 74
3.1.1 Structure and Functions of Leaf 74
3.1.2 Structure and Functions of Trunks 75
3.1.3 Structure and Functions of Roots 76
3.2 Functional Trait Relationships Among Different Plant Organs 76
3.3 Hypotheses on the Relationships of Functional Traits Among Different Plant Organs 77
3.3.1 Convergent Evolution Hypothesis 78
3.3.2 Divergent Evolution Hypothesis 80
3.4 Studies on the Relationships Between Functional Traits Among Different Plant Organs 83
3.4.1 Variations and Influencing Factors of Functional Elements in Leaves and Roots 83
3.4.2 Variations and Evolutionary History of C:N Ratio in Leaves, Branches, Trunks, and Roots 87
3.4.3 Comparison of Trait Economic Spectrum Between Leaves and Fine Roots 91
3.5 Summary 94
References 95

4 Functional Trait Diversity and Its Relationship to Plant Adaptation and Functional Optimization 101
4.1 Definition of Functional Trait Diversity 103
4.2 Classification of Functional Trait Diversity 104
4.3 Calculation of Functional Trait Diversity 105
4.3.1 Calculation of Functional Richness 107
4.3.2 Calculation of Functional Evenness 108
4.3.3 Calculation of Functional Dispersion 109
4.4 Major Advances in Studying the Diversity of Functional Trait 112
4.4.1 Advances in Functional Trait Diversity in Plant Communities 112
4.4.2 Spatial Variations in Functional Trait Diversity and Linking to Productivity 114
4.4.3 Functional Trait Diversity at the Regional Scale and Its Association with Interannual Variability in Productivity 115
4.4.4 Effect of the Number of Functional Traits on the Stability of Functional Trait Diversity 118
4.5 Perspectives of Plant Functional Trait Diversity 122
4.6 Summary 123
References 124

5 Plant Trait Networks and Multidimensional Adaptive Mechanisms 129
5.1 Germination of Network Ideology on Plant Functional Traits 132
5.2 Development of Network Technologies in Plant Functional Trait Research 134
5.3 Definition and Implications of PTNs 137
5.4 Construction Methods and Key Parameters of PTNs 141
5.4.1 Methods for Constructing Functional Trait Networks in Plants 141
5.4.2 Important Parameters of PTNs 143
5.5 Application of Studying Plant Functional Traits 147
5.5.1 Spatial Variation and Mechanism of LTNs in Forest Plants in Eastern China 147
5.5.2 Spatial Variation of LTNs and Adaptation Mechanisms 148
5.5.3 Other Applications of PTNs 152
5.5.4 Potential Applications of Plant Functional Trait Networks 156
5.6 Summary 159
References 159

6 Plant Community Traits and Their Spatial Variation and Influencing Factors 165
6.1 Main Research Progresses on Plant Functional Traits 168
6.1.1 Biogeographical Patterns of Plant Functional Traits and Their Influencing Factors 168
6.1.2 Relationships Between Plant Functional Traits and the Environment 169
6.1.3 Requirements and Theoretical Basis for the Expansion of Functional Traits from Organs to Communities 170
6.2 Definition and Connotation of Plant Community Traits 172
6.2.1 Definition and Specifications of Plant Community Traits 172
6.2.2 Scaling-Up Methods for Functional Traits from Organs to Communities 174
6.2.3 Two-Dimensionality of Plant Community Traits 175
6.3 Spatial Variations in Plant Community Traits and Its Influencing Factors at the Biome Scale 176
6.3.1 Spatial Variation and Regulation Mechanism of Functional Traits in Chinese Forests 177
6.3.2 Spatial Variations in Nitrogen and Its Influencing Factors in Qinghai–Tibet Plateau Natural Communities . . . 179
6.3.3 Spatial Variations in Phosphorus and Its Influencing Factors in the Plant Communities of Qinghai–Tibet Plateau 182

6.3.4 Spatial Variations in Sulfur in Natural Communities and Its Influencing Factors in China 185
6.4 Summary 188
References 189

7 Approach to Predict Ecosystem Productivity on Plant Community Traits 193
7.1 Definition of Ecosystem Primary Productivity and Its Classification 195
7.2 Gross and Net Primary Productivity Mechanisms, Processes, and Classical Prediction Approaches 196
7.3 Theoretical Basis and Approaches for Predicting Productivity Using Plant Community Traits 198
7.3.1 Theoretical Basis 198
7.3.2 Developing Trait-Based Productivity Based on the Two-Dimensionality of Community Traits 199
7.4 Trait-Based Productivity Application in Predicting Natural Ecosystem Productivity 205
7.4.1 Prediction of Gross Spatial Variations in Gross Primary Productivity Using Plant Community Trait Associated with Nitrogen 205
7.4.2 Predicting the Spatial Variations in Grassland Productivity Using Plant Community Traits Associated with Chlorophyll 208
7.4.3 Integrating Multiple Community Traits to Predict Ecosystem Productivity 210
7.4.4 Predicting Ecosystem Multifunctionality on the Two-Dimensionality of Plant Community Traits . . . 212
7.5 Prospects and Significance of the Trait-Based Productivity Model 213
7.6 Summary 217
References 217

8 Root Community Traits and Their Correlations with Ecosystem Productivity 223
8.1 Major Progress in Root Trait Variation 226
8.2 Necessity of Scaling Root Traits from Organ to Community Levels 228
8.3 Definition of Root Community Traits 230
8.4 Parameter System of Root Community Traits 231
8.5 Spatial Variations in Root Community Traits and Their Relationships with Grassland Productivity 232
8.6 Conclusions and Perspectives 241
References 242

9 Ecosystem Traits: Bridging Traditional Functional Traits and Macroecological Research 247
9.1 The Urgent Requirement and Theoretical Basis to Develop Ecosystem Traits 251
9.2 What Are Ecosystem Traits? 253
9.2.1 Definition of Ecosystem Traits 253
9.2.2 Development of Methods for Expanding Ecosystem Functional Traits 255
9.3 Research on Ecosystem Traits 258
9.3.1 Application of Ecosystem Traits in Typical Ecosystems 259
9.3.2 Spatial Variations in Ecosystem Traits at Regional or National Scales 260
9.3.3 Using Ecosystem Traits to Estimate Productivity Across Scales 261
9.4 Prospects of Using Ecosystem Traits in Macroecological Research 262
9.4.1 New Ecosystem Ecology Framework Considering Functional Traits 262
9.4.2 Enriching Ecosystem Ecology Research Approaches 263
9.4.3 Promoting the Integrative Development of Macroecology and Geography 265
9.5 Major Challenges in the Development of Ecosystem Traits 267
9.5.1 Standardization of the Terms of Ecosystem Traits Across Multiple Disciplines 267
9.5.2 Resolving the Nonlinear Effect During Scaling-Up Processes 268
9.5.3 Rapid Expansion of the Field-Measured Data That Can Be Used to Derive Ecosystem Traits 269
9.6 Summary 270
References 270

About the Authors

Nianpeng He is a Distinguished Professor at Northeast Forestry University. His researches center on plant functional traits, ecosystem carbon sequestration, and biogeochemistry. He has made pivotal contributions to the development of conceptual frameworks such as community functional traits, ecosystem functional traits, plant functional trait networks, and trait-based productivity prediction models. These innovations have brought new insights to the fields of community ecology and ecosystem ecology. (ORCID iD: 0000-0002-0458-5953)

Ying Li is an Associate Professor at Beijing Forestry University and a recipient of the Youth Talent Support Projects from the Beijing Association for Science and Technology and the Ecological Society of China. Her research focuses on plant functional trait networks, with related manuscripts published in top-tier journals, including *Ecology Letters*, *Trends in Ecology & Evolution*, and *New Phytologist*. (ORCID iD: 0000-0002-7253-6598)

Congcong Liu is an Associate Professor at Minzu University of China and a recipient of the Special Award of the President of the Chinese Academy of Sciences. His researches specialize in trait-environment relationship, as well as the linkages between plant traits and ecosystem functions. He has led several research projects, including those supported by the National Natural Science Foundation and the Postdoctoral General Program. (ORCID iD: 0000-0003-3949-4194)

Ruili Wang is an Associate Professor at Northwest A&F University. Her interest focuses on plant functional traits, including the variability and coupling relationships of leaf and root functional traits, as well as mechanisms for maintaining ecosystem functions. She has led ten research projects, including those funded by the Shaanxi Provincial Association for Science and Technology's Youth Talent Support Program, the National Natural Science Foundation of China, and the Postdoctoral Science Foundation. (ORCID iD: 0000-0002-0384-7739)

Guirui Yu is a Principal Research Scientist at the Institute of Geographic Sciences and Natural Resources Research, Chinese Academy of Sciences, and an Academician of the Chinese Academy of Sciences. He is the founder of ChinaFLUX, a leading research network dedicated to the study of ecosystem carbon, nitrogen, and water fluxes in China. His researches focus on the development of advanced observation techniques and the exploration of the coupling mechanisms among carbon, nitrogen, and water cycles in ecosystems. (ORCID iD: 0009-0006-2250-4842)

Chapter 1
Current and Future Development of Plant Functional Traits in Macroecology

Plant functional traits are generally defined as the relatively stable and measurable attributes of plants, such as their morphological, physiological, and phenological characteristics, which can directly or indirectly impact their growth, reproduction, and survival. Over the past few decades, the definition and measurement of plant functional traits have been gradually standardized. This has led to thorough exploration of intraspecific and interspecific variation, biogeographic patterns, coordination, and evolution of plant functional traits from various perspectives and scales, particularly at species level. With the establishment of global databases on plant functional traits in the late 1990s, studies are no longer limited to local scales. On the one hand, research on the regional and global biogeography of plant functional traits is booming, and, on the other hand, studies have gradually been extended to explore the coexistence mechanisms of multiple species in communities and the optimization of ecosystem functions. Significant progress has been made in understanding and applying plant functional traits in macroecology. This includes (1) exploring the coordination of functional traits among different plant organs and how plants respond to environmental changes; (2) investigating the multidimensional response and adaptation of various plant functional traits, particularly prompted by the emerging concept of the plant trait network; (3) utilizing plant functional diversity and trait frequency distributions to study community assembly and understand the complexity of community structure; and (4) refining scaling methods from organs to species, communities, and ecosystems levels, and developing plant community traits (or ecosystem traits) as a means of bridging traditional plant traits and macroecology. These new developments are prompting the application of traditional functional traits concepts to natural, social, and economic systems, promoting the rapid development of trait-based ecology to address future ecological problems.

N. He et al., *Plant Functional Ecology*,
https://doi.org/10.1007/978-981-96-8264-5_1

1.1 Introduction

Traits are relatively stable and measurable characteristics of plants, animals, and microorganisms that have developed through long-term adaptation and evolution to the external environment. Characteristics such as leaf size, leaf depth, specific leaf area, root diameter, and root length have important ecological significance and serve as valuable indicators to study the response and adaptation of organisms to their environment. Currently, plant functional traits are defined as the relatively stable and measurable characteristics of plants, such as morphology, physiology, and phenology, which can directly or indirectly impact plant growth, reproduction, and survival at various scales (Violle et al. 2007). In most studies, plant traits and plant functional traits are almost considered as equivalent (He et al. 2019a, b). However, in practice, plant traits have been emphasized for a long time and widely used to explain ecological processes at different scales. Theoretically, plant functional traits are determined not only by genetic factors but also by external environment, at least to some extent. Therefore, spatiotemporal variations in plant functional traits may reflect the responses of plants to changes in resources and the environment. Currently, plant functional traits are applied in botany, genetics, and agronomy, serving as important tools for addressing ecological problems across various spatial and temporal scales.

In recent decades, with the rapid development of trait-based approaches, plant functional traits have been widely used to elucidate complex ecological processes at various spatial and temporal scales. Numerous studies have extensively examined research progress on plant functional traits from various perspectives (Hanisch et al. 2020; Martin and Isaac 2015), significantly contributing to the development of related fields (He et al. 2019a, b, 2020a, b, c). Based on previous studies, we briefly introduce the concepts of "traits" and "functional traits" and summarize research progress on plant functional traits, including intraspecific and interspecific variations, trait–trait relationships, and the impact of traits on community assembly and ecosystem functions in nature. Furthermore, we highlight the increasing use of plant functional traits in macroecology.

1.2 Origin and Development of Traits

1.2.1 Traits Are Closely Related to the History of Human Development

In ancient times, humans learned to make bows and arrows for hunting using hard sticks and flexible branches of plants. Additionally, they used leaves or bark with a low ignition point as materials for rapid ignition; burned tree twigs and stems with high caloric value for warmth and protection from wild animals at night; and used soft foliage and other plant tissues for sleeping surfaces. These practices involved the initial understanding of the "traits" of plants or animals. Centuries ago, the renowned traditional Chinese medical scientist, Shen Nong, tasted hundreds of herbs and documented various tastes, such as sour, sweet, bitter, and spicy. He used these bitter and sour herbs to treat cough and gastrointestinal disorders, respectively. The different tastes of plants were once considered as representative of their unique chemical traits. In agriculture, humans have used these "traits" to improve production, for example, by selecting seeds with full grains to increase yields. Since 200 BC, the 24 divisions of the solar year in the traditional Chinese calendar have been used to guide agricultural activities, based on the relatively stable phenological traits of crops. While ancient humans did not formally define "trait" as a scientific term, nor did they quantify what "traits" were, even indirectly, traits have played a significant role in human development throughout history, particularly in agricultural practices.

1.3 Standardization of Traits

1.3.1 Defining "Traits" and "Functional Traits"

In 1859, Charles Darwin completed his famous book *On the Origin of Species*, which extensively studied the length and shape of the beaks of an ecological tribe of birds known as Coerebini. Darwin introduced the concept of traits in this iconic book, highlighting the significance of traits as key indicators of organism behavior. To the best of our knowledge, this was the first specific ecological application of the traits concept. With the subsequent development of community ecology and ecosystem ecology, the concept of traits has continued to evolve, and trait-based approaches have proven to be essential for addressing ecological problems at various levels, from individual species level to the ecosystem level.

In the history of ecology, there has been a period of confusion regarding the meaning of traits. For example, Petchey et al. (2004) used the term to explain variations in biomass production; however, of the traits referred to, eight were measured at the individual plant level (i.e., four leaf traits, three traits describing plant stature, and seed mass), while three other traits were measured at the plot level (i.e., standing biomass, vegetation cover, and canopy height). Eviner (2004) explored the influence of plant standing biomass, soil conditions, and microbial phosphorus, all categorized as "plant traits" on ecosystem properties. Thus, "traits" have been used to describe the characteristics of individuals, communities, and even soils; in practice, "traits" encompassed the different components of ecosystems, leading to confusion when trying to understand the underlying mechanisms of community assembly and ecosystem functions. To resolve these problems, Violle et al. (2007) published a manuscript titled *Let the Concept of Trait Be Functional!*, in which they proposed that traits should only be considered at the individual level. According to their definition, a trait is any morphological, physiological, or phenological characteristic at the individual level, from the cell to the whole-organism

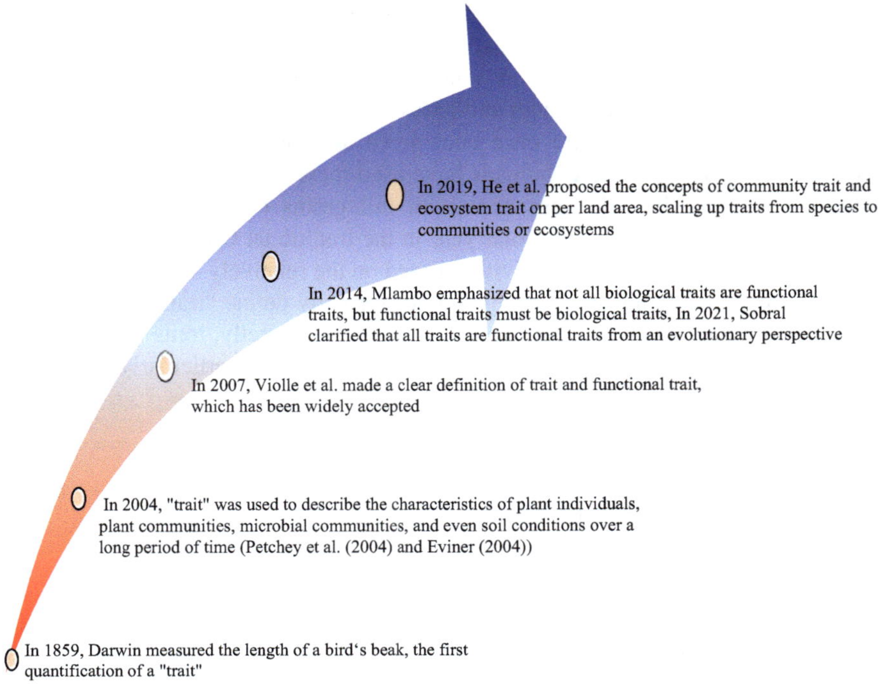

Fig. 1.1 Developmental history of definitions of "trait" and "functional trait"

level, without considering the environment or any other organization level. The term "functional trait" is widely used in the field of plant ecology, although its interpretation and application vary among scientists. It is typically considered as a substitute term for a specific function [e.g., specific leaf area (SLA)] or as a function in itself (e.g., photosynthesis). Functional traits may also be considered as traits that have a significant impact on an organism's performance and can be related to ecosystem functioning. To simplify this concept, Violle et al. (2007) proposed the term "functional traits" to encompass any morpho-physio-phenological trait that affects fitness either directly or indirectly through its impact on performance traits (Fig. 1.1).

Mlambo (2014) strongly opposed the word "any" in the definition of functional traits by Violle et al. (2007). He argued that no trait is unrelated to the fitness of organisms, suggesting that this concept only renames "trait" into "functional trait." Using "functional traits" to represent "physiological traits," "behavioral traits," and "life history traits" could lead to a significant loss of effective information. For example, functional traits should have an impact on ecosystem processes, not fitness, and if a trait does not prove to be "functional," it should be classified as a "biological trait." In other words, all functional traits are biological traits but not all biological traits are functional traits. Functional traits are biological traits that affect ecosystem processes; however, this needs to be tested using the experimental data on biodiversity and ecosystem functioning (Mlambo 2014).

Considering plant phenotypic integration, temporal and spatial changes in selection pressure, and genotype–environment interaction, all traits have theoretical potential effects on plant fitness. Therefore, all plant traits are functional from an evolutionary perspective (Sobral 2021). Currently, the definition of functional traits by Violle et al. (2007) has been widely accepted; however, the distinction between traits and functional traits remains ambiguous. Herein, we define plant functional traits as heritable, relatively stable, and measurable characteristics of morphology, anatomical structure, biochemistry, physiology, and phenology at the individual level (He et al. 2019a, b). Some studies have demonstrated that the growth, survival, and reproduction of plants are simultaneously affected by various functional traits and their complex relationships. Therefore, we should explore these adaptation strategies for plants considering as many functional traits as possible.

1.3.2 Standardization of Trait Measurement

Because of differences in the disciplinary backgrounds and experimental conditions, researchers typically use different methods to measure plant traits. This discrepancy in measuring the same functional traits, owing to different experimental conditions and procedures, not only complicates the study of functional traits but also limits their application in the fields of ecology. To reduce these biases, Cornelissen et al. (2003) summarized 28 measurement methods for plant functional traits, marking the beginning of a global standardized plant functional trait measurement system. As emerging evidence has shown that variations in plant traits and the relationships between them significantly influence ecosystem processes and functions, there is an urgent need to further standardize trait measurements. Pérez-Harguindeguy et al. (2013) provided a review of measurement methods for key whole-plant, leaf, root, stem, and reproductive traits, emphasizing their ecological significance. Furthermore, Wigley et al. (2020) described the measurement methods and possible challenges associated with 34 plant functional traits, and Freschet et al. (2021a, b) compiled a manual for measuring root functional traits and systematically summarized their ecological significance. These studies have greatly helped in standardizing the measurement criteria of plant functional traits at a global level, improving the comparability and integration of data between studies, and expanding the research field of plant functional traits to complex communities and large spatial scales (Table 1.1).

Owing to differences in language, disciplinary backgrounds, and other factors, scientists may differ in their identification and naming of the same functional traits. For example, leaf size could be represented by leaf area (LA), leaf mass, or leaf length. The lack of consistency in naming conventions and their interpretation has greatly hindered the integration and rapid development of plant functional traits. To help address this, Garnier et al. (2017) proposed the naming conventions for plant traits and created the website Plant Trait Thesaurus (www.top-thesaurus.org). In addition, Sack et al. (2010) created PrometheusWiki (http://prometheuswiki.publish.

Table 1.1 Papers related to standardized measurement and nomenclature of plant functional traits

	Title	Journal	Author
1	A handbook of protocols for standardised and easy measurement of plant functional traits worldwide	Australian Journal of Botany	Cornelissen et al. (2003)
2	New handbook for standardised measurement of plant functional traits worldwide	Australian Journal of Botany	Pérez-Harguindeguy et al. (2013)
3	A handbook for the standardised sampling of plant functional traits in disturbance-prone ecosystems, with a focus on open ecosystems	Australian Journal of Botany	Wigley et al. (2020)
4	A starting guide to root ecology: Strengthening ecological concepts and standardising root classification, sampling, processing and trait measurements	New Phytologist	Freschet et al. (2021a, b)
5	Towards a thesaurus of plant characteristics: An ecological contribution	Journal of Ecology	Garnier et al. (2017)
6	A unique web resource for physiology, ecology, and the environmental sciences: PrometheusWiki.	Functional Plant Biology	Sack et al. (2010)

csiro.au/tiki-custom_home.php) to standardize the nomenclature and measurement methods used for plant functional traits, enabling researchers to modify and continuously update entries as required. To date, many scientists have contributed standardized names and precise measurement methods for the plant functional traits relevant to their research.

1.4 Bibliometric Analysis of Advances in Plant Functional Traits

Taking "plant traits" and "plant functional traits" as keywords, we searched the literature from the Web of Science and found that the number of papers related to plant functional traits has rapidly increased since 2000. In 2001, 499 papers were published, and this number increased to over 3000 in 2020 (Fig. 1.2), indicating the growing interest in the study of plant functional traits. Research on plant functional traits primarily involves botany, ecology, and genetics, with publications in major international journals such as *New Phytologist*, *Annals of Botany*, *American Journal of Botany*, and *Functional Ecology*. The development of research on plant functional traits is the result of collaborative efforts among scientists, with foundational works regularly making significant contributions to the advancement of the concept of plant traits. For example, Garnier et al. (2004) first scaled plant traits from species to community level using community-weighted mean methods and linked plant traits to ecosystem functions; Wright et al. (2004a, b) introduced the leaf economic spectrum; Cornwell and Ackerly (2009) first explored community assembly from the perspective of plant traits; Violle et al. (2007) conceptually defined the terms

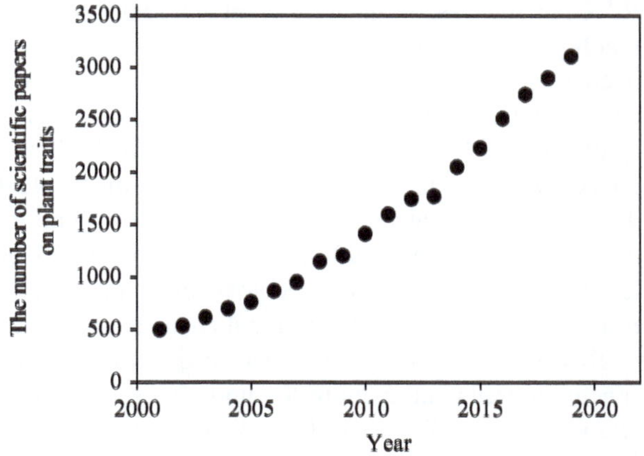

Fig. 1.2 Changes in the number of scientific papers on plant traits published annually from 2000 to 2020

"functional traits" and "traits"; and Pérez-Harguindeguy et al. (2013) developed a standard for measuring plant functional traits.

1.5 Major Advances in Plant Functional Trait Research

After rapid development since the late 1990s, studies on plant functional traits have extended beyond the traditional level of individual or species to encompass community and ecosystem levels and even to various aspects of ecological research. Scientific topics within macroecology related to functional traits include, but are not limited to, (1) variation of functional traits and trait–environment relationships across different scales, (2) relationships between different functional traits at varying scales, (3) functional traits and community assembly, and (4) functional traits and ecosystem functions.

1.5.1 Intraspecific Variations of Plant Functional Traits

Plant functional traits show strong intraspecific variations, reflecting their genetic variations and phenotypic plasticity, which can affect the interactions among plant species as well as between plants and their environment, thereby affecting the processes of community assembly and ecosystem functions (Bolnick et al. 2011; Siefert et al. 2015). Therefore, the intraspecific variations of plant functional traits have attracted considerable interest from scientists worldwide and have been applied to explore community assembly, species interaction, and the response of plants to climate change at different spatiotemporal scales (Luo et al. 2023). By integrating the SLA, LA, nitrogen content per unit area (N_{area}), nitrogen content per unit mass (N_{mass}), phosphorus content per unit mass (P_{mass}), and carbon isotope composition

($\delta^{13}C$) of 109 species under different altitudinal gradients, Midolo et al. (2019) found that N_{area}, N_{mass}, and $\delta^{13}C$ were significantly increased, while SLA was significantly decreased with increasing altitude. Furthermore, in experiments simulating global change, nitrogen addition resulted in increases in leaf nitrogen content and photosynthetic rates by 18.4% and 12.6%, respectively (Liang et al. 2020). Increasing CO_2 concentrations can also significantly promote plant photosynthetic rates and decrease plant stomatal conductance, thereby significantly improving plant water-use efficiency (Wang and Wang 2021). Liu et al. (2022b) found that the plasticity of leaf functional traits, such as photosynthetic rate, SLA, and leaf nitrogen content, in response to factors related to global change was not limited by phylogeny.

Researchers have recently tried to quantify the intraspecific variations of plant functional traits and explore the underlying influencing factors at different spatio-temporal scales. Based on 629 plant communities and 36 plant functional traits, Siefert et al. (2015) demonstrated that intraspecific functional trait variation can account for 25% of the total variation of traits in a specific community and for 32% of the total variation among different communities. Furthermore, there is greater intraspecific variation in size-related traits, such as tree height, than in other functional traits (e.g., LA and thickness) and leaf element content (e.g., leaf nitrogen content and phosphorus content). Thomas et al. (2020) analyzed six plant functional traits in the tundra and found that intraspecific variation accounted for 23.2% of the total trait variation, with leaf nitrogen content showing the largest variation, accounting for 55% of the total variation.

In natural communities, each plant species typically has a specific distribution range in space, and plant individuals can adjust their functional traits to achieve optimal adaptation in different environments. For example, the leaf thickness, SLA, K^+ content, and Na^+ content of *Leymus chinensis* is modified by mean annual temperature and mean annual precipitation (Wang et al. 2011a, b, c). The size and relative growth rate of *Arabidopsis thaliana* also show a latitudinal pattern, primarily influenced by climate (Li et al. 1998). The leaf morphology and elemental content of *Quercus* are significantly correlated with mean annual precipitation (Castro-Diez et al. 1997).

It is unclear when intraspecific trait variation should be considered in studies (Bolnick et al. 2011); while it can promote the development of trait-based approaches, measuring it in individual ecosystems and across various environments is almost impossible. Therefore, the decision to include intraspecific variation in functional traits in future studies should depend on their specific purpose, scale, and available resources.

1.5.2 Interspecific Variation of Plant Functional Traits

The interspecific variation of functional traits and their differences among different functional groups and plant communities have been well studied. Kattge et al. (2011) reported the interspecific variation of 52 functional traits and quantitatively assessed their statistical characteristics, such as the maximum, minimum, and mean values of each functional trait, and revealed that most were log-normally distributed at regional and global scales.

The differences in plant functional traits among different functional groups have also been explored. For example, the SLA, leaf nitrogen content, and photosynthetic rate of deciduous plants are significantly higher than those of evergreen plants (Wright et al. 2004a, b); the stomatal density and maximum stomatal conductance of angiosperms are significantly higher than those of gymnosperms (Liu et al. 2018a, b); and the root tissue density of woody plants is significantly higher that of non-woody plants .

Recently, the investigation of the spatial distribution of functional traits and their driving factors at regional and global scales has become a hot topic in the fields of biogeography and macroecology. Wang et al. (2016) measured LA, thickness, and SLA for 847 plant species in a north–south transect across Eastern China and found that the LA and thickness decreased with latitude, while SLA increased. Han et al. (2011) identified the relative contributions of climate, soil, and plant functional groups to 11 necessary elements in plants at a national scale and proposed stability in the limiting elements hypothesis. Paroshy et al. (2021) found that the global pattern of trunk carbon content is primarily affected by mean annual precipitation, and He et al. (2020c) found that temperature and precipitation during the growing season independently drive spatial variation in xylem hydraulic conductivity on a global scale. Freschet et al. (2017) revealed the global spatial variation of plant root traits and found that climate, soil, and plant functional types were the main driving factors. In addition, seed mass has been shown to decrease with increasing latitude on a global scale (Moles et al. 2009). Song et al. (2022) also found that floral longevity increases with increasing latitude, with mean annual temperature being the main climatic driver for flowering season.

With the accumulation of functional trait databases and the rapid development of machine learning, particularly neural networks, scientists have been gradually expanding studies on plant functional traits from the local to regional and global scales with the help of the explosion in available environmental and phylogenetic information (Boonman et al. 2020). Furthermore, an increasing number of global functional trait datasets, such as SLA, tree height, wood density, and leaf nitrogen content, have been made publicly available and freely shared, thereby promoting our understanding of plant functional traits on a broad scale.

1.5.3 Trait–Trait Relationships

Trait–trait relationships are essential and have attracted considerable research attention over the last few decades. First, some trait–trait relationships are directly related to mechanistic (i.e., physiological structure–function relationships), assuming that the size or number of a given structure may determine the physiological output in a specific process. Second, traits are co-selected for optimal design, that is, each trait independently contributes structurally to one or several functions. In addition, leaf hydraulic traits, such as stomatal morphological traits, are not only closely related to water use and leaf temperature maintenance but are also typically correlated to venation traits to maintain water balance in leaves. Lastly, concerted convergence occurs in some functional traits, where each trait contributes independently to an advantage in the specific environment (Liu et al. 2020; Sack and Scoffoni 2013).

Researchers have developed the classic "leaf economics spectrum" by exploring the correlations among six leaf functional traits: leaf nitrogen content, leaf phosphorus content, leaf lifespan, SLA, photosynthetic rate, and respiratory rate (Wright et al. 2004a, b). The leaf economics spectrum summarizes two approaches: one end of the spectrum represents plant species that grow fast and live for short periods, and the other end represents those plant species that grow slowly and live for long periods. Studies have also confirmed that the relationships among various leaf economic traits are less influenced by local climate, and the leaf economic spectrum is relatively stable in the tundra, the Tibetan Plateau region, in wetlands, and even in simulation experiments related to global change (Cui et al. 2020). Subsequently, researchers have gradually developed the economic spectra of wood (Chave et al. 2009), roots (Weemstra et al. 2016), and flowers (Roddy et al. 2021). Among these, the root economic spectrum classifies root respiratory rate, specific root length, and root nitrogen content as traits closely related to resource acquisition, while root dry matter content, root diameter, and lignin are considered as conservative traits (Weemstra et al. 2016). Recently, numerous studies have demonstrated that root–fungi collaboration breaks the traditional economic spectrum and expands the coordination and trade-off relationships of functional traits into multidimensional space. Through comprehensive analyses of root functional traits from 1810 species worldwide, scientists have confirmed that there is a conservation gradient and a mycorrhizal collaboration gradient in root traits (Bergmann et al. 2020). Of these, one end of the conservative trait axis may represent plant species characterized by slow return on investment, long lifespans, and high tissue density, while the other end may characterize plant species with a fast return on investment, short lifespans, high nitrogen content, and high metabolic rate. Simultaneously, one end of the collaboration gradient represents plant species with high specific root lengths and thin roots, while the other end represents those with thick roots and close cooperation with mycorrhizas (Bergmann et al. 2020).

Trait–trait relationships between leaves and roots have also received extensive attention. Wang et al. (2017a, b, c) found that leaves and roots are coordinated in tissue nitrogen concentrations but divergent in morphology, and Zhao et al. (2016a, b) reported that the elemental contents of leaves and roots have similar control mechanisms. By integrating global root and leaf traits, Valverde-Barrantes et al. (2020) found that vein density is closely related to fine root diameter, root tissue density, and specific root length. Weigelt et al. (2021) opined that the conservation gradient of roots and leaves should be coupled, and there is a significant positive correlation between plant height and root depth. In addition to the trait–trait relationships between roots and leaves, relationships among different plant organs have attracted attention. Based on 15 functional traits, such as the carbon, nitrogen, and phosphorus contents in leaves, stems, and roots of *Artemisia*, Liu et al. (2021b) found that the trait–trait relationships among different organs were relatively stable, even if the environment changed. Based on six bark traits, seven stem traits, and twelve leaf traits, Li et al. (2022a) reported a strong relationship between traits across various plant organs, leading to the formation of a whole plant economic spectrum. Zhang et al. (2020a, b) explored the coevolutionary patterns of C:N ratios among leaves, branches, stems, and roots, along with their climatic drivers at a regional scale, and Zhao et al. (2020a, b) tested the allometric rule of multiple functional elements among different organs, showing that the allometry of these functional elements was conservative among different organs.

1.5.4 Plant Traits and Community Assembly

The mechanism of species coexistence within natural communities is always at the core of ecological research. Theoretically, functional traits are closely related to species distributions and interactions among different plant species; therefore, it is useful for us to reveal the underlying mechanism of plant community assembly from the perspective of functional traits. Species with different functional traits in a community tend to utilize different resources or the same resources over different timescales to ensure coexistence. Furthermore, habitat filtering and limiting similarity are the main processes that determine community assembly. Owing to environmental filtering, species with extreme values in functional trait may be inhibited or even disappear in a long run, which gradually reduces the distribution range of functional traits in the community, thereby resulting in trait convergence. However, limiting similarity may reduce resource competition among species by increasing differences among functional traits, leading to trait divergence. Based on trait data for the leaves, roots, and seeds for 102 perennial plants, Kraft et al. (2015) established that differences in functional traits indeed promote the coexistence of multiple species in natural communities.

Generally, habitat filtering and limiting similarity would synergistically affect the distribution of plant traits (e.g., mean, variance, range, skewness, and kurtosis) within a community. For forest communities in California, Cornwell and Ackerly (2009) showed that habitat filtering narrowed the variability of SLA and tree height, while limiting similarity resulted in a more uniform distribution of SLA and leaf nitrogen content within the community. Based on the community-weighted mean and variance of functional traits, Bernard-Verdier et al. (2012) found that the distribution of leaf dry matter content was more diverse in areas with shallow soil layers, while seed mass was more diverse in areas with deep soil layers. Using the mathematical relationships between skewness (S) and kurtosis (K), where $K \geq S^2 + 1$, Gross et al. (2017) demonstrated that the skewness and kurtosis of tree height and SLA in arid lands are distributed along the straight line $K = S^2 + 1$, indicating that the distribution of tree height and SLA within the community is almost uniform. Liu et al. (2020) also found that water availability and temperature determine the evenness of leaf hydraulics and economic traits within the community, respectively, and the assembly processes of functional traits within natural communities improve ecosystem productivity at large scales.

1.5.5 Plant Functional Traits and Ecosystem Functions

Violle et al. (2007) divided plant functional traits into response traits and effect traits, where response traits describe the response of plants to environmental change and effect traits describe the influence of plants on ecosystem functions. Thus, a specific functional trait can reflect a plant's adaptive strategy or its impact on ecosystem function. Theoretically, plant functional traits can track environmental changes and play an important role in determining ecosystem functions, such as productivity, water-use efficiency, and nitrogen-use efficiency. Therefore, how the relationships between plant functional traits and ecosystem functions can be established is of critical importance in ecological research. Although there is a clear distinction between response traits and effect traits in the definitions of Violle et al. (2007), they are difficult to distinguish in practice owing to the direct or indirect relationships among multiple functional traits, with many traits serving as both response traits and effect traits simultaneously.

Plant functional traits may affect ecosystem functions in various ways and further affect ecosystem services (Fig. 1.3). Changes in external environments and biological conditions lead to changes in community structure, which influences effect trait and has a certain impact on specific ecosystem function. Moreover, changes in external environments and biological conditions influence response traits, and because species respond differently to such changes, community structure is altered, which ultimately affects ecosystem functions by regulating these traits. There is a relationship between response traits and effect traits that leads to changes in effect traits, further affecting ecosystem functions (Liu et al. 2021b).

There are two main approaches to identify the relationships between plant functional traits and ecosystem functions: one is based on the community-weighted mean values of functional traits through species relative abundance or community biomass, which corresponds to the selection effect. The other approach is based on functional diversity, which corresponds to the niche complementary effect. Influenced by the "mass ratio" hypothesis, the traditional perspective is that dominant species play a crucial role in ecosystem function, and the community-weighted mean approach primarily depends on the functional traits of these dominant species. Studies have found strong relationships between community-weighted mean and ecosystem functions. For example, Liu et al. (2018a, b) found that the community-weighted mean of stomatal density is positively correlated with ecosystem water-use efficiency. With the development of niche theory, people have gradually realized that nondominant species can impact ecosystem functions through complementary effects. For example, Li et al. (2022b) found that ecosystem productivity is strongly influenced by the functional diversity index, and Tang et al. (2022) demonstrated

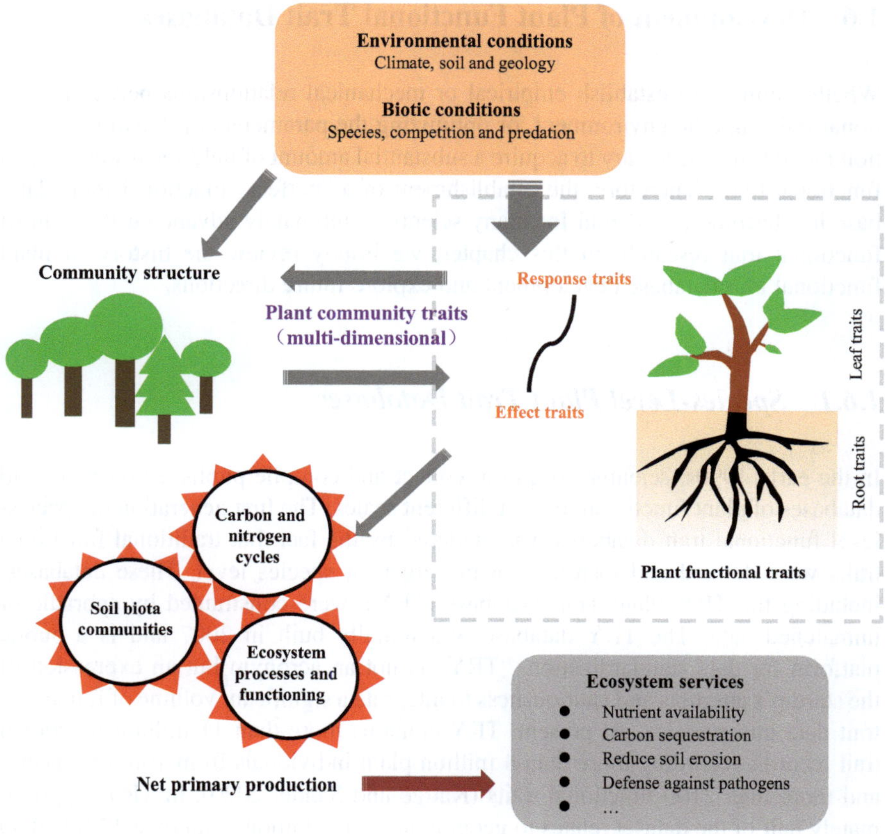

Fig. 1.3 Integration of effect traits and response traits at a community scale to demonstrate how plant functional traits determine ecosystem functioning

that the impact of tree diversity and genetic diversity on productivity is primarily achieved through functional diversity. It can be confusing how the impacts of selection effects and complementary effects on ecosystem functions may vary with time and external environment. For example, Bongers et al. (2021) found that the effect of complementary effects on forest ecosystem productivity gradually exceeds the effect of selection as the forest ages. However, simultaneously considering the effects of both selection and complementary effects on ecosystem function can offer a comprehensive understanding of community assembly, although this remains a complex and difficult task in practice.

1.6 Development of Plant Functional Trait Databases

Whether aiming to establish empirical or mechanical relationships between functional traits and the environment, or optimizing the parameters of dynamic vegetation models, it is necessary to acquire a substantial amount of field-measured data on functional traits. Therefore, the establishment of a "perfect" functional traits database has become a key goal for many scientists, ultimately advancing the field of functional trait research. In this chapter, we briefly review the history of plant functional trait database development and explore future directions.

1.6.1 Species-Level Plant Trait Databases

In the early 1990s, scientists began to collect and compile published data to build databases of plant functional traits at different scales. The first generation of species-level functional trait databases was hindered by the fact that traditional functional traits were defined and measured at the organ or species level. These databases, including the TRY Plant Trait Database (TRY), were constrained by sporadic or unmatched data. The TRY database was initially built in 2007 and is a public platform for data standardization. "TRY" is not an acronym but an expression of the startup's attempts and cautiousness to integrate a significant volume of functional trait data into practice. At present, TRY contains more than 11 million functional trait records, recording more than 4 million plant individuals from 160,000 species and more than 2100 functional traits (Kattge and Sandel 2020). In TRY, approximately half of the dataset relates to geographical distribution, with over 15,000 field measurement sites worldwide. Researchers have recently published a series of high-impact studies using TRY data. For example, Diaz et al. (2016) used data for 46,085 vascular plant species from 423 families within the TRY database (which contained a total of 500,000 individuals and 800,000 functional traits) to analyze the relationships between traits related to plant growth, survival, and reproduction on a global scale, including adult plant height, wood density, LA, SLA, and seed mass. Kunstler et al. (2016) integrated data on wood density, SLA, and maximum height from the TRY database with growth information for over 3 million trees in 140,000 monitoring plots worldwide to assess the impact of plant functional traits on the strength of community competition. Tian et al. (2019) compiled a database of leaf nitrogen content and leaf phosphorus content for 3227 plant species worldwide; Wang et al. (2018a, b, c) collected functional trait data such as plant morphology, chemical characteristics, and photosynthesis rates of 1215 plant species from 122 field-samples in China; and Wang et al. (2022a, b) updated the Chinese plant functional trait database, which includes the morphological, physical, chemical, photosynthetic, and hydraulic traits of 1529 species from 140 sites.

1.6.2 Plant Functional Trait Databases Combined with Community Structure

The distribution of plants varies among communities, and some species become extinct if they cannot adapt to a changing environment. Most existing plant function databases are created by a single researcher or small research team. To explore the underlying mechanisms affecting the geographical distribution of species diversity and to use the increasing amount of data available for research, the Botanical Information and Ecology Network (BIEN, https://bien.nceas.ucsb.edu/bien/) began to establish a botanical research foundation database in 2008. The BIEN database has been constantly updated, and the latest version, BIEN 4.1, contains more than 200 million observations, 364,477 quadrats, and 485,902 species. In addition, the BIEN has also developed the "RBIEN" package, which allows researchers to freely access all data and data products online, including species distribution ranges, evolutionary trees, functional traits, and quadrat data, and has become a cutting-edge database for current studies on plant functional traits (Maitner et al. 2018).

Recently, scientists have developed sPlot as a tool for global vegetation analysis. sPlot integrates global vegetation plot data to analyze the spatial and temporal variations of global plant taxonomic diversity, functional diversity, and phylogenetic diversity (Bruelheide et al. 2019). Currently, sPlot contains 1,121,244 plant plots, in which the coverage and abundance of 23,586,216 plant species are recorded. In combination with the TRY database, sPlot has been used to scale key plant functional traits from the species level to the community level (Bruelheide et al. 2018).

1.6.3 Plant Functional Trait Databases Based on in Situ Community Surveys and Multi-trait Measurement

The TRY database is essential for studying functional traits at a global scale, with an impressive 279,000 species and 11.85 million traits recorded from published data. This database has been extensively utilized across various scales, playing a crucial role in advancing research on plant functional traits (Bjorkman et al. 2018a, b; Bruelheide et al. 2018). However, as studies delve into complex communities found in nature, the limitations of the TRY database become more evident. Thus, plant functional trait databases with community structure information have been developed, such as the combination of the TRY database with sPlot. While these databases are beneficial, they have limitations in terms of the data they contain. Notably, current databases primarily contain trait data that are easy to measure, such as leaf size and thickness, nitrogen and phosphorous contents, and individual size. However, there is a lack of matched leaf–branch–trunk–root functional trait data, as well as matched functional trait data for plant species within complex communities.

Recently, the principles of a systematic approach, multidimensionality, and compatibility have been proposed (Table 1.2) as a foundation for a next-generation

Table 1.2 Principles of systematicness, multidimensionality, and compatibility in functional trait databases

	Unique property[a]	Object	Target
1	Systematicness	Covering all functional traits of plant structures (i.e., leaf, stem, trunk, and root), animal, and microbe and involving supplementary data such as plant community structure, ecosystem C and H_2O flux, and ecosystem productivity	Systematically obtain all kinds of data related to whole ecosystems
2	Multidimensionality	Including three dimensions: Spatial, temporal, and research objectives. For example, organ, species, community, ecosystem, landscape, region, and global	Can be applied at different scales and disciplines
3	Compatibility	Taking different technologies (e.g., ground surveys, remote sensing, and eddy-flux observations) and adapting to different disciplines (e.g., plant, animal, microbe, ecology, geoscience, and remote sensing)	Compatible for different technologies and disciplines

[a]Functional trait databases on ecosystems, such as China_Traits, should possess these three properties because they can adapt to the requirements of multiple organisms, targets, and scales within ecosystems. However, there is a need for other functional trait databases, each with its own set of strengths and weaknesses

plant functional trait database at the community scale (He et al. 2019a, b). For example, by conducting in situ community investigations and measuring multiple traits from the leaves, branches, trunks, and roots of all plant species in natural communities, a database of plant functional traits within Chinese ecosystems has been established (China_Traits; Fig. 1.4), consisting of 112 plant functional traits (He et al. 2019a, b). This database includes detailed information of plant community structure, soil and soil microbial properties, and the functional traits of leaves, branches, trunks, and roots of all plant species alongside the contents of 16 key elements including C, N, P, and K. Functional trait data for small soil animals, such as locusts, soil nematodes, beetles, hoppers, and earthworms, have also been incorporated into the China_Traits database, although data are currently missing for seeds, flowers, and other propagators. However, any database on plant functional traits will have its own advantages and limitations owing to its multi-target and multi-pathway characteristics.

The China_Traits database is groundbreaking and valuable for promoting research on functional traits in complex natural communities; it matches plant functional traits with data on community structure, soil microbes, soil physicochemical properties, and other factors, helping researchers extend functional traits measured at the organ level to higher levels of organization. It can also open new research avenues in functional traits and support the development of novel approaches to reveal multidimensional plant adaptation mechanisms (Fig. 1.5). Overall, the China_Traits database addresses the current gap in supporting trait

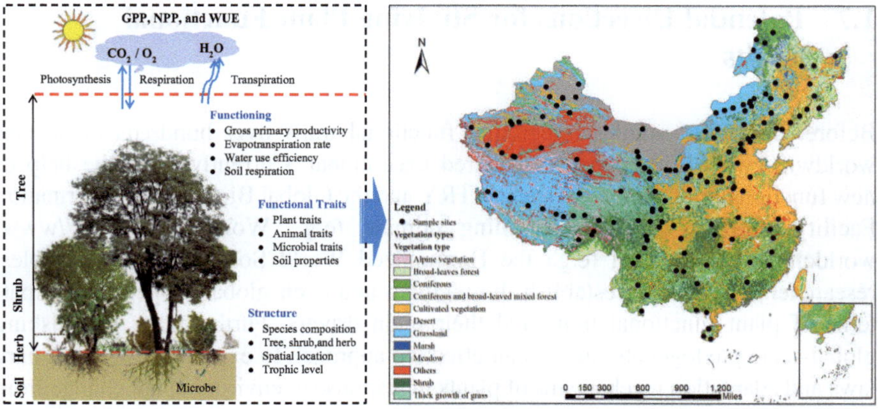

Fig. 1.4 Database of plant functional traits of Chinese ecosystems, "China_traits" containing various functional traits obtained under the principles of systematic and matched measurement. To date, the China_Traits database includes more than 120 typical ecosystems covering most of the ecosystems in China

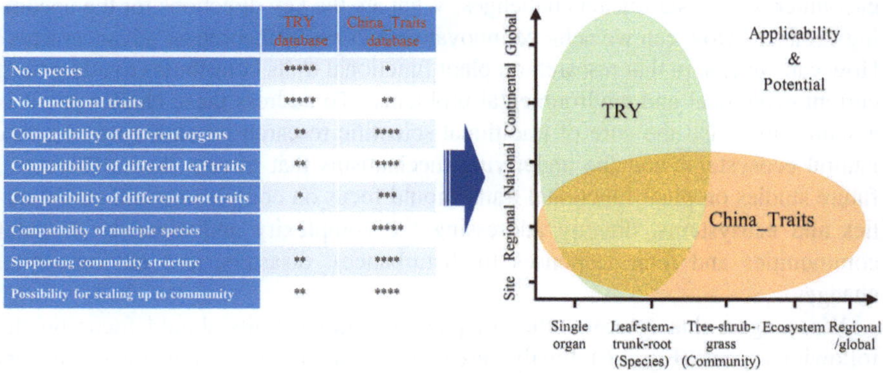

Fig. 1.5 Comparison of the advantages, applicability, and research potential for the TRY database and China_Traits database. The number of asterisks (*) for each database feature indicates relative advantage (high number) and disadvantage (low number)

data and community structure data, allowing for multidimensional research on species, functional groups, and communities (He et al. 2020b; He et al. 2019a, b).

The China_Traits database enables a novel perspective on multidimensional trait research on species, functional groups, and communities that can help reveal the essential law of nature. With the development and improvement of matched databases, in situ community surveys and multi-trait measurements are becoming increasingly recognized by scientists globally. Thus, research on functional traits in complex natural ecosystems should become more common, offering a better understanding to provide new solutions to regional ecological problems as well as the response of ecosystem structure and function to global change.

1.7 Potential Directions for Studying Plant Functional Traits

Before the 1990s, studies integrating functional traits over hundreds of species worldwide would have been considered exceptional. Currently, with the help of new functional trait databases, such as TRY and the Global Biodiversity Information Facility (GBIF), and corresponding climate (e.g., WorldClim, http://www.worldclim.org/) and soil (e.g., the Harmonized World Soil Database) variables, researchers can quickly establish the regional and even global biogeographic patterns of plant functional traits and their main drivers. Furthermore, establishing global-scale phylogenetic trees is an effective approach to explore the evolutionary laws and adaptation mechanisms of plants in response to environmental change from a long-term perspective (Jin and Qian 2023).

Through a rapid period of development over the last 30 years, research on plant functional traits has made significant progress, expanding to various spatiotemporal scales and establishing a new field known as "trait-based ecology." While acknowledging and honoring the achievements of earlier researchers, current scholars encounter several significant challenges: What are the key directions for the upcoming decades? How can we achieve innovative and major theoretical advancements? How can we ensure that research on plant functional traits contributes to addressing current ecological and environmental problems? To address these challenges, it is essential to revisit the core of traditional scientific research by exploring complex natural ecosystems and the underlying mechanisms that govern them. Therefore, future studies on plant functional traits should focus on complex natural communities and ecosystems, directly addressing the complexity and diversity of plant communities and their responses to disturbances, disasters, and environmental changes.

We suggest that future studies of plant functional traits should focus on the following aspects (Fig. 1.6). Firstly, research should further develop new parameters that integrate microgenomics, remote sensing technology, and new algorithms supported by Big Data. Secondly, we should move beyond traditional concepts of plant functional traits at the organ or species level and innovatively develop new theoretical systems to effectively utilize the growing number of data sources available. This will allow us to conduct comprehensive multidimensional research on plant functional traits, transitioning from studying individual traits or organs to considering multiple traits and organs. Third, by advancing theory, methods, and technologies, research can better understand the response mechanisms that govern natural community structure based on functional traits. Lastly, it is crucial to explore how the parameters of structure interact with functional traits to achieve optimal and stable ecosystem functions at various spatiotemporal scales, rather than just focusing on the traditional understanding that structure determines function. Systematic and matched plant functional trait databases will be the cornerstone for these future research directions.

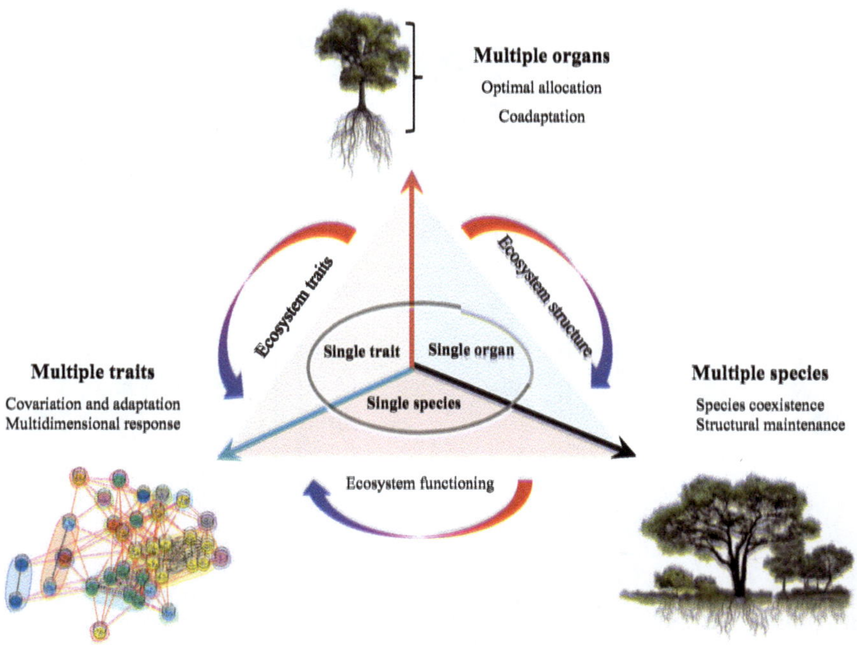

Fig. 1.6 Potential directions for research on plant functional traits

1.7.1 Development and Application New Plant Functional Trait Parameters

In the past few decades, various plant functional trait parameters have been applied; however, most of them are easily measurable at regional and global scales. With the deepening of our understanding of plant functional traits and the rapid development of observation techniques, additional parameters closely related to plant responses and adaptation to the environment may emerge. For example, when considering stomata, traditional studies on stomatal morphological traits have primarily focused on density and size. Theoretically, under the same stomatal density and size, the spatial arrangement or combination of stomata can vary and significantly affect the maximum stomatal conductance, thereby affecting the response mechanism of leaves. To explore this, Liu et al. (2021a) introduced new parameters, such as the stomatal evenness index, divergence index, and aggregation index, based on the distance between stomata (Fig. 1.7). These parameters not only improve upon traditional parameters but also open up new research avenues for crop cultivation, plant adaptation, and evolution. Similarly, leaf pH is an emerging parameter in plant functional trait research (Cornelissen et al. 2011). The variation in leaf pH is high among plant species and is governed by relatively stable protoplastid pH and

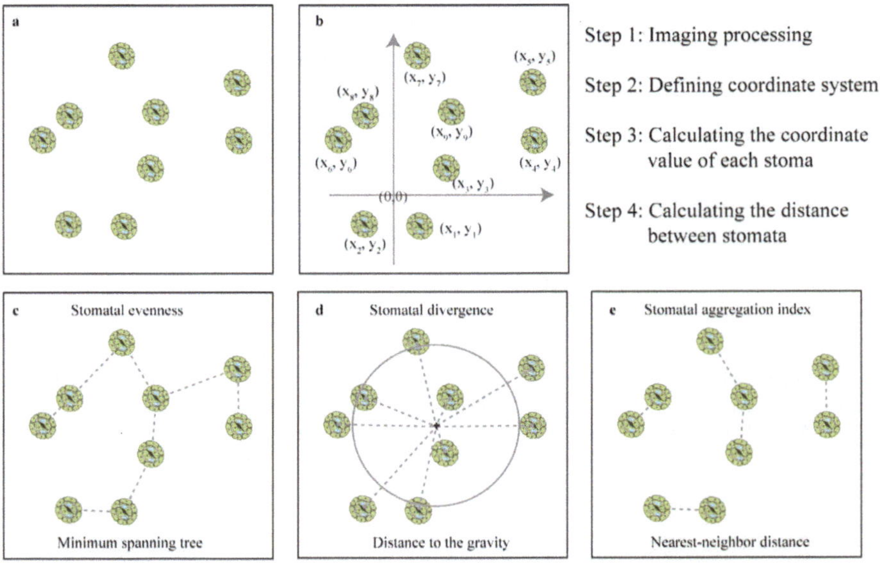

Fig. 1.7 Quantifying stoma spatial assembly based on stomatal distance. (**a**) Number of stomata in each panel is nine, and stomatal density can be calculated as the ratio of stomatal number to the measurement area. (**b**) A coordinate is established, and the locations of stomata and the distances between them can be quantified. (**c**) A minimum spanning tree (MST, dashed line) links the stomata, and stomatal evenness represents the regularity of points along this tree. (**d**) Stomatal divergence quantifies how stomata diverge in their distances from their center of gravity, which is marked with black cross. The gray dashed lines represent the distance of each stoma to the center of gravity, and the gray circle represents the mean distance to the center of gravity. (**e**) A stomatal aggregation index is used to classify the stomatal distribution type (i.e., clustered, random, or regular distribution) on the leaf surface based on the nearest-neighbor distance

relatively sensitive exoplasmic pH. As leaf pH measurement is simple and plays a significant role in nutrient cycling and physiological processes, it should be integrated into ecological process models (Liu et al. 2019). In addition to developing new parameters at the organ level, scientists may also develop new parameters to translate plant functional traits from individual species to communities (He et al. 2019a, b). Overall, the development and application of these new trait parameters are crucial in promoting the ongoing rapid development of plant functional trait research.

1.7.2 Cooperative Adaptation and Optimal Allocation of Functional Traits Among Different Organs

All higher plants are composed of different connected organs, such as leaves, branches, trunks, and roots. Different organs perform different roles; however, they also cooperate and influence each other and jointly regulate the life activities of plants in nature. One of the most important functions of leaves is photosynthesis, where leaves absorb carbon dioxide from the atmosphere, release oxygen, and provide energy by producing carbohydrates to support the material and energy requirements of the plant, as well as other organisms in the ecosystem. Branches and stems have the important functions of support and storage, with the phloem and xylem transporting nutrients and water, respectively. Roots are critical because they absorb water and nutrients from the soil. Although these plant organs are different in terms of their primary function, they complement each other. They are structurally connected and have the same source of nutrients and energy. Therefore, understanding the coevolution and adaptation mechanisms of different plant organs is crucial for scientifically assessing the response and adaptation of plants and communities to changes in resources, disturbance, and environmental change.

Although researchers have recognized the universality and scientific importance of synergistic and optimal allocation of leaves, branches, trunks, and roots in the context of plant functional traits, there is limited research conducted on multispecies at regional and global scales. For example, Zhang et al. (2020a, b) first demonstrated that, at a regional scale, the coordination and trade-offs in elemental contents among different plant organs are essential strategies for plant evolution and adaptation. Understanding the distribution and regulation mechanisms of functional elements among different organs can further reveal the underlying mechanisms of material circulation and energy flow in plants. This understanding is crucial for optimizing ecological process models. During the process of growth and development, plants require approximately 16 functional elements, which participate in various physiological and biochemical reactions (Zhao et al. 2016a, b). The specific functions and requirements of each element and the specific requirements of plants for them are reflected in the content of each element, with a large number of elements involved in essential physio-biochemical processes (Fernández-Martínez 2022). According to the theory of minimum factor restriction, any supply of element that is below the optimal balance of the lowest demand could potentially limit plant growth. Therefore, the contents and variations of different plant elements can effectively reflect the regulation and distribution strategies related to plant nutrition and, to some extent, adaptation mechanisms in response to a changing environment. Researchers should be cautious in separating different organs, rather than focusing solely on a single organ, integrating them as a proxy of the "leaf–branch–trunk–root"; this approach helps prevent one-sided results or drawing unexpected and wrong conclusions (Zhang et al. 2018a, b, 2020a, b). Different research conclusions derived from studies on various organs or single organ traits contribute to the confusion surrounding plant response and adaptation mechanisms.

1.7.3 Multidimensional Adaptation and Response Functional Traits in Plants

Most studies are limited to exploring the variation of single functional traits and the relationships between two or several functional traits, although each process of plant growth, development, and reproduction requires the coordination and trade-offs between multiple functional traits. Therefore, scientists can systematically and comprehensively examine the adaptation strategies of plants in response to changing resources and the external environment only by comprehensively considering the complex relationships among multiple functional traits (He et al. 2020a; Li et al. 2022c). In practice, most researchers adopt "dimensionality reduction" in studies involving various functional traits. Theoretically, if multiple functional traits show a strong correlation, they can be simplified using methods such as principal component analysis. For example, the economic spectrum of plant leaves is considered to vary along a single axis, with one end representing fast-growing, short-lived species and the another representing slow-growing, long-lived species (Wright et al. 2004a, b). An analysis based on dimensionality reduction, using principal component analysis, is intuitive in this case; however, the limitation to this approach is that the complex relationships between various functional traits are obscured, and important ecological information may be lost.

It is common for plants to adapt to the external environment in multidimensional ways. He et al. (2020a) defined plant trait networks (PTNs), in which plant functional traits are regarded as the "node" of the network, and the general relationships between traits are regarded as the "edge" of the network. This new concept and parameters of PTNs help us identify key plant traits. As a new concept, PTNs provide novel perspectives for research, such as the biogeography of plant functional traits, trait dynamics in the succession of plant communities, the relationships between the traits of different plant organs, and the role of traits in the response of plant communities to global changes. For example, based on 35 leaf functional traits, Li et al. (2022c) found that leaf economic traits were "hub traits" in the PTN. Liu et al. (2022a) also reported that the trait networks of lianas are more modularized than those of trees in both wet and dry forests, indicating that these changing environments were less influential on lianas than on trees. Zhang et al. (2021a) further demonstrated that the effects of anthropogenic nitrogen addition on grassland community structure in Inner Mongolia depend on the plasticity of plant functional element networks. They indicated that the use of PTN approach to study multidimensional traits can help us reveal the responses of plants to changing resources, environments, and even human disturbance in nature, representing a new research horizon in plant functional trait research.

1.7.4 Plant Trait Frequency and Community Structure

Trait driver theory suggests that the changes in the interactions between biological, abiotic, and neutral processes may lead to changes in functional trait distributions within plant communities, significantly affecting ecosystem structure and functions (Enquist et al. 2015). The mass ratio hypothesis further predicts that the extent to which the functional traits of a species may affect ecosystem characteristics depends on the relative contribution of that species to total community biomass. However, using the community-weighted mean to scale-up functional traits from organ to communities has its inherent disadvantages, such as potentially exacerbating the role of dominant species and masking other important information on the distributions of functional traits within the community. In contrast, the "niche complementarity hypothesis" emphasizes the importance of nondominant and even rare species within natural communities. However, the roles of dominant species, nondominant species, and rare species should be considered when analyzing frequent distributions of plant traits. To date, four parameters of plant trait frequency have been widely used: the community-weighted mean, variance, skewness, and kurtosis. The community-weighted mean primarily reflects the optimal results of plant communities adapting to environmental change. Variance represents the scale or dispersion of trait values in plant communities; skewness represents the level of left- or right-tailed asymmetry in the trait distribution. Highly absolute skewness indicates that dominant traits are located at the extremities of the trait distribution, which is typically the result of asymmetrical competition and/or rapid changes in the environment, such as strong disturbance events. Lastly, kurtosis reflects the relative extremes throughout the distributions of functional traits of plant communities. Commonly, high values of kurtosis occur when plant communities have species with high abundance and similar trait values, whereas low values of kurtosis typically occur when plant communities have species with even abundance but varying trait values (i.e., high diversity).

In practice, the analysis of plant trait distributions aids in examining the maintenance mechanisms of natural community structure across various scales. At the population scale, plant trait distributions reflect the response of populations to different selection pressures and other evolutionary factors. At the community scale, plant trait distributions reflect the differences in local selection, species interactions, and ecological drivers associated with species coexistence, as well as historical disturbances, species migration, extinction, and environmental change. The theoretical basis of plant trait distributions has been well developed. However, owing to a lack of systematic and matched functional trait data, scientists can only use easily measurable parameters, such as plant height and SLA, in the analyses of plant function trait distributions to infer the rules of community assembly and the response of population, community, and even ecosystems to climate change (Bagousse-Pinguet et al. 2021). With new developments in functional traits databases, Liu et al. (2020) first calculated the community-weighted mean, variance, skewness, and kurtosis of stomatal morphological traits and explored their influence on ecosystem productivity at a regional scale. In summary, plant trait distributions consider both the selection effect of dominant species and the niche complementary effect of nondominant species, highlighting the importance of including intraspecific trait variation in future research.

1.7.5 Plant Functional Traits and Ecosystem Functions

"Structure determines function" has been considered a primary law of ecosystem ecology. However, in the face of complex communities or complex ecosystems in nature, it is more important and practical for us to understand *how* "structure determines function." Theoretically, the changes in plant functional traits and their complex relationships are key to determine the structure and function of ecosystems in response to global change. For historical reasons, the methods of plant functional type, which have been widely used, do not focus on plant functional traits but, instead, plant species. This hinders the development of global vegetation dynamic models to a certain extent. In nature, plant functional traits can not only reflect the response and adaptation of plants to environmental changes but also are closely related to the structure and function of natural ecosystems. Thus, trait-based approaches may significantly improve the predictions of global vegetation dynamic models in the future by integrating these functional traits into ecosystem processes.

For all ecosystem functions, productivity is of fundamental importance, and, therefore, how the spatiotemporal changes in natural ecosystems can be accurately predicted, at regional and even global scales, is one of the most important scientific questions in ecosystem ecology. Scientists have developed the "big-leaf" model to predict the spatiotemporal variation of ecosystem productivity. The approach is based on leaf-tissue-level photosynthesis and respiration, through which the unity principle simulates ecosystem productivity and predicts spatiotemporal variations in productivity primarily based on environmental variables. In practice, this approach has high prediction accuracy at the local scale but high uncertainty in regional and global predictions owing to the difficulty in obtaining physiological parameters across different sites. Researchers have recently pointed out that plant functional traits, such as leaf size, SLA, and C:N ratio, may affect or even determine productivity at multiple scales, which is also important for future model improvement and development. For example, based on the classical engine model of physics and bidimensional plant community functional traits, He et al. (2022) proposed a new framework to predict ecosystem productivity from a system of quantitative traits, efficiency traits, and growth period (Fig. 1.8). In this new framework, the effects of environmental factors, such as temperature, precipitation, and soil nutrients, on ecosystem productivity indirectly affect the spatiotemporal variation of ecosystem productivity by affecting community structure, species composition, and intraspecific functional traits. For example, the standardized quantitative and efficiency traits of chlorophyll per unit land area explained 60% of the spatial variation in grassland productivity in northern China (Zhang et al. 2021b). Along a forest successional gradient in temperate forests in northeast China, 78% of the variation in ecosystem productivity could be explained using the quantitative and efficiency traits of SLA

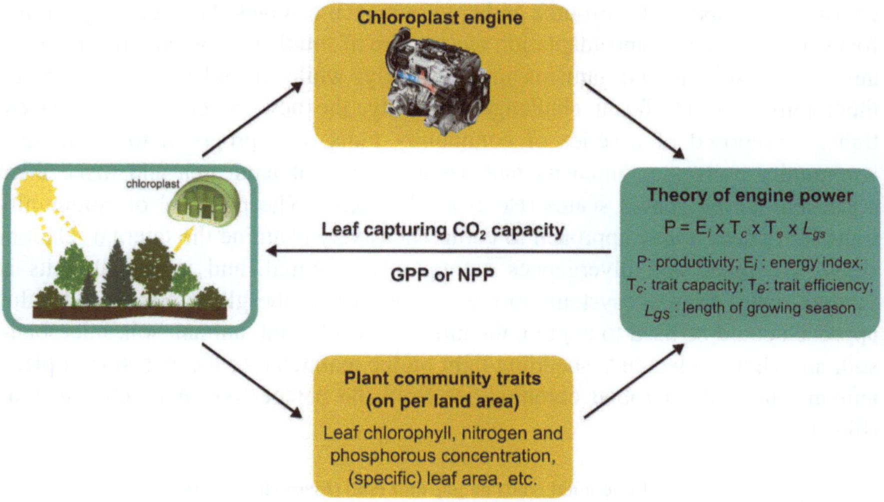

Fig. 1.8 Theoretical framework for predicting ecosystem productivity from plant community traits

(Li et al. 2021a, b, c). Furthermore, the quantitative and efficiency traits of leaf nitrogen content jointly explained 81% of the spatial variation in natural ecosystem productivity in China (Yan et al. 2022). Thus, with the growth in the availability of plant functional trait data and the development of more advanced technologies, particularly high-resolution remote sensing, plant community traits offer an important way to predict ecosystem productivity. This is expected to support the application of quantitative traits, efficiency traits, and growth period in predicting productivity using new-generation ecological process models.

1.7.6 Multidisciplinary Integration of Plant Functional Traits

The traditional definition of functional traits includes all biological groups, such as plants, animals, and, microorganisms (Violle et al. 2007). In complex natural ecosystems, plant, animal, and microbial functional traits should be interrelated, independent, or even unique. Therefore, theoretically, the rapid development of new concepts, methods, and technologies in the field of plant functional traits offers a valuable reference for other organisms.

Natural ecosystems are composed of plants, animals, microorganisms combined with abiotic factors in a specific space, all of which being connected by material circulation and energy flows. In recent years, researchers have gradually collected and compiled functional trait databases related to vertebrates (such as reptiles, fish, birds, amphibians, and mammals), invertebrates, corals, and fungi (Gallagher et al. 2020). However, it is very rare for the relationships between the functional traits of plants, animals, and microbes to be studies at the ecosystem scale, let alone their coordinated response to resource and environmental changes. Most existing studies focus on the variation and adaptation mechanism of simple functional traits in plants, animals, or soil microorganisms independently, while the relationships between them remain a significant challenge. Recently, the new concept of "ecosystem traits," composed of a series of community traits, was proposed to standardize community traits (i.e., intensity and density) per unit land area and make them consistent across spatial scales (He et al. 2019a, b). The proposal of community traits introduces a new approach to comprehensively examine the internal relationships, synergies, and divergences among plant, animal, and microbial traits at various scales from ecosystems to regions and across the globe. Specifically, this approach could be used to explore the interactions of plant–animal–soil, microbial–soil, and climate elements, shedding light on the maintenance mechanisms of plant, animal, and soil microbial communities from the perspective of functional traits (Fig. 1.9).

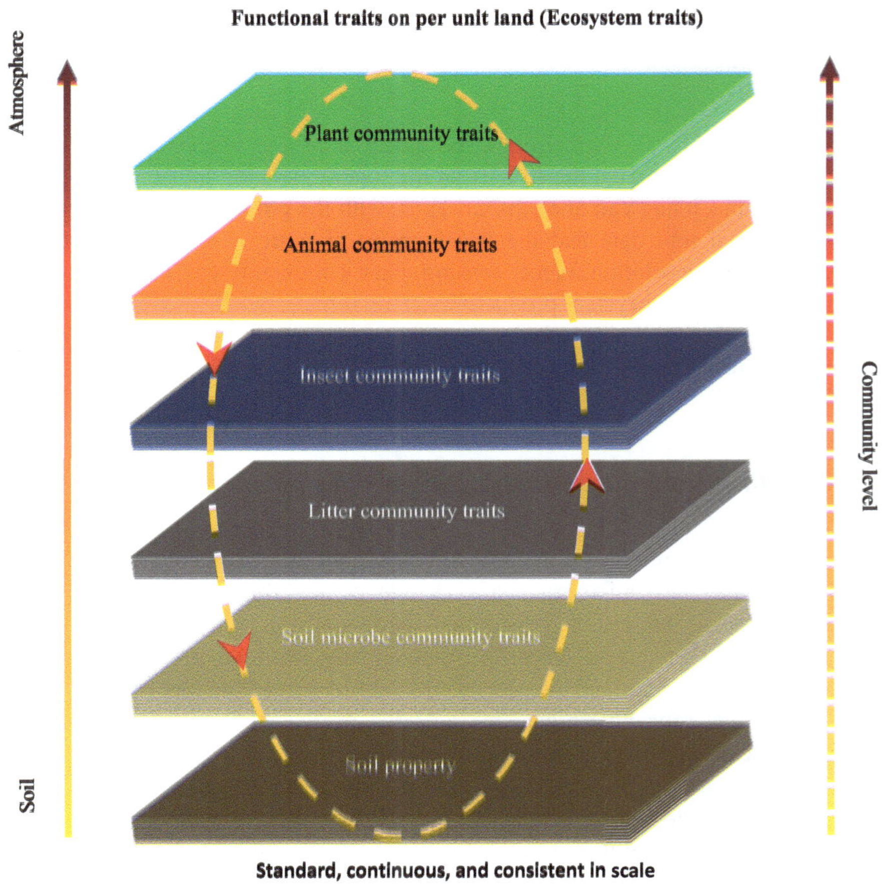

Fig. 1.9 Framework to explore plant–animal–microorganism interactions through community traits. Ecosystem traits in natural ecosystems comprise all community traits

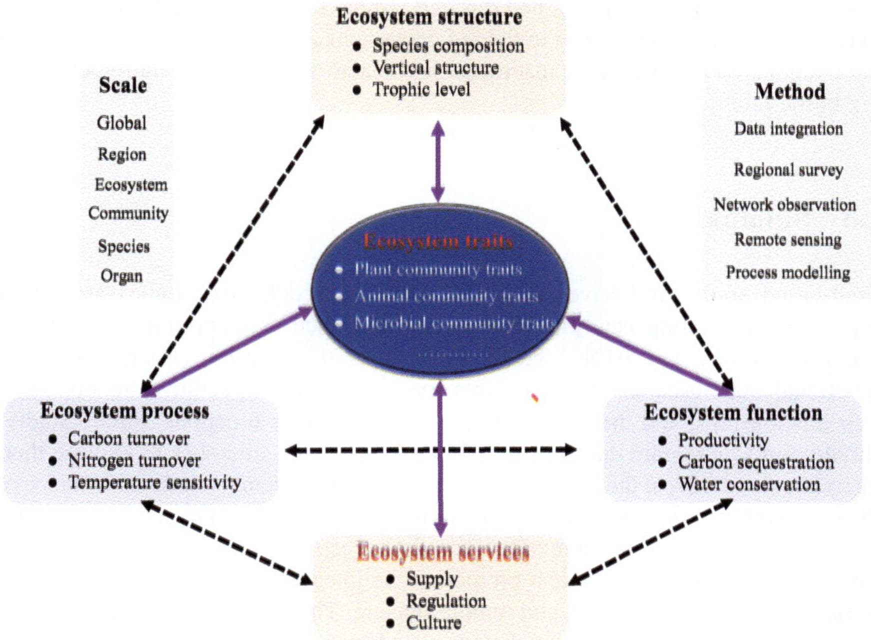

Fig. 1.10 Theoretical framework of integrative ecosystem ecology based on ecosystem traits or a combination of community traits

Furthermore, ecosystem traits or ecosystem functional traits are combinations of community traits (plant, animal, microbe, and other community traits) per unit area of land, which addresses the mismatch of spatial scales among different functional trait data. At the same time, frameworks for predicting ecosystem productivity from plant community traits, along with recent theoretical developments in PTNs and ecosystem traits, are breaking through the traditional barriers of plant functional trait research focused at the species and organ levels. Thus, these approaches lay the foundations for expanding trait-based ecology to encompass multiple species and scales.

Under the new framework shown in Fig. 1.10, researchers can make full use of various technologies, particularly remote sensing, to further develop an "Integrative Ecosystem Ecology of Traits" and establish the underlying relationships among ecosystem components, structure, process, function, and services at different scales. With the rapid development of macroscale technologies, such as remote sensing and flux observation, additional parameters at different spatiotemporal scales, including LA index, SLA, and fluorescence parameters, could be used to explain ecosystem structure. These parameters will undoubtedly be new growth areas of future research. Overall, the innovative concept of ecosystem traits, which encompass plant community traits, animal community traits, soil microbial community traits, and others, has formed a strong foundation for connecting traditional ground-survey data with macroscale data obtained through remote sensing and eddy-flux technologies. This not only advances the field of ecosystem ecology but also contributes to the integrated development of macroecology, macro-geosciences, and remote sensing in the future.

1.8 Summary

Trait-based approaches have been widely used in ecological studies at various scales. By integrating newly developed concepts such as plant community traits, ecosystem traits, and PTNs, we have reviewed the research progress in plant functional trait research, including intraspecific and interspecific variation, trait–trait relationships, and the impact of traits on ecosystem functions. Future studies should aim to integrate the traits of animals, plants, and microbes to explore their interactions and reveal the mechanisms underlying the assembly and maintenance of plant, animal, and soil microbial communities. We thus recommend developing a theoretical framework for integrated ecosystem ecology specifically focusing on functional traits in the future and promoting the integration of multiple fields, such as macroecology, geoscience, and remote sensing. This approach will allow for the effective application of functional traits to address current and future ecological problems at regional and global scales.

References

Bagousse-Pinguet YL, Gross N, Saiz H et al (2021) Functional rarity and evenness are key facets of biodiversity to boost multifunctionality. Proc Natl Acad Sci USA 118:e2019355118

Bergmann J, Weigelt A, Plas FVD et al (2020) The fungal collaboration gradient dominates the root economics space in plants. Science. Advances 6:eaba3756

Bernard-Verdier M, Navas M-L, Vellend M et al (2012) Community assembly along a soil depth gradient: contrasting patterns of plant trait convergence and divergence in a Mediterranean rangeland. J Ecol 100:1422–1433

Bjorkman AD, Myers-Smith IH, Elmendorf SC et al (2018a) Plant functional trait change across a warming tundra biome. Nature 562:57

Bjorkman AD, Myers-Smith IH, Elmendorf SC et al (2018b) Tundra trait team: a database of plant traits spanning the tundra biome. Glob Ecol Biogeogr 27:1402–1411

Bolnick DI, Amarasekare P, Araújo MS et al (2011) Why intraspecific trait variation matters in community ecology. Trends Ecol Evol 26:183–192

Bongers FJ, Schmid B, Bruelheide H et al (2021) Functional diversity effects on productivity increase with age in a forest biodiversity experiment. Nat Ecol Evol 5:1594–1603

Boonman CCF, Benítez-López A, Schipper AM et al (2020) Assessing the reliability of predicted plant trait distributions at the global scale. Glob Ecol Biogeogr 29:1034–1051

Bruelheide H, Dengler J, Purschke O et al (2018) Global trait-environment relationships of plant communities. Nat Ecol Evol 2:1906–1917

Bruelheide H, Dengler J, Jimenez-Alfaro B et al (2019) sPlot – a new tool for global vegetation analyses. J Veg Sci 30:161–186

Castro-Diez P, Villar-Salvador P, Perez-Rontome C et al (1997) Leaf morphology and leaf chemical composition in three Quercus (Fagaceae) species along a rainfall gradient in NE Spain. Trees 11: 127–134

Chave J, Coomes DA, Jansen S et al (2009) Towards a worldwide wood economics spectrum. Ecol Lett 12:351–366

Cornelissen JHC, Lavorel S, Garnier E et al (2003) A handbook of protocols for standardised and easy measurement of plant functional traits worldwide. Aust J Bot 51:335–380

Cornelissen JHC, Sibma F, Van Logtestijn RSP et al (2011) Leaf pH as a plant trait: species-driven rather than soil-driven variation. Funct Ecol 25:449–455

Cornwell WK, Ackerly DD (2009) Community assembly and shifts in plant trait distributions across an environmental gradient in coastal California. Ecol Monogr 79:109–126

Cui E, Weng E, Yan E et al (2020) Robust leaf trait relationships across species under global environmental changes. Nat Commun 11:2999

Díaz S, Kattge J, Cornelissen JHC et al (2016) The global spectrum of plant form and function. Nature 529:167–171

Enquist BJ, Norberg J, Bonser SP et al (2015) Chapter nine – scaling from traits to ecosystems: developing a general trait driver theory via integrating trait-based and metabolic scaling theories. In: Pawar S, Woodward G, Dell AI (eds) Advances in ecological research, vol 52. Academic Press, pp 249–318

Eviner VT (2004) Plant traits that influence ecosystem processes vary independently among species. Ecology 85:2215–2229

Fernández-Martínez M (2022) From atoms to ecosystems: elementome diversity meets ecosystem functioning. New Phytol 234:35–42

Freschet GT, Valverde-Barrantes OJ, Tucker CM et al (2017) Climate, soil and plant functional types as drivers of global fine-root trait variation. J Ecol 105:1182–1196

Freschet GT, Pagès L, Iversen CM et al (2021a) A starting guide to root ecology: strengthening ecological concepts and standardising root classification, sampling, processing and trait measurements. New Phytol 232:973–1122

Freschet GT, Roumet C, Comas LH et al (2021b) Root traits as drivers of plant and ecosystem functioning: current understanding, pitfalls and future research needs. New Phytol 232:1123–1158

Gallagher RV, Falster DS, Maitner BS et al (2020) Open Science principles for accelerating trait-based science across the tree of life. Nat Ecol Evol 4:294–303

Garnier E, Cortez J, Billes G et al (2004) Plant functional markers capture ecosystem properties during secondary succession. Ecology 85:2630–2637

Garnier E, Stahl U, Laporte M-A et al (2017) Towards a thesaurus of plant characteristics: an ecological contribution. J Ecol 105:298–309

Gross N, Bagousse-Pinguet YL, Liancourt P et al (2017) Functional trait diversity maximizes ecosystem multifunctionality. Nat Ecol Evol 1:0132

Han WX, Fang JY, Reich PB et al (2011) Biogeography and variability of eleven mineral elements in plant leaves across gradients of climate, soil and plant functional type in China. Ecol Lett 14: 788–796

Hanisch M, Schweiger O, Cord AF et al (2020) Plant functional traits shape multiple ecosystem services, their trade-offs and synergies in grasslands. J Appl Ecol 57:1535–1550

He NP, Liu CC, Piao SL et al (2019a) Ecosystem traits linking functional traits to macroecology. Trends Ecol Evol 34:200–210

He L, Chen JM, Liu J et al (2019b) Diverse photosynthetic capacity of global ecosystems mapped by satellite chlorophyll fluorescence measurements. Remote Sens Environ 232:111344

He NP, Li Y, Liu CC et al (2020a) Plant trait networks: improved resolution of the dimensionality of adaptation. Trends Ecol Evol 35:908–918

He NP, Liu CC, Xu L et al (2020b) Ecosystem traits open new insights into macroecology: opportunity and challenge. Acta Ecol Sin 40(8):2507–2522

He NP,Yan P, Liu CC et al (2022) Predicting ecosystem productivity based on plant community traits. Trends Plant Sci 28:34–53

He P, Gleason SM, Wright IJ et al (2020c) Growing-season temperature and precipitation are independent drivers of global variation in xylem hydraulic conductivity. Glob Chang Biol 26: 1833–1841

Jin Y, Qian H (2023) U. PhyloMaker: an R package that can generate large phylogenetic trees for plants and animals. Plant Divers 45:347–352

Kattge J, Sandel B (2020) TRY plant trait database-enhanced coverage and open access. Glob Chang Biol 26:5343–5343

Kattge J, Diaz S, Lavorel S et al (2011) TRY – a global database of plant traits. Glob Chang Biol 17: 2905–2935

Kraft NJB, Godoy O, Levine JM (2015) Plant functional traits and the multidimensional nature of species coexistence. Proc Natl Acad Sci 112:797–802

Kunstler G, Falster DS, Coomes DA et al (2016) Plant functional traits have globally consistent effects on competition. Nature 529:204–207

Li B, Suzuki JI, Hara T (1998) Latitudinal variation in plant size and relative growth rate in Arabidopsis thaliana. Oecologia 115:293–301

Li X, He N, Xu L et al (2021a) Spatial variation in leaf potassium concentrations and its role in plant adaptation strategies. Ecol Indic 130:108063

Li Y, Li Q, Xu L et al (2021b) Plant community traits can explain variation in productivity of selective logging forests after different restoration times. Ecol Indic 131:108181

Li Y, Liu CC, Xu L et al (2021c) Leaf trait networks based on global data: representing variation and adaptation in plants. Front Plant Sci 12:710530

Li X, Li MX, Xu L et al (2022a) Allometry and distribution of nitrogen in natural plant communities of the Tibetan plateau. Front Plant Sci 13:845813

Li Y, Hou J, Xu L et al (2022b) Variation in functional trait diversity from tropical to cold-temperate forests and linkage to productivity. Ecol Indic 138:108864

Li Y, Liu CC, Sack L et al (2022c) Leaf trait network architecture shifts with species-richness and climate across forests at continental scale. Ecol Lett 25:1442–1457

Liang X, Zhang T, Lu X et al (2020) Global response patterns of plant photosynthesis to nitrogen addition: a meta-analysis. Glob Chang Biol 26:3585–3600

Liu CC, He NP, Zhang JH et al (2018a) Variation of stomatal traits from cold-temperate to tropical forests and association with water use efficiency. Funct Ecol 32:20–28

Liu J, Milne RI, Cadotte MW et al (2018b) Protect third pole's fragile ecosystem. Science 362: 1368–1368

Liu S, Yan Z, Chen Y et al (2019) Foliar pH, an emerging plant functional trait: biogeography and variability across northern China. Glob Ecol Biogeogr 28:386–397

Liu CC, Li Y, Zhang JH et al (2020) Optimal community assembly related to leaf economic-hydraulic-anatomical traits. Front Plant Sci 11:2222341

Liu CC, Li Y, Yan P et al (2021a) How to improve the predictions of plant functional traits on ecosystem functioning? Front Plant Sci 12:622260

Liu XR, Chen HX, Sun TY et al (2021b) Variation in woody leaf anatomical traits along the altitudinal gradient in Taibai Mountain, China. Glob Ecol Conserv 26:e01523

Liu CC, Li Y, He NP et al (2022a) Differential adaptation of lianas and trees in wet and dry forests revealed by trait correlation networks. Ecol Indic 135:108564

Liu CC, Sack L, Li Y et al (2022b) Contrasting adaptation and optimization of stomatal traits across communities at continental-scale. J Exp Bot 73:6405–6416

Luo W, Griffin-Nolan RJ, Song LT et al (2023) Interspecific and intraspecific trait variability differentially affect community-weighted trait responses to and recovery from long-term drought. Funt Ecol 37:504–512

Maitner BS, Boyle B, Casler N et al (2018) The BIEN R package: a tool to access the botanical information and ecology network (BIEN) database. Methods Ecol Evol 9:373–379

Martin AR, Isaac ME (2015) REVIEW: plant functional traits in agroecosystems: a blueprint for research. J Appl Ecol 52:1425–1435
Midolo G, De Frenne P, Hölzel N, Wellstein C (2019) Global patterns of intraspecific leaf trait responses to elevation. Glob Chang Biol 25:2485–2498
Mlambo MC (2014) Not all traits are 'functional': insights from taxonomy and biodiversity-ecosystem functioning research. Biodivers Conserv 23:781–790
Moles AT, Warton DI, Warman L et al (2009) Global patterns in plant height. J Ecol 97:923–932
Paroshy NJ, Doraisami M, Kish R et al (2021) Carbon concentration in the world's trees across climatic gradients. New Phytol 232:123–133
Pérez-Harguindeguy N, Díaz S, Garnier E et al (2013) New handbook for standardised measurement of plant functional traits worldwide. Aust J Bot 61:167–234
Petchey OL, Hector A, Gaston KJ (2004) How do different measures of functional diversity perform? Ecology 85:847–857
Roddy AB, Martínez-Perez C, Teixido AL et al (2021) Towards the flower economics spectrum. New Phytol 229:665–672
Sack L, Scoffoni CJ (2013) Leaf venation: structure, function, development, evolution, ecology and applications in the past, present and future. New Phytol 198:983–1000
Sack L, Cornwell WK, Santiago LS, Barbour MM, Choat B, Evans JR, Munns R, Nicotra A (2010) A unique web resource for physiology, ecology and the environmental sciences: PrometheusWiki. Funct Plant Biol 37:687–693
Siefert A, Violle C, Chalmandrier L et al (2015) A global meta-analysis of the relative extent of intraspecific trait variation in plant communities. Ecol Lett 18:1406–1419
Sobral M (2021) All traits are functional: an evolutionary viewpoint. Trends Plant Sci 26:674–676
Song B, Sun L, Barrett SCH et al (2022) Global analysis of floral longevity reveals latitudinal gradients and biotic and abiotic correlates. New Phytol 235:2054–2065
Tang T, Zhang N, Bongers FJ et al (2022) Tree species and genetic diversity increase productivity via functional diversity and trophic feedbacks. elife 11:e78703
Thomas HJD, Bjorkman AD, Myers-Smith IH et al (2020) Global plant trait relationships extend to the climatic extremes of the tundra biome. Nat Commun 11:1351
Tian D, Kattge J, Chen YH et al (2019) A global database of paired leaf nitrogen and phosphorus concentrations of terrestrial plants. Ecology 100:e02812
Valverde-Barrantes OJ, Maherali H, Baraloto C et al (2020) Independent evolutionary changes in fine-root traits among main clades during the diversification of seed plants. New Phytol 228: 541–553
Violle C, Navas ML, Vile D et al (2007) Let the concept of trait be functional! Oikos 116:882–892
Wang Z, Wang C (2021) Responses of tree leaf gas exchange to elevated CO2 combined with changes in temperature and water availability: a global synthesis. Glob Ecol Biogeogr 30:2500–2512
Wang J, Mo H, Wang F et al (2011a) Exploring the network structure and nodal centrality of China's air transport network: a complex network approach. J Transp Geogr 19:712–721
Wang L, Wang N, You X et al (2011b) Analysis of chain saw selective felling operations on damage rate of residual trees during winter time in a mixed conifer-broad-leaved forest in China. For Prod J 61:283–289
Wang RZ, Huang WW, Chen L et al (2011c) Anatomical and physiological plasticity in Leymus chinensis (Poaceae) along large-scale longitudinal gradient in Northeast China. PLoS One 6: e26209
Wang RL, Yu GR, He NP et al (2016) Latitudinal variation of leaf morphological traits from species to communities along a forest transect in eastern China. J Geogr Sci 26:15–26
Wang RL, Wang QF, Zhao N et al (2017a) Complex trait relationships between leaves and absorptive roots: coordination in tissue N concentration but divergence in morphology. Ecol Evol 7:2697–2705

Wang R, Gibson CD, Berry TD et al (2017b) Photooxidation of pyrogenic organic matter reduces its reactive, labile C pool and the apparent soil oxidative microbial enzyme response. Geoderma 293:10–18
Wang H, Prentice IC, Davis TW et al (2017c) Photosynthetic responses to altitude: an explanation based on optimality principles. New Phytol 213:976–982
Wang H, Harrison SP, Prentice IC et al (2018a) The China plant trait database: toward a comprehensive regional compilation of functional traits for land plants. Ecology 99:500–500
Wang S, Wang X, Han X et al (2018b) Higher precipitation strengthens the microbial interactions in semi-arid grassland soils. Glob Ecol Biogeogr 27:570–580
Wang RL, Wang Q, Liu CC et al (2018c) Changes in trait and phylogenetic diversity of leaves and absorptive roots from tropical to boreal forests. Plant Soil 432:389–401
Wang H, Harrison SP, Li M et al (2022a) The China plant trait database version 2. Scientific Data 9: 769
Wang R, Li M, Xu L et al (2022b) Scaling-up methods influence on the spatial variation in plant community traits: evidence based on leaf nitrogen content. Journal of geophysical research. Biogeosciences 127:e2021JG006653
Weemstra M, Mommer L, Visser EJW et al (2016) Towards a multidimensional root trait framework: a tree root review. New Phytol 211:1159–1169
Weigelt A, Mommer L, Andraczek K et al (2021) An integrated framework of plant form and function: the belowground perspective. New Phytol 232:42–59
Wigley BJ, Charles-Dominique T, Hempson GP et al (2020) A handbook for the standardised sampling of plant functional traits in disturbance-prone ecosystems, with a focus on open ecosystems. Aust J Bot 68:473–531
Wright IJ, Reich PB, Westoby M et al (2004a) The worldwide leaf economics spectrum. Nature 428:821–827
Wright IJ, Groom PK, Lamont BB et al (2004b) Leaf trait relationships in Australian plant species. Funct Plant Biol 31:551–558
Yan P, Li MX, Yu GR et al (2022) Plant community traits associated with nitrogen can predict spatial variability in productivity. Ecol Indic 140:109001
Zhang JH, Zhao N, Liu CC et al (2018a) C:N:P stoichiometry in China's forests: From organs to ecosystems. Funct Ecol 32:50–60
Zhang JH, He NP, Liu CC et al (2018b) Allocation strategies for nitrogen and phosphorus in forest plants. Oikos 127:1506–1514
Zhang JH, He NP, Liu CC et al (2020a) Variation and evolution of C:N ratio among different organs enable plants to adapt to N-limited environments. Glob Chang Biol 26:2534–2543
Zhang Y, Li Y, Wang RM et al (2020b) Spatial variation of leaf chlorophyll in northern hemisphere grasslands. Front Plant Sci 11:1244
Zhang JH, Li MX, Xu L et al (2021a) C:N:P stoichiometry in terrestrial ecosystems in China. Sci Total Environ 795:148849
Zhang JH, Ren TT, Yang JJ et al (2021b) Leaf multi-element network reveals the change of species dominance under nitrogen deposition. Front Plant Sci 12:580340
Zhao N, Yu GR, He NP et al (2016a) Coordinated pattern of multi-element variability in leaves and roots across Chinese forest biomes. Glob Ecol Biogeogr 25:359–367
Zhao WL, Chen YJ, Brodribb TJ et al (2016b) Weak co-ordination between vein and stomatal densities in 105 angiosperm tree species along altitudinal gradients in Southwest China. Funct Plant Biol 43:1126–1133

Chapter 2
Variation, Adaptation, and Optimization Mechanism of Plant Functional Traits at the Organ Level

Plant functional traits play a crucial role in linking morphology, structure, and function of plants, making their variation, adaptations, and optimization mechanisms in response to changing environments long-term hot topics in ecology. Research on plant functional traits has recently expanded from small-scale, single-organ, and individual functional traits to large-scale, multiorgan, and multifunctional traits, thereby improving the development of biogeography, ecosystem ecology, and global change ecology. In this chapter, we discuss the research progress on plant functional traits at the organ level to connect with other chapters and develop a continuous understanding from the organ to the ecosystem level. First, as the primary organ responsible for plant energy production through photosynthesis, leaf functional traits may reflect photosynthetic capacity, resource utilization strategy, and adaptability of plants to the external environment. In recent decades, numerous studies on leaf functional traits have examined the spatiotemporal variations in morphological traits, functional elements, anatomical traits, vein and stomatal traits, and the key influencing factors. The main factors influencing the spatiotemporal variation in leaf functional traits are species-specific differences, soil conditions, and climate. Furthermore, researchers have extensively investigated the geographical patterns of leaf functional elements, studied the stoichiometric properties of functional elements and their responses to global change factors, and proposed relevant theoretical hypotheses, such as the stability of limiting elements hypothesis, temperature–plant physiology hypothesis, biogeochemistry hypothesis, and soil matrix age hypothesis. Studies on root functional traits have lagged their aboveground counterparts, such as leaves. Currently, root ecology focuses on the adaptive strategies of root morphological traits, functional elements, and mycorrhizal traits to the fluctuating environment, multiple dimensions of trait relationships, and the underlying influence on ecosystem processes and functions. Although related research has expanded to all aspects of ecology and promoted the rapid development of plant ecology, there is considerable scope for theoretical advances. For example, there is an urgent need to

N. He et al., *Plant Functional Ecology*,
https://doi.org/10.1007/978-981-96-8264-5_2

explore the synergies of multiple organs or functional traits, the effects of intraspecific variation and phylogeny on plant functional traits, and the adaptive mechanisms of plant functional traits under extreme environmental conditions. Furthermore, organ-level functional trait studies should be combined with these control experiments to effectively reveal the plant response mechanism to changing environmental conditions.

As important components of the terrestrial ecosystem, plants can adapt to the environment through changes in morphological traits, internal adaptation of physiological and biochemical processes, and even life history strategies. Therefore, they are considered ideal for simulating the response of plants to climate change (Cornelissen et al. 2003; Reichstein et al. 2014). Global changes, such as "climate change," significantly impact plant growth and distribution at regional and global scales, as well as the structure and function of ecosystems, leading to subsequent feedback on global climate change at various biological levels (IPCC 2021). How and which plants adapt to climate change and how this adaptation influences ecosystem processes and functions are still hot topics today, presenting a big challenge for scientists.

Plant functional traits have become a hot topic in ecology as an important bridge between plants and the external environment. The study of plant functional traits has a long history that can be traced back to Raunkiaer's life form classification system in the 1930s. Violle et al. (2007) defined plant traits as a set of characteristics which have a significant impact on the growth, reproduction, and survival of plants. Therefore, they can, individually or jointly, represent the response of plants to environmental changes and have a strong impact on ecosystem processes at different scales. He et al. (2019a, b) added a new connotation to plant functional traits, defining them as "heritable, relatively stable, and measurable" in contrast to the traditional perspective, which further improves the scientific nature and comparability of plant functional traits. Over the past two decades, ecologists have been using plant functional traits to reexamine ecological processes from a new perspective (Fig. 2.1). Moreover, plant functional traits have proven to be a valuable tool in exploring various frontier scientific problems in ecology. With the gradual improvement of the parameters of plant functional traits and research methods, related research has gradually transitioned from isolated investigations to large-scale geographical patterns and their underlying mechanisms (Wright et al. 2004a, b; Moles et al. 2014). Furthermore, researchers aim to study the adaptation and feedback processes of organisms at various levels, from individual, population, community, and ecosystem to global changes, through multiple approaches (van Bodegom et al. 2014; Li et al. 2020).

As the main organs of plants, leaves and fine roots play a key role in plant growth strategies and resource utilization. Thus, the functional traits of leaves and fine roots, as described in Fig. 2.1, have attracted widespread attention. In this chapter, we focus on the variations in leaf and fine root anatomical features and functional elements along environmental gradients and their influencing factors.

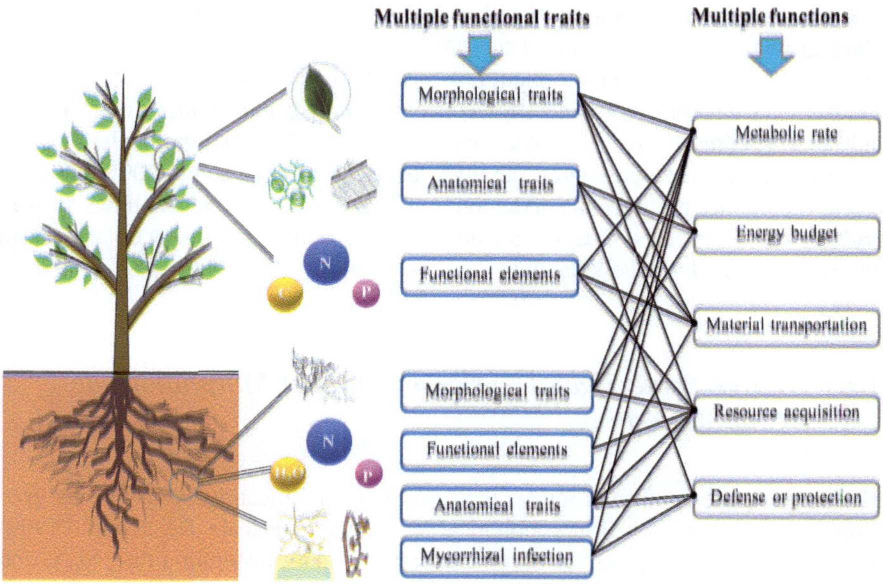

Fig. 2.1 Brief description of the traits of leaf and fine root and their related functions

2.1 Variations in Leaf Morphological Traits Along the Environmental Gradient

The leaf morphological traits, such as specific leaf area (SLA), leaf dry matter content (LDMC), leaf tissue density (LTD), leaf thickness, and leaf area, are simple to measure and are relatively stable in specific environments compared to other traits. Therefore, these traits are frequently used in studying trait–environment relationships, community building, and model optimization in recent decades (Violle et al. 2014; Bjorkman et al. 2018a, b; Li et al. 2020). Of these leaf traits, SLA or specific leaf weight (LMA) is generally considered to be closely related to photosynthesis or plant growth rate and is therefore typically used to explore various plant strategies (Reich 2014). Different species are found at opposite ends of the leaf economics spectrum, and two opposing functional strategies have been identified: conservative resource strategies and acquisitive resource strategies (Wright et al. 2004a, b; Reich 2014). At one end of the leaf economics spectrum, plant species with conservative resource strategies tend to have high LMA, high LTD, low nitrogen content, and long leaf lifespan, enabling them to enhance their competitive ability in arid or nutrient-poor environments. At another end, plant species with acquisitive resource strategies tend to be associated with rapid resource acquisition, as these species typically have high relative growth rates and therefore could be dominant in wet and nutrient-rich habitats. Theoretically, LDMC and LTD can reflect the construction cost of leaves, which is considered a conservative strategy. High LDMC typically indicates high construction costs and stress tolerance and low metabolic rate and plant growth rate (Cornelissen et al. 2003). LT plays an important role in plant growth and is closely linked to plant resource acquisition and utilization strategies. Thick leaves generally contain a high concentration of photosynthetic enzymes per unit area, leading to high N content per unit area and CO_2 requirements. Furthermore, physical resistance is strong in thick leaves (Schemske et al. 2009). Leaf area can also influence the thermal conductivity of leaf boundaries, and small leaf sizes in arid regions result in a reduced boundary effect, which can help cool leaves in extremely hot and windless habitats.

2.1.1 *Large-Scale Spatial Variations in Leaf Morphological Characteristics*

The spatial pattern of environmental factors may significantly influence the geographical variations in leaf morphological traits on a large scale. Reich et al. (1999) revealed that leaf lifespan, photosynthetic capacity, dark respiration rate, SLA, and leaf N content were significantly different in six different biomes from tropical to alpine tundra. It has long been hypothesized that high herbivore grazing pressure in tropical species results in leaves that are generally more physically resistant than those of temperate species (Schemske et al. 2009). However, Onoda et al. (2011) showed that tropical leaves did not display stronger physical resistance than the leaves of temperate plants. Moreover, no similarity was observed on a global scale, although a negative correlation was observed between leaf thickness and precipitation at the local scale (Onoda et al. 2011). Variations in leaf morphological traits are also affected by the heterogeneity of soil resources and other environmental stresses (Ordoñez et al. 2009; Onoda et al. 2011). Wang et al. (2016) analyzed the leaf traits of 847 plant species in nine forests along the north–south transect of eastern China and revealed that SLA increased and LDMC decreased with increasing latitude; however, leaf size and leaf thickness did not significantly vary. These studies were primarily attributed to the fact that the greatest variation in leaf area and leaf thickness occurred among coexisting species within a site, thereby weakening their geographic variation at large scales.

Similar to the latitudinal gradient, the altitudinal gradient is ideal for studying plants' response to climate change, because it brings drastic changes in many environmental factors, such as temperature, humidity, light, and soil nutrients (Korner 2007). Leaf length, leaf area, and SLA decreased, while leaf thickness, leaf N content, and water storage increased with increasing elevation (Luo et al. 2005). These results may be attributed to the fact that leaf palisade tissue thickness increased as an adaptation to low temperatures at high altitudes, and small leaves may reduce water loss through transpiration. Meanwhile, leaves with low SLA have high physical resistance and could adapt to severe weather in high altitudes, particularly strong wind and rapid frost (Saura-Mas and Lloret 2007).

2.1.2 *Factors Affecting Spatial Variations in Leaf Morphological Traits*

2.1.2.1 Species-Specific Differences

Spatial variations in leaf functional traits may be affected by species differences, climatic factors, soil nutrients, and soil environment (He et al. 2010; Wang et al. 2016). At large or global scales, the geographical patterns of various leaf functional traits along environmental gradients are not readily discernible, and a great amount of variation was observed among coexisting species within the community (Freschet et al. 2011; Moles et al. 2014; Wang et al. 2016). For example, global studies have revealed that 20–67% of the spatial variation in leaf economic traits can be attributed to these coexisting species (Wright et al. 2004a, b; Freschet et al. 2011; Moles et al. 2014). These results suggest that the plant functional traits are influenced by their phylogenetic conservatism and that the spatial patterns of large-scale plant functional traits are primarily derived from these differences in species composition within natural communities (Wang et al. 2015, 2016).

It is widely accepted that plant functional types (PFTs), including phylogenetic type (angiosperms and gymnosperms), growth type (herbs, grasses, shrubs, and woody plants), and leaf habits (evergreen and deciduous plants), can explain more spatial variations in the morphological, nutrient, and physiological traits of leaves than environmental factors at regional or global scale (Reich et al. 2007; Wang et al. 2016). Based on the leaf economics traits of 2021 plant species worldwide, Reich et al. (2007) found that PFTs can explain 33–67% of the variability in SLA, photosynthetic rate, and leaf N and P contents, while climatic factors explained only 5–20% of the variations. Wang et al. (2016) analyzed the leaf traits of 847 species from 9 forests in eastern China and observed that 19.4–41.6% of the spatial variation in leaf morphological traits was explained by differences in PFTs, while climate and soil factors only accounted for weak effect on the spatial variation of leaf morphological traits.

Differences in leaf morphological traits among different PFTs are the result of a combination of genetic differences of the species and its adaptation to the habitat (Poorter et al. 2009). In nutrient-poor or severely stressed ecosystems, plants can preferentially allocate carbon to storage and defense organs with high LDMC and low SLA, typically resulting in long leaf lifespans. Therefore, these plants are mostly sclerophyllous or evergreen (Poorter et al. 2009). Through the extensive use of fossil data, plant phylogenetic history has been shown to significantly influence leaf functional traits. Most leaf functional traits, such as leaf thickness and SLA, are phylogenetically conservative (Li et al. 2015; Wang et al. 2017a, b, c). In Chinese grasslands, phylogeny could explain 27% of the spatial variation in LMA, photosynthesis rate, and leaf N content in 171 species, while climate and soil factors together explained 11% of the total variations (He et al. 2010).

2.1.2.2 Climatic Factors

Temperature could influence leaf energy balance, metabolic rate, and plant growth rate to some extent (Moles et al. 2014). In general, LMA increases with increasing temperature to reduce plant water loss through transpiration (Poorter et al. 2009).

Water is a critical factor for plant growth, particularly in arid and semiarid regions. Plant species growing in arid regions tend to have high water-use efficiency (Wright et al. 2003), which may reduce tissue turnover by increasing leaf lifespan and reducing SLA (Reich et al. 1999). Leaf thickness generally increases with decreasing annual mean precipitation owing to the low area-to-volume ratio of thick leaves, which may increase the carbon assimilation rate per unit of water loss (Wright et al. 2003; Onoda et al. 2011). Thus, plants with thick leaves have a high growth advantage in arid or adverse environments. Furthermore, plant species in arid habitats typically have small leaves and high leaf potassium and phosphorus contents (Tomlinson et al. 2013).

The effect of light on leaves is also important. Under low light conditions, plants tend to have large leaves and SLA (Ackerly et al. 2002; Poorter et al. 2009) to improve light interception. Studies on canopy development in the tropical rainforests of French Guiana revealed that canopy growth rate slowed, canopy widened, leaf layers decreased, leaf elongation costs decreased, and SLA increased with the decrease in light radiation intensity (Sterck and Bongers 2001).

2.1.2.3 Soil Factors

Changes in leaf morphological and chemical traits are closely related to the availability of soil nutrients and soil microenvironment (Ordoñez et al. 2009; He et al. 2010). For example, SLA and leaf N and P contents increase with increasing soil fertility (Ordoñez et al. 2009). Leaf N and P contents of plants in resource-poor soils tended to be lower than those in fertile soils (Poorter et al. 2014). At a global scale, soil nutrients could explain more variation in leaf traits than climate, and SLA and leaf N content per unit area are affected by a combination of climatic and soil factors. Furthermore, climatic variables explain relatively small variations in spatial patterns of leaf functional traits (Moles et al. 2014; Wang et al. 2016); however, they have direct and indirect effects on the spatial variation of leaf traits. On the one hand, climatic factors could have a direct influence on the morphological structure of plant leaves and the carbon distribution associated with metabolic activities (Moles et al. 2014). On the other hand, climatic conditions shape the large-scale geographical patterns of vegetation types and regulate the availability of soil nutrients (Chapin et al. 2002).

2.2 Spatial Variations in Leaf Anatomical Traits Along Environmental Gradients

2.2.1 Geographical Patterns of Leaf Anatomical Traits

The anatomical structures of leaves include epidermis, stomata, palisade tissue, and spongy tissue. These traits might affect leaf photosynthesis rate and gas exchange processes. For example, palisade tissue and spongy tissue are considered to be the major limiting factors for light absorption and gas exchange during leaf photosynthesis (Terashima et al. 2011). Plant species can evolve different types of leaf anatomical traits in response to various environmental pressures, and plant species might adjust the relative proportion of different leaf anatomical traits to ensure their survival when environmental stresses change (Royer et al. 2008; Li and Bao 2014). He et al. (2018a, b, c) found that the ratio of palisade tissue to spongy tissue thickness and the ratio of palisade tissue to leaf thickness increase with increasing latitude, while the ratio of spongy tissue to leaf thickness decreases with increasing latitude from tropical to cold temperate forests. Abaxial epidermis thickness, leaf thickness, palisade tissue thickness, and spongy tissue thickness first increase and then decrease with increasing latitude (He et al. 2018a, b, c).

In addition, plants could modulate different leaf anatomical traits to adapt to changing environmental conditions, such as temperature, humidity, and oxygen with increasing altitude. In high-altitude areas, the thickness of epidermis is increased to improve the ability to regulate temperatures (Kuster et al. 2016). Furthermore, low temperatures at high altitudes could weaken the enzyme activity associated with photosynthesis and slow down the enzymatic reaction, leading to a decrease in the photosynthetic rate (Terashima et al. 2011). Therefore, by increasing the thickness of palisade tissue, plant species could increase the number of chloroplasts (Terashima et al. 2011) and enhance light absorption, thereby improving photosynthetic rates and compensating for reduced photosynthesis in limited growth periods at low temperatures (Chen et al. 2010). However, the increase in sponge tissue thickness results in an increased distance between mesophyll cells, creating increased space for gas exchange and promoting the diffusion of CO_2. Thus, these changes provide more raw materials for leaf photosynthesis and help maintain normal plant growth at high altitudes.

2.2.2 Factors Influencing the Spatial Variations in the Anatomical Traits of the Leaves

2.2.2.1 Climatic Factors

In general, temperature and moisture are the most important factors affecting functional traits when plant leaves are exposed to the environment. Low- or high-temperature stress would lead to a reduction in plant photosynthetic rate.

Leaves could modulate their internal structures in the long term to adapt to harsh environments and maintain their normal physiological functions (Mathur et al. 2014). For example, under low-temperature conditions, plants tend to develop narrow and thick leaves with thickened epidermis and palisade tissue for defense performance (Jankowski et al. 2017; Liu et al. 2021a, b, c, d). Additionally, the increase in leaf palisade tissue thickness could promote the ability to capture light energy in low-temperature environments (He et al. 2018a, b, c; Liu et al. 2021a, b, c, d).

As a key factor for plant growth, water availability and water conservation are also important with regard to leaf anatomical traits. The anatomical structure of leaves may undergo adaptive changes in response to prolonged drought stress. Leaf structural traits are closely related to plant drought tolerance (Ma et al. 2020). Cuticle is a lipid layer on the outer surface of the plant surface cell wall. Their main function is to reduce leaf water loss, improve energy reflection, and reduce plant transpiration. Therefore, a well-developed cuticle can increase the drought resistance of plants. In arid and semiarid environments, plants have thick palisade tissue and a high ratio of palisade tissue thickness to spongy tissue, which can reduce mechanical damage during plant wilting (Tian et al. 2016; He et al. 2018a, b, c).

2.2.2.2 Soil Factors

Soils are the sources of nutrients, water, and root-fixing substrate to support plant growth, which represent an important factor affecting the spatial variation of leaf traits. For example, N, as one of the major elements for establishing plant morphology, is an important limiting element in regulating plant growth and development in natural ecosystems. Moreover, N is involved in many physiological processes that accompanies plant growth and development (Högberg et al. 2017). In the palisade tissue of the leaves, there are many chloroplasts, which represent an important site for photosynthesis for plants. To maintain normal plant growth, the thickness of palisade tissue and spongy tissue increases under N-deficiency situations in the Loess Plateau. Furthermore, plants increase leaf thickness and palisade tissue layers or change mesophyll types to adapt to soil drought and water shortage (Chen et al. 2010).

2.2.2.3 Species-Specific Differences

There are significant differences in leaf anatomical traits among different PFTs. For example, the palisade tissue–sponge tissue thickness ratio and the palisade tissue–leaf thickness ratio in woody species decrease with increasing drought (He et al. 2018a, b, c). Liu et al. (2021a, b, c, d) studied anatomical leaf traits in different vertical vegetation zones in the Taibai Mountain and found that plant communities dominated by these deciduous broad-leaved trees exhibit thin leaves and low anatomical trait values, while evergreen coniferous forests have thick leaves and high anatomical traits. Compared to evergreen conifers, deciduous conifers have smaller leaf thickness, palisade tissue thickness, and spongy tissue thickness. These results indicate that deciduous conifers shed foliage as an adaptive strategy to cold winter environments, instead of maintaining their physiological activities by increasing anatomical leaf thickness (Gower and Richards 1990; Liu et al. 2021a, b, c, d). Additionally, the anatomical traits of leaves are generally phylogenetically conservative, and their variation due to species difference is greater than that due to climatic and soil factors (Liu et al. 2021a, b, c, d).

2.3 Spatial Variation in Leaf Vein Traits Along Environmental Gradients

Leaf veins generally comprise xylem and phloem, which are responsible for transporting water from the petiole through the mesophyll layer and transporting photosynthesis products and signaling molecules from the leaves to other plant organs (Sack et al. 2012). According to the distribution characteristics of leaf veins, leaves can be divided into reticular venation, parallel venation, bifurcate venation, and single venation. Generally, reticular venation is common in dicotyledons, parallel venation is typically found in monocots, bifurcate venation is the most common in ferns, and single-vein structures are found in conifers. In reticular venation, the major vein in the middle of the leaf is typically called the first-order vein, and the branches of the first-order vein are called the second-order vein, and the third-order vein and higher are called minor veins. However, these minor veins account for more than 80% of the total leaf veins and play an important role in leaf water transport (Sack and Scoffoni 2013). Vein density and vein diameter (VD) are the two most commonly used parameters. Vein density is defined as the length of the vein per unit area (VLA), which may reflect the efficiency of material transport. VD is the thickness of the vein, which can affect the distribution of substances among different leaf veins. Sack et al. (2012) recently defined vein volume per unit area (VV) to represent plant investment in veins. A relatively stable negative correlation exists between VLA and VD (Sack et al. 2012). Such a trade-off relationship could be explained by the return on investment hypothesis and the space restriction hypothesis as described in Box 2.1.

Box 2.1: Theoretical Basis for the Trade-off Relationships Between Vein Density and Vein Diameter

At present, the theoretical foundations for the trade-off relationship between vein density and vein diameter are primarily attributed to the space limitation hypothesis and return on investment hypothesis. The spatial limitation hypothesis emphasizes that there must be a minimum permissible distance between

(continued)

Box 2.1 (continued)
minor veins owing to the requirement for minimal continuity of CO_2 diffusion between stomata and photosynthetic tissue (Brodribb et al. 2007). If the leaf veins are densely distributed and have a large diameter, the vertical connection between the stomata and the photosynthetic tissue is cut off, and CO_2 cannot pass through the xylem (Blonder et al. 2017). Therefore, the only way to increase vein density without making it intersect with adjacent veins is to reduce vein diameter (Feild and Brodribb 2013). Large leaf vein density can achieve high leaf hydraulic conductivity, high stomatal conductivity, and high gas exchange rate per unit leaf area. However, high vein density also implies high construction costs, because the veins are rich in lignin and have high carbon costs (Beerling and Franks 2010). From a return on investment perspective, thinning and elongating leaf veins are the most cost-effective strategy, because thin and long leaf veins can achieve high leaf hydraulic conductivity and fast photosynthesis rates for a given carbon input, particularly with poor or severely stressed situations.

2.3.1 Large-Scale Geographical Pattern of Leaf Vein Traits

The values of VLA typically tend to increase from humid to dry regions, while the values of VD decrease (Sack and Scoffoni 2013). With increasing longitude, the VLA of minor veins decreases, whereas the VD increases (Sack and Scoffoni 2013). However, the spatial variations in leaf vein characteristics remain controversial till date. For example, some plant species in tropical rainforests have high VLA (Sack and Frole 2006). Zhang et al. (2022a, b) reported that the values of VLA on the Loess Plateau did not change significantly as the longitude increased, while the VD decreased. Moreover, in nature, there are considerable differences in leaf vein traits among different vein types. Therefore, studying how the spatial variation of leaf vein traits varies between different leaf vein types along environmental gradients is crucial.

As expected, there is no consistent conclusion about the altitudinal variation of leaf vein traits. In tropical and subtropical montane forests, Zhao et al. (2016a, b) revealed no significant altitudinal patterns of leaf vein traits. In contrast, Blonder and Enquist (2014) found a negative correlation between the mean minor VLA of dicotyledonous woody plants and altitude in tropical and subtropical forests. Wang et al. (2020a, b) reported that the altitudinal patterns of vein traits were different between trees and shrubs in Taibai Mountain. Specifically, the VLA of trees increases and VD decreases with increasing altitude, while the altitudinal trends are reversed for shrubs.

2.3.2 *Influencing Factors of Spatial Variation in Leaf Vein Traits*

2.3.2.1 Species-Specific Differences

The evolution of leaf vein structure should be considered as one of the important mechanisms through which angiosperms have a high photosynthetic gas exchange and are superior to other plant functional groups (Boyce et al. 2009). During radiation development of angiosperms, leaf veins become dense and thin; however, basal angiosperms and non-angiosperms generally have a relatively low VLA (Feild and Brodribb 2013). Recent studies have revealed that vein traits have significant phylogenetic signals (Zhang et al. 2014). Wang et al. (2020a, b) hypothesized the phylogenetic conservatism of leaf vein traits. That is, differences in leaf vein networks can be primarily attributed to intrinsic evolutionary adaptations between major phylogenetic clades, and their variability is conservative and less regulated by environmental plasticity. Furthermore, significant differences in leaf vein traits are observed among different growth forms. For example, trees have the highest vein density, followed by shrubs, and herbs have the lowest. In nature, deciduous plants typically have a higher vein density than evergreen plant species (Sack and Scoffoni 2013).

2.3.2.2 Climatic Factors

Among all these factors, water availability or moisture is critical for the spatial variation of leaf veins traits. Uhl and Mosbrugger (1999) reported that VLA was high in arid environments, which could contribute to high plant transpiration or low water-use efficiency. To quickly absorb large amounts of water during a relatively short precipitation period, plants in arid areas typically have a high VLA to improve their water supply and maintain strong photosynthesis and transpiration, thereby ensuring sufficient water availability and nutrient supply for plants during the drought period. The negative correlations between VLA and annual mean precipitation or humidity have been verified at different spatial scales (Sack and Scoffoni 2013). However, tropical plants typically have strong leaf transpiration and high VLA owing to strong light and high temperatures, although high humidity and precipitation occur frequently (Beerling and Franks 2010).

Temperature can influence the transpiration rate in plants and lead to an adaptive change in leaf vein traits. As the temperature increases, more water is released from the leaves. Therefore, plants will increase their investment in leaf vein traits, particularly VLA and VD, to compensate for water loss through plant transpiration. Temperature increases may cause an increase in VLA, according to both global field data and warming experiment results. This could be because plants in warm environments require increased transpiration to cool down, thereby reducing the damage caused by high temperatures and maintaining a stable, reasonably appropriate temperature (Sack and Scoffoni 2013).

Changes in atmospheric CO_2 concentration play an important role in the development of leaf vein traits at the long-term evolutionary scale. During the early Cretaceous period, high atmospheric CO_2 concentration allowed for reduced gas exchange which maintained a high rate of photosynthesis, resulting in a few leaf stomata and low transpiration rates but efficient photosynthesis. Consequently, it is advantageous for plants to have low VLA under less water transport stress (Franks and Beerling 2009). Over the next 130 million years, long-term decreases in atmospheric CO_2 concentration forced plants to increase leaf stomatal conductance and maintain gas exchange, resulting in high rates of transpiration and water loss. To compensate for this additional water loss, plants needed to improve their investment in their vascular systems, creating a competitive hydraulic environment among plant species. This CO_2-driven change in vein structure allowed for high gas exchange capacity. Moreover, this change must be coordinated with the increased hydraulic flow (Brodribb et al. 2010).

2.3.2.3 Soil Nutrients

In general, leaf veins are dense in regions with poor soil nutrients (Uhl and Mosbrugger 1999). In nutrient-poor areas, plants can obtain large nutrient transport capacity by increasing the density of minor veins, thereby improving nutrient uptake and nutrient-use efficiency and allowing them to survive in harsh environments. Zhu (2013) found that the minor vein density of *Quercus variabilis* in East Asia was significantly negatively correlated with soil potassium and calcium contents. When soil is deficient in certain nutrients, plants increase VLA to improve the ability to transport nutrients. For grass plants on the Loess Plateau, Zhang et al. (2022a, b) reported that minor VLA increased with the decrease in soil nutrients, and soil water and organic matter contents are the main factors affecting the large-scale spatial variation of leaf vein traits.

2.4 Spatial Variations in Leaf Stomatal Traits Along the Environmental Gradient

Stomata are important for gas exchange between plants and the external environment. They are typically composed of two guard cells. Their morphological traits and opening–closing behavior are closely related to photosynthesis and transpiration of plants (Hetherington and Woodward 2003). To some extent, the development of their stomata, cuticles, and roots is considered the cornerstone of the plants' spread from aquatic to terrestrial habitats. Therefore, leaf stomatal traits are one of the most important functional traits for plants. Theoretically, revealing the spatial variation and long-term evolution of stomatal traits is helpful in further exploring the

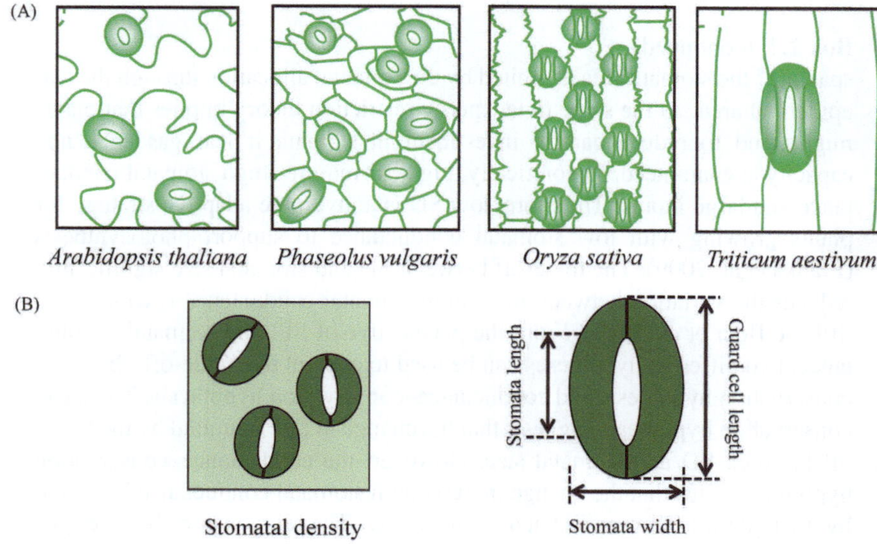

Fig. 2.2 Leaf morphological traits and some common parameters

adaptation strategies of plants to changing environmental conditions. Stomatal density (VD) is a widely used indicator with respect to stomatal morphology, commonly defined as the number of stomata per unit leaf area. Stomatal length (SL), or guard cell length, and stomatal width (SW) are also used to describe the variations in stomatal morphological traits (Fig. 2.2). SD and SL are the two most important indicators, and a relatively stable negative correlation exists between them (Franks et al. 2009). However, the theoretical mechanism and development history of this trade-off are controversial as per the description in Box 2.2. Furthermore, Sack et al. (2013) developed the stomatal pore area index (SPI), which can be defined as the product of SL and SD.

Box 2.2: Theoretical Basis of the Trade-off Relationship Between Stomatal Density and Stomatal Size

It is generally believed that the trade-off between stomata density (SD) and stomata size can be explained through physical space limit theory and energy restriction theory (Franks et al. 2009). According to the theory of physical space restriction, plants can achieve the best stomatal conductance by adjusting the size and number of stomata on their leaves; however, there must be a certain proportion of stomatal and non-stomatal epidermal cells. As stomatal development in most plants follows the principle of "single cell

(continued)

Box 2.2 (continued)
spacing," the stomata area is limited by the stomatal allocation through the leaf epidermal area. At the same time, energy restriction theory implies that plants might tend to reduce carbon investment in stomata if leaf gas exchange capacity is guaranteed. Theoretically, high SD requires high stomatal conductance and large inputs. Therefore, low SD is a favorable adaptive strategy for plants growing with low stomatal conductance to support photosynthesis (Franks et al. 2009). The trade-off between SD and stomatal size significantly reduces the variation between maximum stomatal conductance (gwmax) and SPI (de Boer et al. 2016). From the perspective of SPI and stomatal conductance, two different hypotheses can be used to explain this trade-off: the area–conservation hypothesis and conductance–conservation hypothesis. The area–conservation hypothesis suggests that the change in SPI is limited by the trade-off between SD and stomatal size. However, the conductance–conservation hypothesis states that the change in maximum stomatal conductance is limited by the trade-off between SD and stomatal size. The relationships between SD and stomatal size are not directly verified by the environment but are the result of checking the maximum stomatal conductance and SPI. The area–conservation hypothesis indicates that an abnormally high SPI is extremely unfavorable for plants because a high SPI implies that the leaf epidermis has to provide more space for the stomata, and therefore, the maintenance and operation of the stomata require more energy (de Boer et al. 2016). Thus, if the trade-off between SD and stomata size conforms to the area–conservation hypothesis, the maximum stomatal conductance is constrained by SPI, and the former is the result of the latter adaptation to the environment. According to the conductance–conservation hypothesis, the unusually high maximum stomata conductance should be unfavorable for plants, because high stomata conductance requires high water consumption. The magnitude of SPI is determined by the variation in maximum stomatal conductance under the conductance–conservation hypothesis.

Traditionally, variation in plant maximum stomatal conductance is restricted by the stomatal density on the leaf surface (i.e., area–conservation hypothesis). We analyzed stomata data from 2408 species in global forests through field investigation and literature collection and found that the trade-off between SD and stomata size completely limited the variation of maximum stomata conductance, supporting the conductance–conservation hypothesis to some extent (Fig. 2.3).

(continued)

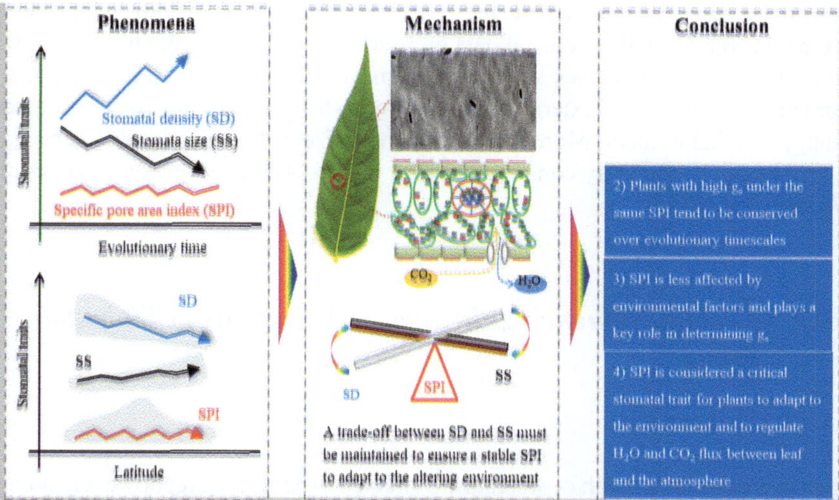

Fig. 2.3 Underlying mechanisms for the trade-off between stomata density and stomata size in the leaves of global forests. Herein, the interaction effect between stomatal density and size ensures that the stomatal pore area index remains relatively stable to better adapt to the environment

2.4.1 Geographic Pattern of Stomatal Morphological Traits in Leaves

Regarding the intraspecific variations in stomatal traits, the SD of *Leymus chinensis* in the Inner Mongolia steppe showed a decreasing trend from east to west. Tian et al. (2016) studied these dominant species in 9 typical forests in eastern China and found that temperature and precipitation were important environmental factors that influence the latitudinal variations in stomatal traits. Sun et al. (2021) studied leaf stomatal traits of 181 species in different grasslands on the Loess Plateau and found that the stomatal parameters of SD were large in meadow grasslands, the SPI was high in these typical grasslands, and the stomatal size was large in desert grasslands. With increasing aridity along the transect of forests in eastern China, the SD decreases, while stomata size increases. Through measured stomatal traits of 760 plant species in the transect, Wang et al. (2015) quantified the differences in SD and stomatal size among different functional groups and demonstrated that stomatal traits were closely related to large-scale forest productivity.

Wen et al. (2018) measured stomatal morphological traits of four *Quercus* species at different altitudes and found that the SL and SW of *Q. variables* and *Q. aliena* significantly decreased with increasing altitude, whereas SD and SPI increased to some extent. The length of stomatal apparatus of *Q. aliena* var. *acuteserrata* initially decreased and then increased from east to west with increasing altitude. The SL of

Q. wutaishansea increased with the decrease in height and width, while the SD decreased. Based on the stomatal data of 36 herbaceous plants, Bucher et al. (2017) found that the plant species with large stomata size tended to change the size of stomata to adapt to the external environment, while plants with large SD tend to change stomata number. Moreover, the stomata size of the upper and lower epidermis of plants is relatively consistent, which may be related to genome size control (Franks et al. 2012).

2.4.2 Factors Influencing the Spatial Variations in Stomatal Traits

2.4.2.1 Temperature

The formation and growth of stomata are affected by temperature; however, stomatal traits of different plant species respond differently to environmental changes. As temperature has a significant impact on leaf area expansion, its impact on SD should be indirect. However, the relatively stable SPI may partially neutralize the effect of leaf expansion, and therefore, the response of stomatal traits to changing temperature is relatively stable. For example, some studies have found that temperature have little effect on stomata morphological traits, primarily because leaves, as organs being responsible for sensing ambient environment temperature, could only develop within an appropriate temperature range. Through the temperature gradient control experiment, Zuo et al. (2005) reported that the stomata index of seven herbaceous and three woody plants was more sensitive to temperature changes than to SD. The relationships between stomatal index and temperature were observed to be positive, negative, or not significant for different plant species, and the relationships between stomatal density and temperature were positive or not relevant for different plant species. Similarly, the length of guard cells in different species responded differently to temperature gradients.

2.4.2.2 Precipitation

SD increase with decreasing precipitation and increasing plant height, plant density, and leaf area (Wang and Gao 2003; Gazanchian et al. 2007). Xu and Zhou (2008) showed that SD first increased and then decreased with increasing drought stress, and there should be a trade-off between stomata formation and leaf growth. Moderate drought conditions have the potential to directly accelerate the increase in SD. However, when water is deficient, leaf growth will be restricted leading to an increase in SD. Therefore, the sharp decrease in CO_2 concentration caused by drought contributes to a significant reduction in SD (Joos et al. 2004). However, a meta-analysis showed that increased CO_2 can only explain 5% of the decrease in SD (Ainsworth and Rogers 2010). Yan and Zhong (2017) collected data from 1854 field-control experiments worldwide to examine the response of SD and stomatal index to the main factors of global change and found that both SD and stomatal index decreased with increasing CO_2 concentration. In addition, the SD's responses to four global change factors, such as CO_2 concentration, warming, nitrogen deposition, and aridity, were different. Specifically, SD increased with increasing atmospheric CO_2 concentration, decreased with global warming, increased with the increasing aridity, and decreased with increasing nitrogen deposition. Overall, the response of SD to future global changes remains unclear and should be explored further in the future.

2.5 Variations in Leaf Functional Elements Along Environmental Gradients

2.5.1 *Geographical Patterns of Leaf Functional Elements*

Currently, 92 elements are known in nature, and approximately 30 elements are involved in plant growth, and at least 17 of them are essential elements for plant growth (Marschner 2011). These elements are essential for the normal growth and life cycle of plants, thereby playing an important role in the growth and physiological metabolism of plants. These essential elements can be divided into four categories according to their physiological functions. The first category consists of basic elements, such as C, H, O, N, and S, which primarily serve as structural substances in plant species. The second category includes the elements P and B that participate in the energy transfer reaction. The third category includes elements that regulate the osmotic potential, electrochemical potential, and activating enzymes of plants, such as K, Na, Ca, Mg, Mn, and Cl. The fourth category includes elements, such as Fe, Cu, Zn, and Mo, which can carry out electron transport and form cogroups (Fig. 2.4).

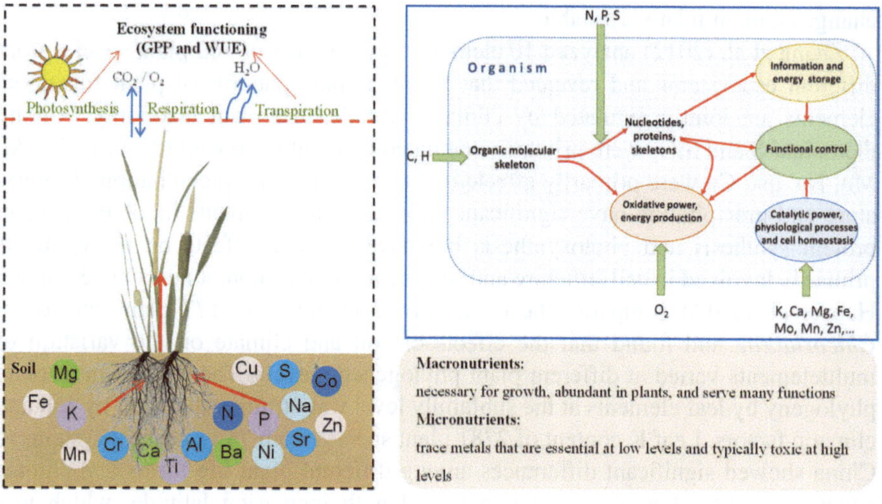

Fig. 2.4 Essential elements for plant growth

The content and variation of these functional elements in different plant organs are an important part of plant functional traits, which can not only reflect the adaptive strategies of plant nutrient uptake and utilization but also serve as important indicators of ecosystem function. For example, K is closely related to water use in plant leaves (Li et al. 2021a, b, c). Ca influences plant resistance to drought and cold (White and Broadley 2003), while Ca and Mg are related to the composition of plant cell walls, which might influence herbivore grazing pressure on plants (Mladkova et al. 2018).

N and P are considered the main limiting elements for plant growth and are the most concerned parameters among plant chemical traits. In recent decades, the biogeographic patterns, major influencing factors, and underlying mechanisms of N, P, and N:P in leaves have been extensively and intensively investigated. For example, Reich et al. (2004) analyzed data from 1280 terrestrial plants worldwide and discovered that leaf N and P contents decreased with decreasing latitude, while the N:P ratio increased with decreasing latitude. Chinese researchers collected the relative data on 753 terrestrial plant species in China (Han et al. 2005) and 654 plant species in north–south transect of eastern China (Ren 2007) and consistently found that leaf N and P contents decreased with decreasing latitude, while N:P ratio showed no significant trend with latitude. Furthermore, data from 213 dominant plants in Chinese grasslands showed that leaf N, P, and N:P did not vary significantly with temperature and precipitation (He et al. 2006; He et al. 2008).

In recent decades, studies on plant stoichiometric traits of C, N, and P have been extended to other elements, even to the trace elements. Han et al. (2011) explored the stoichiometric relationship, large-scale spatial variation, and influencing factors of 11 elements in more than 1900 plants in China. Based on the relative contributions of climate, soil factors, and plant functional groups to plant stoichiometric traits, the "Stability of Limiting Elements Hypothesis" was developed. This hypothesis assumes that owing to physiological and nutrient balance constraints, the content of limiting elements in plants and their response to large-scale environmental changes remain relatively stable.

Zhang et al. (2012) analyzed 10 elements in the leaves of 702 plant species from different ecosystems and revealed that biogeographic patterns of plant functional elements are jointly regulated by climate, latitude, and plant taxonomy. Among them, leaf S and SiO_2 were primarily regulated by plant taxa, while N, P, K, Fe, Al, Mn, Na, and Ca were primarily affected by temperature and precipitation. Temperature and precipitation have significant impacts on the elements involved in plant protein synthesis and photosynthesis but have minimal effects on the elements primarily involved in cell structure and serving as enzyme components or activators. Hao et al. (2014) compared the content of 7 elements in 177 plant species of *Chicoriaceae* and found that the effects of soil and climate on the variation of multielements varied at different plant phylogenetic levels. The regulation of plant phylogeny by leaf elements at the subfamily level was stronger than that by soil and climatic factors. Leaf K content of 2781 plant species from 78 typical ecosystems in China showed significant differences among different plant life forms and climate zones (Fig. 2.5). Leaf K content increased with increasing latitude, which was primarily influenced by climatic variables, particularly temperature; however, the

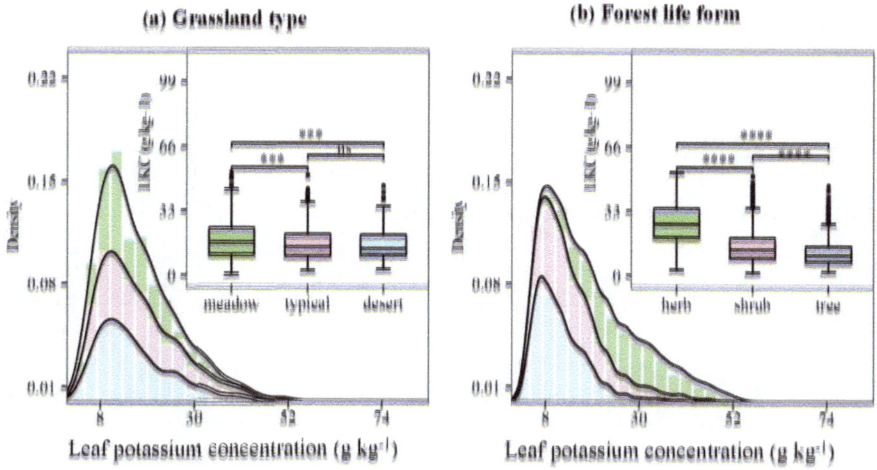

Fig. 2.5 Frequency of leaf potassium (K) concentrations (LKC, g kg^{-1}) in different grassland types (panel **a**) and plant species in forests (panel **b**). LKC was highest in meadow grasslands. LKC significantly differed between different plant life forms in forest ecosystems. **** and *** denote significant differences between different taxa at $P = 0.0001$ and 0.001, respectively; ns denotes no significant difference between different taxa

limiting effect of plant phylogeny was very small. The high leaf K concentration in low-temperature regions may be related to the adaptation of plants to severe cold (Li et al. 2021a, b, c). The leaves of herbaceous plants and those in arid regions have a high S content (Fig. 2.6). Under the conditions of drought, low temperature, and strong ultraviolet radiation, leaf S contents tend to become high. Temperature, precipitation, radiation, and soil S content jointly regulate plant leaf S concentration, partially confirming the adaptive mechanism for the spatial variation of leaf S content (Fig. 2.7). These studies reveal that leaf elements (K and S) play an important role in the adaptation of plants to extreme environments (Zhao et al. 2022a, b).

With increasing altitude, temperature, precipitation, and soil nutrient conditions change drastically, which can lead to a significant change in vegetation types in the vertical zones. These studies have focused on the altitudinal variation of leaf C, N, and P contents (Hultine and Marshall 2000; Kohler et al. 2006). On a global scale, nonstructural carbon, such as starches, lipids, and soluble sugars, increases with elevation, which may aid plants in regulating osmotic pressure in response to low-temperature stress and adapting to difficult habitats by increasing storage materials (Hoch and Korner 2012). With increasing altitude, plants reduce leaf P content and improve P-use efficiency as an adaptation by reducing soil P availability.

Stoichiometric traits of plants have also been considered a limiting factor for species distribution and community assembly. Other studies have shown that alien species typically have low specific leaf weight and high photosynthetic capacity and ability to acquire functional elements, such as macroelements (N, P, and K0) and microelements (Fe, Ni, Cu, and Zn), occupying biogeochemical niches different from those of native species (Penuelas et al. 2010).

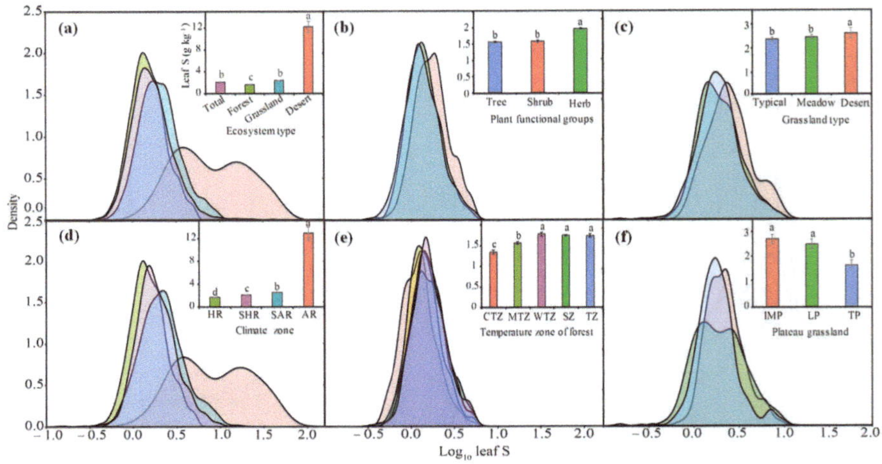

Fig. 2.6 Frequency distribution of leaf sulfur content (LSC) (left panel) and comparison of LSC between different vegetation types and regions (right panel). *HR* humid region, *SHR* semi-humid region, *SAR* semiarid region, *AR* arid region, *CTZ* cold temperate zone, *MTZ* middle temperate zone, *WTZ* warm temperate zone, *SZ* subtropical zone, *TZ* tropical zone, *Typical* typical grassland, *Meadow* meadow grassland, *Desert* desert grassland, *IMP* grasslands in the Inner Mongolia Plateau, *LP* grasslands in the Loess Plateau, *TP* grasslands in the Tibetan Plateau

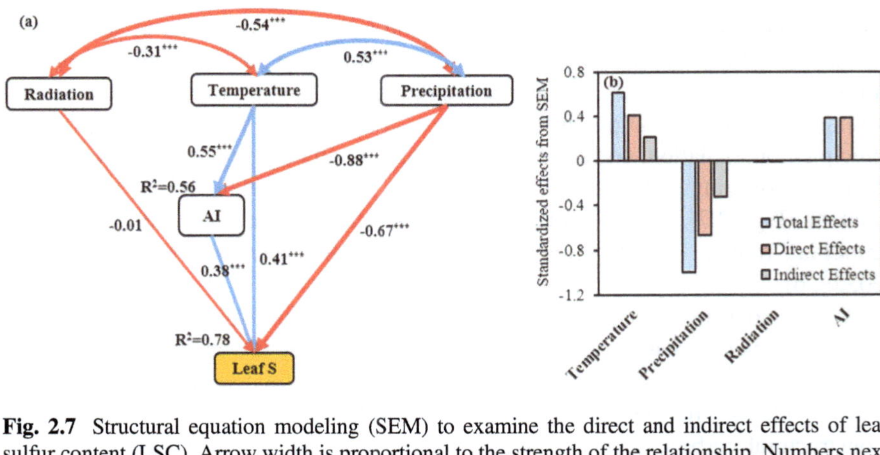

Fig. 2.7 Structural equation modeling (SEM) to examine the direct and indirect effects of leaf sulfur content (LSC). Arrow width is proportional to the strength of the relationship. Numbers next to the arrows are standardized path coefficients. The proportion of variance explained (R^2) appears next to each response variable in the model

2.5.2 Factors Influencing the Spatial Variations in Leaf Functional Elements

2.5.2.1 Climatic Factors

Temperature and precipitation are important factors that drive spatial variations in plant stoichiometric traits. Reich and Oleksyn (2004) opined that temperature was an important driving factor for the latitudinal distribution of leaf N and P and proposed the temperature–plant physiological hypothesis. This hypothesis states that temperature influences the accumulation and transfer of C, N, and P in plant leaves by affecting the physiological metabolic rate of plants. When the temperature is low, plants increase enzyme (N rich) and RNA (P rich) to compensate for the decrease in the enzyme and metabolic activities of plants due to the low temperature. Therefore, plants in cold habitats typically have high N and P contents. Precipitation and temperature have some covariance with latitude, and precipitation may affect the geographical variations in stoichiometric traits among plant communities through changes in vegetation types, species composition, and nutrient distribution of plant species (Reich 2005). Furthermore, leaching caused by precipitation should indirectly affect stoichiometric traits by regulating the supply of mineral elements to soils.

2.5.2.2 Soil Conditions

Except C, most plant functional elements are obtained from the soil. Soil nutrient supply capacity is a non-negligible factor affecting the biogeographical pattern of plant stoichiometric traits. Researchers have proposed a temperature–biogeochemistry hypothesis and soil substrate age hypothesis to elucidate the role of soil nutrient supply in the formation of biogeographic patterns of plant stoichiometric traits (Hedin et al. 2003). Soil nutrients have different sources. For example, available N primarily originates from litter decomposition, and available P is derived from rock weathering. Temperature and precipitation affect soil nutrient patterns by influencing litter decomposition and soil leaching. In low-temperature areas, microbial activity is limited, which reduces the mineralization rate of soil organic matter. In areas with high precipitation, the leaching effect increases, which can lead to a reduction in the N and P contents available in the soil. Moreover, the age of the soil matrix influences the soil nutrient status. Overall, nutrient contents are low in extremely young or old soils but high in middle-aged soils. For example, it was assumed that soils in equatorial regions have a long formation time, high leaching, and low nutrient content (Vitousek et al. 2010), leading to high C:N and C:P ratios for plants at low latitudes. In addition, soil pH is an important factor in influencing the biogeographic pattern of plant stoichiometric traits on a large scale. Mineral elements in plants are closely related to soil pH.

Because leaf stoichiometric traits are closely related to soil nutrient status, leaf N:P ratio has been widely used to diagnose N or P limitation in plant growth. The availability of N and P is a significant factor that restricts plant growth in the majority of ecosystems, and the leaf N:P ratio can indicate the extent to which soil N and P limit plant growth. In general, when the leaf N:P is less than 14, plant growth can be considered to be N limited, and when it is greater than 16, plant growth is considered to be P limited. Plant growth is jointly limited by N and P when $14 < \text{leaf N:P} < 16$ (Koerselman and Meuleman 1996). Based on the fertilization experiments in grassland, Zhang et al. (2003) found that leaf N:P is relatively stable and can be used as an indicator to assess the nutrient limitation of species. They determined the critical N:P index for herbaceous plants in Inner Mongolia grassland; when N:P ratio is <21, plant growth is limited by N availability, and when it is >23, plant growth is considered to be limited by P availability.

2.5.2.3 Vegetation Types

Vegetation type and plant species composition are other important factors that determine the biogeographical pattern of plant stoichiometric traits. Vegetations can be classified into different PFTs according to plant life form and other characteristics. There is ample evidence that the stoichiometric traits of plants significantly differ across different PFTs. For example, herbs typically have higher leaf N:P ratio than woody plants (He et al. 2006, 2008), more angiosperms than gymnosperms, more dicotyledonous plants than monocotyledonous plants, more broadleaf plants than coniferous plants, more seed plants than ferns, more deciduous plants than evergreens, and more C3 plants than C4 plants (Han et al. 2005).

Plant stoichiometric traits also differ among different vegetation types. McGroddy et al. (2004) found significant differences in C, N, and P contents between different plant communities, such as temperate broad-leaved forests, temperate coniferous forests, and tropical rain forests. Using the C: N: P stoichiometric relationships of deciduous broad-leaved forests, evergreen coniferous forests, and evergreen broad-leaved forests in Tiantong Mountain, Zhejiang Province, Yan et al. (2010) revealed significant differences in leaf stoichiometric relationships between the three forest communities.

Researchers have recently recognized that the effects of taxonomy and phylogeny on stoichiometric traits are not negligible. Some plant species can accumulate certain elements (White et al. 2007). For example, legumes typically have high N content. Dicots have higher Ca content in their leaves than monocots (White and Broadley 2003). A study of the stoichiometric relationships of N, P, K, Ca, and Mg in 50 tree species in European forests showed that species differences can explain most of the variations (56.7%) in element content and stoichiometric relationships (Sardans and Penuelas 2014). However, coexisting species in the same genus exhibit greater stoichiometric differences than non-coexisting species, which may be the result of niche differentiation to reduce competition for the same or similar resources.

In conclusion, plant functional elements are affected by the combined influence of climate, soil, and plant types. The effects of these factors are not independent and are subject to complex interactions between them. Climatic factors may affect the geographical distribution of vegetation types and soil nutrients, which could further influence the formation of the geographical pattern of plant functional elements. Plants can also respond to environmental changes through ecological processes, such as photosynthesis and litter decomposition. Therefore, understanding the biogeographic patterns of plant functional elements and their driving mechanisms may provide a theoretical basis for the comprehensive study of the material cycling of ecosystems.

2.6 Variations in Fine Root Traits Along Environmental Gradients

Studies on root functional traits originally started in agronomy, where the focus was on achieving high yields and studying high-quality crops. Since the late 1990s, the study of fundamental functional traits as a link between biology, ecology, earth sciences, and environmental sciences has gradually become a research focus in contemporary ecology. In recent decades, research into fundamental functional traits have developed rapidly and achieved promising results. The role of root traits in plant nutrient utilization strategies, species coexistence mechanisms, and ecosystem functions has attracted extensive attention (Freschet et al. 2021a, b). Moreover, scientists have revealed that fine roots or absorptive roots, the main underground plant organs for absorbing nutrients and water from the soil, have fast growth and turnover rates (Pregitzer et al. 2002; McCormack et al. 2015).

Root morphological and architectural traits, such as root depth, root length, and root density, determine the spatial structure of the entire root system. Root morphological traits or root size grades [such as root diameter (RD), specific root length (SRL), root tissue density (RTD)], physiological traits of fine roots (such as root respiration, root nutrient uptake, and root tissue nutrient content), root exudates, and rhizosphere microorganisms play an important role in nutrient uptake of fine roots (Bardgett et al. 2014; McCormack et al. 2015). These root traits not only show large variation across species and genotypes but also demonstrate high environmental plasticity, enabling plants to effectively adapt to changing environments, particularly changes in soil nutrients and water availability (Fort and Freschet 2020).

2.6.1 *Geographic Pattern of Root Morphological Traits*

Some studies have demonstrated that at the organ level, plants can optimize the root physiological functions by adjusting architectural and structural traits, such as RD, SRL, root lifespan, and mycorrhizal infection rate, to improve the nutrient uptake of fine roots. In general, plants with thick roots typically achieve better nutrient uptake through increased mycorrhizal colonization rate and hyphal density, whereas plants with thin primarily achieve increased root branching and SRL to serve a similar function. Based on the global data on root functional traits of 1115 species, Freschet et al. (2017) found that roots of plants in tropical regions had relatively conservative root traits; however, plants in colder regions had thin and long roots, which are favorable for plants to survive in habitats with relatively low nutrients. Similarly, Ma et al. (2018) examined the first-order root traits of 369 species across the world's seven biomes and revealed that, from tropical rainforest to desert, RD became thin and less dependent on symbiotic fungi. These changes suggest that the efficiency of nutrients obtained through carbon investment in plant units has been optimized, potentially improving the ability of plants to adapt to the changing environment. On an evolutionary scale, absorptive roots have evolved in a more efficient and independent direction, laying the foundation for plant species to colonize new habitats over 400 million years of plant evolution. Steidinger et al. (2019) mapped the symbiotic microbial network on a global scale for the first time using a database of more than 28,000 tree species in more than 70 countries. Their results showed that climatic factors modulated the exchange of key symbiotic organisms between plant roots and microorganisms. There is an apparent transition of forest microbial symbiosis from arbuscular mycorrhiza and nitrogen-fixing bacteria at low latitudes to ectomycorrhizal ecosystems at high latitudes. While arbuscular mycorrhizal trees predominate in seasonally warm tropical forests and the abundance of nitrogen-fixing trees peak around 30°N or in arid regions, ectomycorrhizal symbionts predominate in forests where seasonally cold and dry climates inhibit decomposition.

2.6.2 Spatial Variations in Root Morphological Traits and Their Influencing Factors

2.6.2.1 Species-Specific Differences

Variations in fine root functional traits are jointly affected by phylogeny and their living environment (Fig. 2.8). Most studies have demonstrated that the morphological, chemical, anatomical, and structural traits of fine roots have significant phylogenetic signals (Chen et al. 2013; Comas et al. 2014). At the family level, plants tend to develop thin and long fine roots. Phylogenetic analyses of extant angiosperms have confirmed that basal angiosperm lineages have relatively thicker diameters and smaller SRL than recently diverged angiosperm lineages (Comas et al. 2014; Kong et al. 2014; Wang et al. 2017a, b, c). Chen et al. (2013) reported that the RD of angiosperm families decreased, while SRL increased, with a divergence time of 120–64 million years.

Angiosperms generally have thinner RD, lower RTD, higher SRL, and higher root branching ratio than gymnosperms (Pregitzer et al. 2002; Freschet et al. 2017). Wang et al. (2006) found that RD, length, and tissue density of *Larix gmelinii* were significantly higher than those of *Fraxinus mandshurica*, while the SRL was lower. In the first two root orders, root branching ratio of *Fraxinus mandshurica*'s is significantly larger than that of *Larix gmelinii*. The differences in root morphological and architectural traits between these species reflect their different ecological strategies. The high root length density of *Fraxinus mandshurica* may promote rapid root growth, and low RTD could reduce carbon input for developing dense root systems, thereby improving subsurface competitiveness.

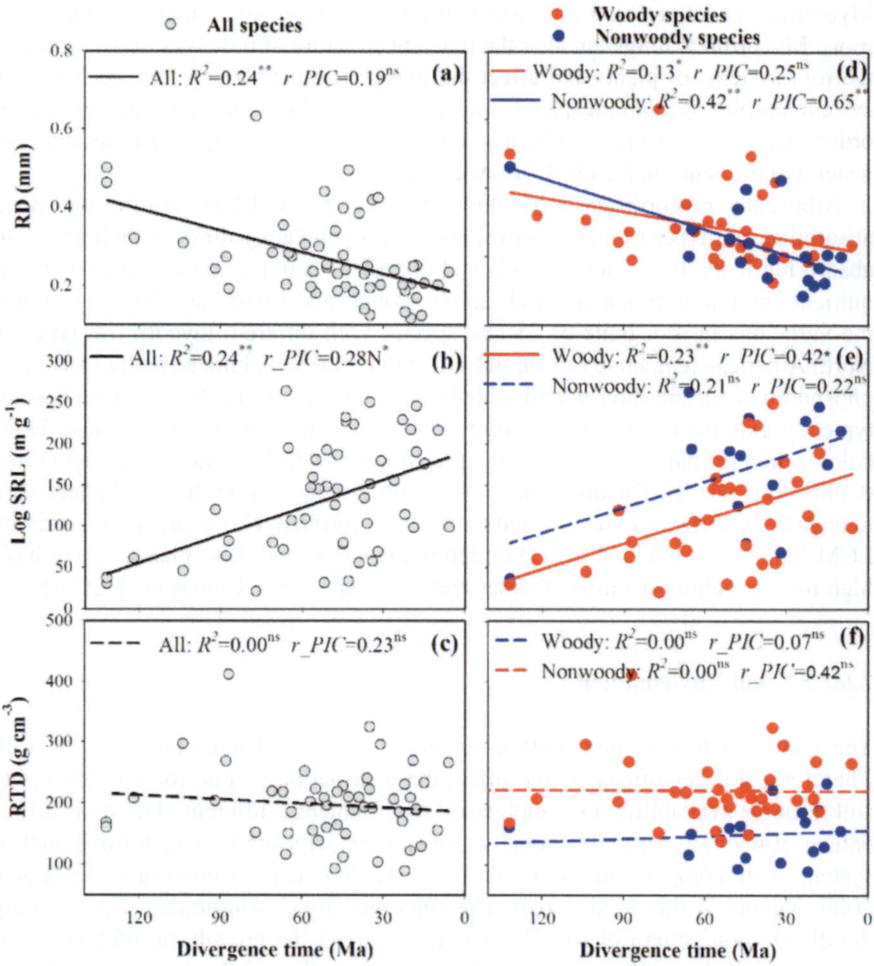

Fig. 2.8 Linear regression analysis and Pearson correlation between phylogenetic independent contrasts (*r_PIC*) of absorptive root traits and divergence time (million years ago, Ma), using the datasets of all 56 angiosperm families, 20 woody families and 42 nonwoody families, respectively. Here some families that included both woody and nonwoody plants were analyzed separately. Dashed lines indicate nonsignificant relationships ($P > 0.05$)

2.6.2.2 Rhizosphere Microorganisms

The rhizosphere microbiome, including bacteria, actinomycetes, fungi, and algae, can interact with root systems and affect the structure and composition of roots, which further influence the function of root systems. Most terrestrial plants develop a symbiotic relationship with mycorrhizal fungi (Wang and Qiu 2006), and more than 90% of angiosperms roots are infected by mycorrhizal fungi (Brundrett 2009).

Mycorrhiza is defined as a symbiont that arises from the combination of fungi and roots. Mycorrhizal fungi can alter the inherent structure of root systems by infecting the root systems of plants and affect the function of the root system, such as root system morphology and lifespan. Compared to the higher root orders, the distal first-order roots are most likely to be infected by fungi, which helps the roots improve water and nutrient uptake (McCormack et al. 2015).

Arbuscular mycorrhizas (AM) and ectomycorrhizas (ECM) are the two main absorptive root types of angiosperms, and they significantly differ in their ability to absorb nutrients. It is generally believed that ECM could dominate under low soil nutrient conditions (Lambers et al. 2008). Comas and Eissenstat (2009) examined the variations in root traits of 25 tree species with different mycorrhizal types in North American temperate forests and found that root branching intensity (SRL) and phosphorus concentration varied widely between species. ECM-colonized species typically exhibited more intense root branching than AM species because ECM colonization primarily occurred at the root tip, while AM colonized the root cortex (Comas et al. 2014). The different colonization strategy implies that AM plants tend to have thick cortexes to be infected by more mycorrhiza. However, colonization by ECM fungi is not affected by RD or cortex thickness, and these plants tend to have high root branching intensity (Comas and Eissenstat 2009; Comas et al. 2014).

2.6.2.3 Soil Environment

The main function of fine roots is to absorb water and nutrients from the soil. Therefore, soil conditions should affect the distribution of fine roots by changing soil resource availability. In complex soil environments, different plant roots differentiate structurally and functionally, making underground roots form branched system with complex structures and diverse functions (Comas and Eissenstat 2009; Kramer-Walter et al. 2016). The heterogeneity of soil nutrients, particularly the distribution pattern of soil N content, can affect the growth, proliferation, and branching structure of fine roots, which can further influence the distribution of plant roots (Fort and Freschet 2020). Holdaway et al. (2011) analyzed the changes in fine root traits of plants grown in different soil age sequences in New Zealand and found that soils with a young site age were typically phosphorus limited, and plants grown on such soils have high SRL, thin RD, high RTD and branching intensity, and low nutrient content.

Soil moisture is also an important factor that influences the growth and vertical distribution of fine roots. The variation in root morphological traits along the water gradient represents an adaptation strategy for plants to drought stress. As soil drought stress increases, RD decreases (West et al. 2004), while SRL and root surface area increase (Metcalfe et al. 2008). Furthermore, soil bulk density can greatly influence the morphological traits of fine roots. In general, high soil bulk density allows for thick, dense root systems, leading to increased penetrating power for plants, which in turn positively impacts plant growth and resource extraction in dense soils.

Notably, there is still controversy on how soil factors influence root traits. For example, roots in poor soils should be conservative, when following the leaf economic spectrum. Some studies have confirmed that thick roots with high RTD are advantageous in infertile soils, and plants with fast strategies dominate in soils with rich nutrients and low bulk density (Reich 2014; Fort and Freschet 2020). However, some studies have also found that soil fertility is negatively correlated with SRL and positively correlated with RD (Kramer-Walter et al. 2016), which may be owing to the positive correlation between RD and mycorrhizal colonization rate (Comas et al. 2014). Based on the fine root trait data of 1115 species, Fort and Freschet (2020) showed that herbaceous plants with rapid resource utilization strategies mostly dominated in favorable soil environments (e.g., high nutrient and water availability); however, woody plants followed an opposite strategy. These findings indicate that it is impossible for us to use a simple, uniform root trait–environment relationship to explain plant adaptation mechanisms.

2.6.2.4 Climatic Factors

Climatic factors could influence root traits by changing soil temperature, moisture, and nutrient content. A meta-analysis indicated that the SRL of woody plants decreased under fertilized conditions and was negatively correlated with light, temperature, and atmospheric CO_2 concentration (Ostonen et al. 2007). Ostonen et al. (2011) found that latitude and N deposition can explain a large portion of the variation in fine root traits of *Picea abies*. With increasing latitude, the root length and RTD of ectomycorrhizae increased linearly; however, N content decreased. Hertel et al. (2013) investigated the growth dynamics of fine roots of *Fagus sylvatica* along a precipitation gradient (820–540 mm $year^{-1}$) in Central Europe and found that the response of fine root morphology to changing precipitation differed among different soil textures. In sandy soil, the SRL and specific root area of *Fagus sylvatica* increased along with decreasing precipitation, and the average RD decreased. By analyzing the variation in fine root traits of 1115 species worldwide, Freschet et al. (2017) revealed that climatic factors, particularly temperature, strongly influenced root traits related to resource acquisition ability, which was positively correlated with RD and negatively correlated with SRL.

Furthermore, some studies have demonstrated that the increased CO_2 concentrations have a strong effect on the growth and function of plant roots (Iversen et al. 2008; Nie et al. 2013). However, the response of fine root traits to increased CO_2 concentrations is highly inconsistent. For example, the relationship between root length and atmospheric CO_2 concentration is found to be positive (Norby et al. 2004), negative (Wan et al. 2004), or without any significant correlation (Higgins et al. 2002). Similarly, no consistent conclusions have been drawn on the effects of elevated CO_2 concentrations on root growth and death. Nie et al. (2013) integrated the results of 110 experiments and found that the morphology and function of root systems considerably changed under the increased CO_2 concentration. Specifically, root length, RD, and root nitrogen content increased by 26.0, 8.4, and 7.1%, respectively. Meanwhile, root respiration, rhizosphere sediment, and mycorrhizal colonization rate increased by 58.9, 37.9, and 3.3%.

2.7 Changes in the Environment That Affect the Functional Elements of Fine Roots

2.7.1 Geographic Patterns of Fine Root Functional Elements

Fine roots are the most important organ in plants for absorbing water and nutrients from the soil. It is estimated that global N reserves in fine roots are approximately 4.8×10^8 Mg, which accounts for 1/7 of the N reserves in terrestrial vegetation. Moreover, carbon and nutrients released by root decomposition processes are of great importance for the carbon and nutrient fluxes of terrestrial ecosystems. In forests, the amount of nutrients released into the soil through the production and decomposition of fine roots should equal or exceed the amount of nutrients released through the decomposition of aboveground litter. Therefore, it is important to clarify the geographical variations in N, P, and other functional elements in fine roots on a large scale.

As N and P are important factors that limit plant growth in most terrestrial ecosystems, the geographical patterns of their contents in roots have long attracted considerable attention. According to the meta-analysis of root data from 211 studies in 51 countries, Yuan et al. (2011) found that the N and P contents in fine roots were lower than those in leaves; however, their N:P ratio was similar to that of green leaves. The exponential decrease in root N:P ratio with increasing latitude was observed at different RDs, which was inconsistent with the previous results of a linear decrease in leaf N:P ratio along the latitudinal gradient. The latitudinal patterns of root N:P ratio are likely to be influenced by various factors, such as vegetation type, climate, soil age, and soil weathering. Meanwhile, studies on plant multielements have also revealed that the content of fine root elements shows significant latitudinal patterns. For example, with increasing latitude, Ca and Na contents in the fine roots significantly increase, while the contents of Mg, S, Al, Fe, Cu, Pb, Ni, and Co decrease; however, the root contents of K, Mn, and Zn do not show any significant latitudinal variation (Zhao et al. 2016a, b).

To clarify the underlying mechanism of the biogeographic patterns of plant functional elements, scientists have proposed several hypotheses, including some hypotheses underlying the logistic relationships between plant growth and the environment. Among them, the growth rate hypothesis states that plants tend to decrease N:P and C:P ratios, while P content tends to increase with increasing growth rate. The hypothesis helps us to understand the regulation mechanism of plant growth rate by the content of functional elements (Sterner and Elser 2002). The temperature–plant physiology hypothesis typically focuses on the influence of temperature on the metabolic rate of plant growth with plants requiring high N and P contents to compensate for the decline in physiological efficiency under low-temperature conditions (Reich and Oleksyn 2004). The temperature–biogeochemistry hypothesis highlights that temperature influences soil nutrient availability by affecting all activities of soil microorganisms, thereby leading to variations in N and P contents of different plant organs (Reich and Oleksyn 2004). The soil substrate age hypothesis suggests that the age of soil development can influence the nutrient supply capacity of soil parent materials and thus the content of plant functional element (Reich and Oleksyn 2004). The stability of limiting elements hypothesis assumes that, owing to the restriction of plant physiology and nutrient balance, the limiting elements are more stable in their contents and responses to environmental changes (Han et al. 2011). These hypotheses attempted to explain the possible mechanisms driving the latitudinal variation of plant functional elements from the perspective of plant physiology, environmental factors, and nutrient requirements. By and large, they all have their own theoretical basis, and there are still significant uncertainties. Therefore, there is potential for the development of general hypotheses in the future using new data from multiple sources, advanced observation technologies, and analytical methods.

2.7.2 Factors Influencing the Spatial Variations in Fine Root Function Elements

The variation in the content of fine root functional elements is affected by vegetation type, RD, survival status, and environmental factors. It is common for N, P, and N:P ratio in fine roots to significantly differ among different vegetation types. For example, Yuan et al. (2011) observed that P content in fine roots increased from tropical forests, savannahs, temperate forests, and temperate grasslands to boreal forests, whereas N:P ratio decreased correspondingly. Furthermore, root N and P contents were affected by root size and survival status. With the increase in root diameter, C contents increased, while N and P contents decreased. N:P ratio was high in coarse roots with diameters >5 mm; however, no significant difference was observed in roots with diameters less than 5 mm, and C: N and C:P ratios increased with increasing root diameters. Compared to those in live roots of the same size, C and P contents were generally lower in dead roots, while overall N contents were similar, which may result in a higher N:P ratio in dead roots than in live roots. The C, N, and P contents and their ratios in fine roots were significantly regulated by temperature, precipitation, and the N and P contents of the soil. Fine root P content significantly decreased with the increase of MAT and MAP, while root N content did not show a significant trend (Yuan et al. 2011).

2.8 Challenges and Prospect

In recent years, scientists worldwide have conducted numerous research studies in the field of plant page: functional traits at the organ level and made substantial progress. However, most studies focus on the dominant species or simply collect species data from different regions. Although these data are suitable for the large-scale study of the relationships between plant functional traits and environmental factors, the complexity of natural vegetation including community structure and species composition has not been considered. Therefore, it represents a major challenge for scientists when using these data to study community composition and predict ecosystem functions on a large scale. Therefore, some aspects should be given priority in the future when investigating plant functional characteristics:

1. Synergistic or divergent patterns of multiple functional traits

The adaptation strategies of plant species to their environment must involve different organs of the plant. The study of plant functional traits cannot focus only on the responsive and adaptive strategy of a single organ to changing environmental conditions. A comprehensive approach to how functional traits of different organs act synergistically or are decoupled from each other should help us to better understand the adaptive strategies of plants to the environment and their functional optimization.

The current global database is dominated by these easily accessible functional traits referred to as soft traits. However, some important functional traits, known as hard traits, such as decomposition rate and root exudates, have not been included owing to the challenges associated with their measurement. Despite this, these traits are theoretically closely related to multiple functions. In the future, more hard traits related to plant metabolism and physiology need to be measured to compensate for the current deficits in large-scale functional trait data. Furthermore, studies on belowground roots continue to lag compared to those on leaves, primarily owing to challenges associated with sampling root systems and the lack of standard measurement methods. For example, in most studies to date, the suction roots or first-order roots are the main organs for water and nutrient absorption by plants. However, the absorbing roots or first-order roots are easily damaged or lost during field sampling, leading to high uncertainty in various studies. Therefore, future research should not only analyze the intraspecific, interspecific, and regional variation of different traits but also explore the synergistic variations of multiple traits among different organs (such as leaf, branch, stem, root, flower, and seed) and comprehensively analyze the responsive and adaptive mechanisms of plant species to changing environment.

The recently proposed "plant trait networks" should provide a novel perspective to reveal the response and adaptation strategies of plants to environmental or resource changes, from the perspectives of multidimensionality (He et al. 2020a, b, c; Li et al. 2022a, b, c, d).

2. Intraspecific variations in traits and its relationship to phylogenetics

The differences in functional traits among different plant species contribute to the coexistence of species in natural communities, and the variations in specific functional traits between individuals within a species can play an important role in species coexistence and the resilience of plant species to external dispersal. There is growing evidence that interspecific trait variation is relatively small but cannot be ignored. The responses of different plant species within specific communities to environmental changes and resource competition can only be better reflected if variation in intraspecific and interspecific traits is comprehensively considered. However, many studies on plant functional traits have paid little attention to within-species variation of individual traits because they are limited by field sampling and study scale. Meanwhile, the variability of plant functional traits is closely related to the phylogenetic background, as they are the result of a combination of genetic factors and environmental adaptations. Previous findings have demonstrated that most functional traits are phylogenetically conservative (Chen et al. 2013; Wang et al. 2017a, b, c), and their variability is primarily owing to the differences among species rather than the regulatory mechanisms under changing environment. Incorporating intraspecific variation and phylogenetic history into pertinent ecological models is essential for future research on plant functional traits and adaptation strategies.

3. Adaptive mechanisms of plant functional traits under extreme environment

Under extreme environments, such as the alpine grasslands of the Qinghai–Tibet Plateau and the desert areas of Xinjiang, where drought stress and disturbance coexist, plant species have their own unique strategy for survival in nature. In these regions, further emphasis should be placed on how different plant organs collectively adapt to the harsh environment and whether their variation and adaptation mechanisms differ at different levels. Studies on the spatial variation of plant functional traits under extreme environments will help us better understand the adaptive strategies and optimization mechanisms of plants under future global changes.

References

Ackerly DD, Knight CA, Weiss SB et al (2002) Leaf size, specific leaf area and microhabitat distribution of chaparral woody plants: contrasting patterns in species level and community level analyses. Oecologia 130:449–457

Ainsworth EA, Rogers AJ (2010) The response of photosynthesis and stomatal conductance to rising CO_2: mechanisms and environmental interactions. Plant Cell Environ 30:258–270

Bardgett RD, Mommer L, De Vries FT (2014) Going underground: root traits as drivers of ecosystem processes. Trends Ecol Evol 29:692–699

Beerling DJ, Franks PJ (2010) The hidden cost of transpiration. Nature 464:495–496

Bjorkman AD, Myers-Smith IH, Elmendorf SC et al (2018a) Plant functional trait change across a warming tundra biome. Nature 562:57

Bjorkman AD, Myers-Smith IH, Elmendorf SC et al (2018b) Tundra trait team: a database of plant traits spanning the tundra biome. Glob Ecol Biogeogr 27:1402–1411

Blonder B, Enquist BJ (2014) Inferring climate from angiosperm leaf venation networks. New Phytol 204:116–126

Blonder B, Salinas N, Bentley PL et al (2017) Predicting trait-environment relationships for venation networks along an Andes-Amazon elevation gradient. Ecology 98:1239–1255

Boyce CK, Brodribb TJ, Feild TS et al (2009) Angiosperm leaf vein evolution was physiologically and environmentally transformative. Proc R Soc B Biol Sci 276:1771–1776

Brodribb TJ, Field TS, Jordan GJ (2007) Leaf maximum photosynthetic rate and venation are linked by hydraulics. Plant Physiol 144:1890–1898

Brodribb TJ, Feild TS, Sack LJ (2010) Viewing leaf structure and evolution from a hydraulic perspective. Funct Plant Biol 37:488–498

Brundrett MC (2009) Mycorrhizal associations and other means of nutrition of vascular plants: understanding the global diversity of host plants by resolving conflicting information and developing reliable means of diagnosis. Plant Soil 320:37–77

Bucher SF, Auerswald K, Grün-Wenzel C et al (2017) Stomatal traits relate to habitat preferences of herbaceous species in a temperate climate. Flora 229:107–115

Chapin FSIII, Matson PA, Mooney HA (2002) Principles of terrestrial ecosystem ecology. Springer, New York

Chen FS, Zeng DH, Fahey TJ et al (2010) Response of leaf anatomy of *Chenopodium acuminatum* to soil resource availability in a semi-arid grassland. Plant Ecol 209:375–382

Chen WL, Zeng H, Eissenstat DM et al (2013) Variation of first-order root traits across climatic gradients and evolutionary trends in geological time. Glob Ecol Biogeogr 22:846–856

Comas LH, Eissenstat DM (2009) Patterns in root trait variation among 25 co-existing north American forest species. New Phytol 182:919–928

Comas LH, Callahan HS, Midford PE (2014) Patterns in root traits of woody species hosting arbuscular and ectomycorrhizas: implications for the evolution of belowground strategies. Ecol Evol 4:2979–2990

Cornelissen JHC, Lavorel S, Garnier E et al (2003) A handbook of protocols for standardised and easy measurement of plant functional traits worldwide. Aust J Bot 51:335–380

de Boer HJ, Price CA, Wagner-Cremer F et al (2016) Optimal allocation of leaf epidermal area for gas exchange. New Phytol 210:1219–1228

Feild TS, Brodribb TJ (2013) Hydraulic tuning of vein cell microstructure in the evolution of angiosperm venation networks. New Phytol 199:720–726

Fort F, Freschet GT (2020) Plant ecological indicator values as predictors of fine-root trait variations. J Ecol 108:1565–1577

Franks PJ, Beerling DJ (2009) Maximum leaf conductance driven by CO_2 effects on stomatal size and density over geologic time. Proc Natl Acad Sci 106:10343–10347

Franks PJ, Drake PL, Beerling DJ (2009) Plasticity in maximum stomatal conductance constrained by negative correlation between stomatal size and density: an analysis using Eucalyptus globulus. Plant Cell Environ 32:1737–1748

Franks PJ, Freckleton RP, Beaulieu JM et al (2012) Megacycles of atmospheric carbon dioxide concentration correlate with fossil plant genome size. Philos Trans R Soc B Biol Sci 367:556–564

Freschet GT, Dias ATC, Ackerly DD et al (2011) Global to community scale differences in the prevalence of convergent over divergent leaf trait distributions in plant assemblages. Glob Ecol Biogeogr 20:755–765

Freschet GT, Valverde-Barrantes OJ, Tucker CM et al (2017) Climate, soil and plant functional types as drivers of global fine-root trait variation. J Ecol 105:1182–1196

Freschet GT, Pagès L, Iversen CM et al (2021a) A starting guide to root ecology: strengthening ecological concepts and standardising root classification, sampling, processing and trait measurements. New Phytol 232:973–1122

Freschet GT, Roumet C, Comas LH et al (2021b) Root traits as drivers of plant and ecosystem functioning: current understanding, pitfalls and future research needs. New Phytol 232:1123–1158

Gazanchian A, Hajheidari M, Sima NK et al (2007) Proteome response of Elymus elongatum to severe water stress and recovery. J Exp Bot 58:291–300

Gower ST, Richards JH (1990) Larches: deciduous conifers in an evergreen world. Bioscience 40: 818–826

Han WX, Fang JY, Guo DL et al (2005) Leaf nitrogen and phosphorus stoichiometry across 753 terrestrial plant species in China. New Phytol 168:377–385

Han WX, Fang JY, Reich PB et al (2011) Biogeography and variability of eleven mineral elements in plant leaves across gradients of climate, soil and plant functional type in China. Ecol Lett 14: 788–796

Hao Z, Kuang Y, Kang M (2014) Untangling the influence of phylogeny, soil and climate on leaf element concentrations in a biodiversity hotspot. Funct Ecol 29:165–176

He JS, Fang JY, Wang ZH et al (2006) Stoichiometry and large-scale patterns of leaf carbon and nitrogen in the grassland biomes of China. Oecologia 149:115–122

He JS, Wang L, Flynn DFB et al (2008) Leaf nitrogen: phosphorus stoichiometry across Chinese grassland biomes. Oecologia 155:301–310

He JS, Wang XP, Schmid B et al (2010) Taxonomic identity, phylogeny, climate and soil fertility as drivers of leaf traits across Chinese grassland biomes. J Plant Res 123:551–561

He NP, Liu CC, Tian M et al (2018a) Variation in leaf anatomical traits from tropical to cold-temperate forests and linkage to ecosystem functions. Funct Ecol 32:10–19

He NP, Liu CC, Zhang JH et al (2018b) Opportunities and challenges for plant trait research: from organs to communities. Acta Ecologic Sinica 38:6787–6796

He NP, Zhang JH, Liu CC et al (2018c) Patterns and influencing factors of traits in forest ecosystems: synthesis and perspectives on the synthetic investigation from the north-east transect of eastern China (NETEC). Acta Ecol Sin 38(18):6359–6382

He NP, Liu CC, Piao SL et al (2019a) Ecosystem traits linking functional traits to macroecology. Trends Ecol Evol 34:200–210

He L, Chen JM, Liu J et al (2019b) Diverse photosynthetic capacity of global ecosystems mapped by satellite chlorophyll fluorescence measurements. Remote Sens Environ 232:111344

He NP, Li Y, Liu CC et al (2020a) Plant trait networks: improved resolution of the dimensionality of adaptation. Trends Ecol Evol 35:908–918

He NP, Liu CC, Xu L et al (2020b) Ecosystem traits open new insights into macroecology: opportunity and challenge. Acta Ecol Sin 40(8):2507–2522

He P, Gleason SM, Wright IJ et al (2020c) Growing-season temperature and precipitation are independent drivers of global variation in xylem hydraulic conductivity. Glob Chang Biol 26: 1833–1841

Hedin LO, Vitousek PM, Matson PA (2003) Nutrient losses over four million years of tropical forest development. Ecology 84:2231–2255

Hertel D, Strecker T, Muller-Haubold H et al (2013) Fine root biomass and dynamics in beech forests across a precipitation gradient – is optimal resource partitioning theory applicable to water-limited mature trees? J Ecol 101:1183–1200

Hetherington AM, Woodward FI (2003) The role of stomata in sensing and driving environmental change. Nature 424:901–908

Higgins PAT, Jackson RB, Des Rosiers JM et al (2002) Root production and demography in a California annual grassland under elevated atmospheric carbon dioxide. Glob Chang Biol 8: 841–850

Hoch G, Körner C (2012) Global patterns of mobile carbon stores in trees at the high-elevation tree line. Glob Ecol Biogeogr 21:861–871

Högberg P, Nasholm T, Franklin O et al (2017) Tamm review: on the nature of the nitrogen limitation to plant growth in Fennoscandian boreal forests. For Ecol Manag 403:161–185

Holdaway RJ, Richardson SJ, Dickie IA et al (2011) Species- and community-level patterns in fine root traits along a 120000-year soil chronosequence in temperate rain forest. J Ecol 99:954–963

Hultine KR, Marshall JD (2000) Altitude trends in conifer leaf morphology and stable carbon isotope composition. Oecologia 123:32–40

IPCC (2021) Climate change 2021: the physical science basis. Contribution of working group I to the sixth assessment report of the intergovernmental panel on climate change. Cambridge Press, Cambridge

Iversen CM, Ledford J, Norby RJ (2008) CO_2 enrichment increases carbon and nitrogen input from fine roots in a deciduous forest. New Phytol 179:837–847

Jankowski A, Wyka TP, Zytkowiak R et al (2017) Cold adaptation drives variability in needle structure and anatomy in Pinus sylvestris L. along a 1900 km temperate-boreal transect. Funct Ecol 31:2212–2223

Joos F, Gerber S, Prentice I et al (2004) Transient simulations of Holocene atmospheric carbon dioxide and terrestrial carbon since the last glacial maximum. Glob Biogeochem Cycles 18: GB2002

Koerselman W, Meuleman AF (1996) The vegetation N: P ratio: a new tool to detect the nature of nutrient limitation. J Appl Ecol 33:1441–1450

Köhler L, Gieger T, Leuschner C (2006) Altitudinal change in soil and foliar nutrient concentrations and in microclimate across the tree line on the subtropical Island mountain Mt. Teide (Canary Islands). Flora 201:202–214

Kong D, Ma C, Zhang Q et al (2014) Leading dimensions in absorptive root trait variation across 96 subtropical forest species. New Phytol 203:863–872

Körner C (2007) The use of 'altitude' in ecological research. Trends Ecol Evol 22:569–574

Kramer-Walter KR, Bellingham PJ, Millar TR et al (2016) Root traits are multidimensional: specific root length is independent from root tissue density and the plant economic spectrum. J Ecol 104:1299–1310

Kuster VC, de Castro SAB, Vale FHA (2016) Photosynthetic and anatomical responses of three plant species at two altitudinal levels in the Neotropical savannah. Aust J Bot 64:696–703

Lambers H, Raven JA, Shaver GR et al (2008) Plant nutrient-acquisition strategies change with soil age. Trends Ecol Evol 23:95–103

Li FL, Bao WK (2014) Elevational trends in leaf size of Campylotropis polyantha in the arid Minjiang River valley, SW China. J Arid Environ 108:1–9

Li L, McCormack ML, Ma C et al (2015) Leaf economics and hydraulic traits are decoupled in five species-rich tropical-subtropical forests. Ecol Lett 18:899–906

Li YQ, Reich PB, Schmid B et al (2020) Leaf size of woody dicots predicts ecosystem primary productivity. Ecol Lett 23:1003–1013

Li X, He N, Xu L et al (2021a) Spatial variation in leaf potassium concentrations and its role in plant adaptation strategies. Ecol Indic 130:108063

Li Y, Li Q, Xu L et al (2021b) Plant community traits can explain variation in productivity of selective logging forests after different restoration times. Ecol Indic 131:108181

Li Y, Liu CC, Xu L et al (2021c) Leaf trait networks based on global data: representing variation and adaptation in plants. Front Plant Sci 12:710530

Li X, Li MX, Xu L et al (2022a) Allometry and distribution of nitrogen in natural plant communities of the Tibetan plateau. Front Plant Sci 13:845813

Li Y, Hou J, Xu L et al (2022b) Variation in functional trait diversity from tropical to cold-temperate forests and linkage to productivity. Ecol Indic 138:108864

Li Y, Liu CC, Sack L et al (2022c) Leaf trait network architecture shifts with species-richness and climate across forests at continental scale. Ecol Lett 25:1442–1457

Li J, Chen X, Niklas KJ et al (2022d) A whole-plant economics spectrum including bark functional traits for 59 subtropical woody plant species. J Ecol 110:248–261

Liu CC, Li Y, Yan P et al (2021a) How to improve the predictions of plant functional traits on ecosystem functioning? Front Plant Sci 12:622260

Liu XR, Chen HX, Sun TY et al (2021b) Variation in woody leaf anatomical traits along the altitudinal gradient in Taibai Mountain, China. Glob Ecol Conserv 26:e01523

Liu C, Li Y, Xu L et al (2021c) Stomatal arrangement pattern: a new direction to explore plant adaptation and evolution. Front Plant Sci 12:655255

Liu R, Yang X, Gao R et al (2021d) Allometry rather than abiotic drivers explains biomass allocation among leaves, stems and roots of Artemisia across a large environmental gradient in China. J Ecol 109:1026–1040

Luo TX, Luo J, Pan Y (2005) Leaf traits and associated ecosystem characteristics across subtropical and timberline forests in the Gongga Mountains, eastern Tibetan plateau. Oecologia 142:261–273

Ma ZQ, Guo DL, Xu XL et al (2018) Evolutionary history resolves global organization of root functional traits. Nature 555:94–97

Ma HY, Lv XX, Ji YN et al (2020) Leaf anatomical structure and drought resistance of 17 *Caragana* species. Res Soil Water Conserv 138(1):346–352

Marschner H (2011) Marschner's mineral nutrition of higher plants. Academic Press

Mathur S, Agrawal D, Jajoo A (2014) Photosynthesis: response to high temperature stress. J Photochem Photobiol B Biol 137:116–126

McCormack ML, Dickie IA, Eissenstat DM et al (2015) Redefining fine roots improves understanding of below-ground contributions to terrestrial biosphere processes. New Phytol 207:505–518

McGroddy ME, Daufresne T, Hedin LO (2004) Scaling of C: N: P stoichiometry in forests worldwide: implications of terrestrial Redfield-type ratios. Ecology 85:2390–2401

Metcalfe DB, Meir P, Aragão LEO et al (2008) The effects of water availability on root growth and morphology in an Amazon rainforest. Plant Soil 311:189–199

Mládková P, Mládek J, Hejduk S et al (2018) Calcium plus magnesium indicates digestibility: the significance of the second major axis of plant chemical variation for ecological processes. Ecol Lett 21:885–895

Moles AT, Perkins SE, Laffan SW et al (2014) Which is a better predictor of plant traits: temperature or precipitation? J Veg Sci 25:1167–1180

Nie M, Lu M, Bell J et al (2013) Altered root traits due to elevated CO_2: a meta-analysis. Glob Ecol Biogeogr 22:1095–1105

Norby RJ, Ledford J, Reilly CD et al (2004) Fine-root production dominates response of a deciduous forest to atmospheric CO_2 enrichment. Proc Natl Acad Sci USA 101:9689–9693

Onoda Y, Westoby M, Adler PB et al (2011) Global patterns of leaf mechanical properties. Ecol Lett 14:301–312

Ordoñez JC, van Bodegom PM, Witte JPM et al (2009) A global study of relationships between leaf traits, climate and soil measures of nutrient fertility. Glob Ecol Biogeogr 18:137–149

Ostonen I, Puttsepp U, Biel C et al (2007) Specific root length as an indicator of environmental change. Plant Biosyst 141:426–442

Ostonen I, Helmisaari HS, Borken W et al (2011) Fine root foraging strategies in Norway spruce forests across a European climate gradient. Glob Chang Biol 17:3620–3632

Peñuelas J, Sardans J, Llusia J et al (2010) Faster returns on 'leaf economics' and different biogeochemical niche in invasive compared with native plant species. Glob Chang Biol 16: 2171–2185

Poorter H, Niinemets U, Poorter L et al (2009) Causes and consequences of variation in leaf mass per area (LMA): a meta-analysis. New Phytol 182:565–588

Poorter H, Lambers H, Evans JR (2014) Trait correlation networks: a whole-plant perspective on the recently criticized leaf economic spectrum. New Phytol 201:378–382

Pregitzer KS, DeForest JL, Burton AJ et al (2002) Fine root architecture of nine north American trees. Ecol Monogr 72:293–309

Reich PB (2005) Global biogeography of plant chemistry: filling in the blanks. New Phytol 168: 263–266

Reich PB (2014) The world-wide 'fast-slow' plant economics spectrum: a traits manifesto. J Ecol 102:275–301

Reich PB, Oleksyn J (2004) Global patterns of plant leaf N and P in relation to temperature and latitude. Proc Natl Acad Sci USA 101:11001–11006

Reich PB, Ellsworth DS, Walter MBS et al (1999) Generality of leaf trait relationships: a test across six biomes. Ecology 80:1955–1969

Reich PB, Wright IJ, Lusk CH (2007) Predicting leaf physiology from simple plant and climate attributes: a global GLOPNET analysis. Ecol Appl 17:1982–1988

Reich PB, Oleksyn J (2004) Global patterns of plant leaf N and P in relation to temperature and latitude. Proc Natl Acad Sci USA 101:11001–11006

Reichstein M, Bahn M, Mahecha MD et al (2014) Linking plant and ecosystem functional biogeography. Proc Natl Acad Sci USA 111:13697–13702

Ren SJ (2007) Spacial pattern of leaf ecological stoichiometry across forest ecosystem in NSTEC. University of Chinese Academy of Sciences, Beijing

Royer DL, McElwain JC, Adams JM, Wilf P (2008) Sensitivity of leaf size and shape to climate within Acer rubrum and Quercus kelloggii. New Phytol 179:808–817

Sack L, Frole KJE (2006) Leaf structural diversity is related to hydraulic capacity in tropical rain forest trees. Ecology 87:483–491

Sack L, Scoffoni CJ (2013) Leaf venation: structure, function, development, evolution, ecology and applications in the past, present and future. New Phytol 198:983–1000

Sack L, Scoffoni C, McKown AD et al (2012) Developmentally based scaling of leaf venation architecture explains global ecological patterns. Nat Commun 3:1–10

Sack L, Scoffoni C, John GP et al (2013) How do leaf veins influence the worldwide leaf economic spectrum? Review and synthesis. Ecology 4:4053–4080

Sardans J, Peñuelas J (2014) Climate and taxonomy underlie different elemental concentrations and stoichiometries of forest species: the optimum "biogeochemical niche". Plant Ecol 215:441–455

Saura-Mas S, Lloret F (2007) Leaf and shoot water content and leaf dry matter content of Mediterranean woody species with different post-fire regenerative strategies. Ann Bot 99: 545–554

Schemske DW, Mittelbach GG, Cornell HV et al (2009) Is there a latitudinal gradient in the importance of biotic interactions? Ann Rev Ecol Evol Syst 40:245–269
Steidinger BS, Crowther TW, Liang J et al (2019) Climatic controls of decomposition drive the global biogeography of forest-tree symbioses. Nature 569:404
Sterck FJ, Bongers F (2001) Crown development in tropical rain forest trees: patterns with tree height and light availability. J Ecol 89:1–13
Sterner RW, Elser JJ (2002) Ecological stoichiometry: the biology of elements from molecules to the biosphere. Princeton University Press, Princeton
Sun JG, Liu CC, Hou JH et al (2021) Spatial variation of stomatal morphological traits in grassland plants of the loess plateau. Ecol Indic 128:107857
Terashima I, Hanba YT, Tholen D et al (2011) Leaf functional anatomy in relation to photosynthesis. Plant Physiol 155:108–116
Tian M, Yu GR, He NP et al (2016) Leaf morphological and anatomical traits from tropical to temperate coniferous forests: mechanisms and influencing factors. Sci Rep 6:19703
Tomlinson KW, Poorter L, Sterck FJ et al (2013) Leaf adaptations of evergreen and deciduous trees of semi-arid and humid savannas on three continents. J Ecol 101:430–440
Uhl D, Mosbrugger VJP (1999) Leaf venation density as a climate and environmental proxy: a critical review and new data. Palaeoclimatol Palaeoecol 149:15–26
van Bodegom PM, Douma JC, Verheijen LM (2014) A fully traits-based approach to modeling global vegetation distribution. Proc Natl Acad Sci USA 111:13733–13738
Violle C, Navas ML, Vile D et al (2007) Let the concept of trait be functional! Oikos 116:882–892
Violle C, Reich PB, Pacala SW et al (2014) The emergence and promise of functional biogeography. Proc Natl Acad Sci USA 111:13690–13696
Vitousek PM, Porder S, Houlton BZ et al (2010) Terrestrial phosphorus limitation: mechanisms, implications, and nitrogen-phosphorus interactions. Ecol Appl 2010(20):5–15
Wan SQ, Norby RJ, Pregitzer KS et al (2004) CO_2 enrichment and warming of the atmosphere enhance both productivity and mortality of maple tree fine roots. New Phytol 162:437–446
Wang RZ, Gao Q (2003) Climate-driven changes in shoot density and shoot biomass in Leymus chinensis (Poaceae) on the north-East China transect (NECT). Glob Ecol Biogeogr 12:249–259
Wang B, Qiu YL (2006) Phylogenetic distribution and evolution of mycorrhizas in land plants. Mycorrhiza 16:299–363
Wang ZQ, Guo DL, Wang XR et al (2006) Fine root architecture, morphology, and biomass of different branch orders of two Chinese temperate tree species. Plant Soil 288:155–171
Wang RL, Yu GR, He NP et al (2015) Latitudinal variation of leaf stomatal traits from species to community level in forests: linkage with ecosystem productivity. Sci Rep 5:1–11
Wang RL, Yu GR, He NP et al (2016) Latitudinal variation of leaf morphological traits from species to communities along a forest transect in eastern China. J Geogr Sci 26:15–26
Wang RL, Wang QF, Zhao N et al (2017a) Complex trait relationships between leaves and absorptive roots: coordination in tissue N concentration but divergence in morphology. Ecol Evol 7:2697–2705
Wang R, Gibson CD, Berry TD et al (2017b) Photooxidation of pyrogenic organic matter reduces its reactive, labile C pool and the apparent soil oxidative microbial enzyme response. Geoderma 293:10–18
Wang H, Prentice IC, Davis TW et al (2017c) Photosynthetic responses to altitude: an explanation based on optimality principles. New Phytol 213:976–982
Wang R, Chen H, Liu X et al (2020a) Plant phylogeny and growth form as drivers of the altitudinal variation in woody leaf vein traits. Front Plant Sci 10:1735
Wang P, Huang KL, Hu SJ (2020b) Distinct fine-root responses to precipitation changes in herbaceous and woody plants: a meta-analysis. New Phytol 225:1491–1499

Wen JW, Chen HX, Teng YP et al (2018) Variation of leaf stomatal traits in *Quercus* species along the altitudinal gradient in Taibai Mountain, China. Acta Ecol Sin 38(18):6712–6721

West JB, Espeleta JF, Donovan LA (2004) Fine root production and turnover across a complex edaphic gradient of a Pinus palustris-Aristida stricta savanna ecosystem. For Ecol Manag 189: 397–406

White PJ, Broadley MR (2003) Calcium in plants. Ann Bot 92:487–511

White PJ, Bowen HC, Marshall B et al (2007) Extraordinarily high leaf selenium to sulfur ratios define 'se-accumulator'plants. Ann Bot 100:111–118

Wright IJ, Reich PB, Westoby M (2003) Least-cost input mixtures of water and nitrogen for photosynthesis. Am Nat 161:98–111

Wright IJ, Reich PB, Westoby M et al (2004a) The worldwide leaf economics spectrum. Nature 428:821–827

Wright IJ, Groom PK, Lamont BB et al (2004b) Leaf trait relationships in Australian plant species. Funct Plant Biol 31:551–558

Xu Z, Zhou GJ (2008) Responses of leaf stomatal density to water status and its relationship with photosynthesis in a grass. J Exp Bot 59:3317–3325

Yan W, Zhong Y (2017) Contrasting responses of leaf stomatal characteristics to climate change: a considerable challenge to predict carbon and water cycles. Glob Chang Biol 23:3781–3793

Yan ER, Wang XH, Guo M et al (2010) C:N:P stoichiometry across evergreen broad-leaved forests, evergreen coniferous forests and deciduous broad-leaved forests in the Tiantong region, Zhejiang Province, eastern China. Chin J Plant Ecol 34(1):48–57

Yuan ZY, Chen HY, Reich PB (2011) Global-scale latitudinal patterns of plant fine-root nitrogen and phosphorus. Nat Commun 2:1–6

Zhang LX, Bai YF, Han XG (2003) Application of N: P stoichiometry to ecology studies. Acta Bot Sin 45:1009–1018

Zhang SB, Zhang JL, Slik J et al (2012) Leaf element concentrations of terrestrial plants across China are influenced by taxonomy and the environment. Glob Ecol Biogeogr 21:809–818

Zhang Y, Yang S, Sun M et al (2014) Stomatal traits are evolutionarily associated with vein density in basal angiosperms. Plant Sci J 32:320–328

Zhang JH, Hedin LO, Li MX et al (2022a) Leaf N:P ratio does not predict productivity trends across natural terrestrial ecosystems. Ecology 103:e3789

Zhang M, Gao HR, Mo WY et al (2022b) Variation in leaf vein traits along environmental gradient in loess plateau grasslands. Acta Ecol Sin 42(19):8082–8093

Zhao N, Yu GR, He NP et al (2016a) Coordinated pattern of multi-element variability in leaves and roots across Chinese forest biomes. Glob Ecol Biogeogr 25:359–367

Zhao WL, Chen YJ, Brodribb TJ et al (2016b) Weak co-ordination between vein and stomatal densities in 105 angiosperm tree species along altitudinal gradients in Southwest China. Funct Plant Biol 43:1126–1133

Zhao WZ, Xiao CW, Li MX et al (2022a) Variation and adaptation in leaf sulfur content across China. J Plant Ecol 15:743–755

Zhao WZ, Xiao CW, Li MX et al (2022b) Spatial variation and allocation of sulfur among major plant organs in China. Sci Total Environ 844:157155

Zhu YH (2013) Patterns of variations in leaf trait and responses to environmental changes in oriental oak (*Quercus variabilis*) across eastern Asia. Shanghai Jiao Tong University, Shanghai

Zuo WY, He JS, Han M et al (2005) Responses of plant stomata to elevated CO_2 and temperature: observations from 10 plant species grown in temperature and CO_2 gradients. Acta Ecol Sin 25: 565–574

Chapter 3
Coevolution and Optimization of Functional Traits Among Different Plant Organs

Plant organs, such as leaves, branches, trunks, and roots, are believed to have evolved in a coordinated manner over the past billion years as part of an optimization system that supports their various functions of growth, development, and reproduction. In recent decades, there has been significant interest in the functional traits of different plant organs owing to their vital role in plant's adaptation to the environment. While different plant organs are structurally connected, they exhibit significant differences in function, necessitating their collaborative or coordinated adaptation to complex natural environments. Most studies have been conducted separately on leaves, branches, trunks, or roots, aiming to investigate the spatial variation and influencing factors of functional traits at different scales. Furthermore, scientists have attempted to elucidate the quantitative relationships between functional traits of specific plant organs (leaves or roots) and the structure and function of ecosystems. However, how multiple plant organs coordinate to adapt to the external environment on a large scale remains unclear, which limits our understanding of plant adaptation strategies and optimization of functions. To explain the variations in functional traits among various organs, we developed two coexisting and complementary hypotheses, convergent evolution hypothesis and divergent evolution hypothesis, by considering the structural connections and diverse functions of various plant organs. Functional traits, that can "flow" as elements or other chemical substances among different organs, converge in content because of the unique functions different plant organs serve. However, these morphological traits tend to be divergent among different organs to facilitate adaptation, optimize life history, and improve plant fitness. Herein, we further discuss the practical implications of the two hypotheses by examining the variations and influencing factors of multiple functional elements in leaves and roots, as well as the variations and evolutionary history of the C:N ratio of leaves, branches, trunks, and roots, on a large scale. This empirical evidence demonstrates significant differences in the contents of multiple elements between leaves and roots, although their spatial variations are similar. The C:N ratio significantly differs between leaves, branches, trunks, and roots; however, the directions of allometry are convergent. Few studies have examined the covariation of

N. He et al., *Plant Functional Ecology*,
https://doi.org/10.1007/978-981-96-8264-5_3

functional traits among multiple organs and species on a large scale, although it is important for the adaptation of plants to complex environmental conditions in nature. This area of research is promising and should be emphasized in the future, particularly in view of the global change scenarios.

Plants gradually adapt to environmental changes during long-term evolution (Morris et al. 2018). Approximately 3.8 billion years ago, inorganic substances in nature began to form organic matter, and approximately 3.5 billion years ago, cyanobacteria emerged, demonstrating the capability of photosynthesis (Pech et al. 2011). To date, plant communities on land have evolved to be dominated by angiosperms with different organs, such as leaves, branches, trunks, roots, flowers, fruits, and seeds. In brief, the evolutionary history of plants can be categorized into four distinct stages: the thallophyta age, fern age, gymnosperm age, and angiosperm age. In the thallophyta age, autotrophs, such as cyanobacteria, dominated the environment, and they were able to capture the energy of sunlight through photosynthesis to store organic matter and release oxygen, laying important biochemical and climatic foundations for the evolution of plants. Approximately 400–500 million years ago, fungi (or bacteria) and algae became the pioneers of terrestrial ecosystems (Cheng et al. 2019). They interacted with each other and gradually changed the microhabitat, which contributed to the evolution and development of psilophytons and mosses. Unlike plants that were prostrate on the ground, the majority of plants that survived on land in the early stages were small and laid on the ground. These plants with upward tips could grow fast to some extent owing to increased sunlight exposure and efficient gas exchange. These plants evolved into psilophytons, characterized by distinct, upright, dichotomous stems and lack of leaves and true roots (Kenrick and Crane 1997). However, plants with fast growth should have to suffer from insufficient water supply at the top. Therefore, plants of medium height could not only maintain adequate photosynthesis but also reduce the cost of transporting water over long distances, ultimately evolve into bryophytes. Over 30 million years of evolution, plants developed apical growths, vascular tissues, epidermal structures, true roots, and more. Some ferns emerged with giant leaves, consistent with the age of fern-dominated plant communities on land (Willis and McElwain 2014). Approximately 370 million years ago, early gymnosperms emerged and gradually evolved into seed organs. This marked the beginning of the age of seed propagation. Approximately 130 million years ago, angiosperms emerged and evolved into flowers and fruits. Since then, plant structures have undergone significant improvements, becoming highly complex, and the plants' ability to adapt to land has rapidly enhanced. Compared to these gymnosperms, angiosperms have true flowers and develop seeds after fertilization, enabling them to effectively resist environmental stress and facilitate the spread and reproduction of plants, thereby establishing their dominance on land as observed today (Evert and Eichhorn 2013).

The evolutionary history and development of different plant organs may reflect the adaptation strategies of plants to specific environments or the optimization of plant functions (Fig. 3.1). Considering fine roots as an example, the diameter of fine roots has become thin over long-term evolutionary history (Chen et al. 2013), and their dependence on mycorrhiza has also become weak. All these developments

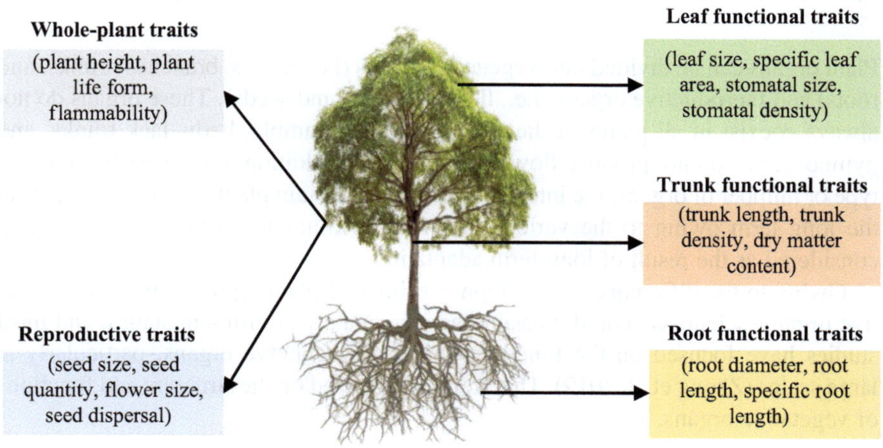

Fig. 3.1 Different plant organs and their functional traits

allow plants to absorb more water and nutrients from soils with less carbon investment (Ma et al. 2018). Similarly, plant stomata have become smaller and denser throughout evolutionary history, which could improve the ability to rapidly regulate carbon and water exchange in arid or semiarid environments, sustain photosynthesis, and maintain water and heat balance in leaves (Liu et al. 2022a, b, c). Moreover, C:N ratios of leaves and branches are expected to decrease over the course of evolution, indicating that plants have a high efficiency of nutrient utilization in the early stages of development when nutrient limitation is more significant. Evolution also enables plants to rapidly accumulate organic matter, enhancing their competitiveness after nutrient limitation is alleviated (Zhang et al. 2020a).

Different plant organs are related to each other and should work together to fulfill the overall functions of the plants, each responsible for unique and essential functions (Fig. 3.1). Notably, these organs exhibit paradoxical structural connectivity and functional diversity, allowing for survival in complex environments. Numerous studies have demonstrated that the spatial variations in functional traits influence specific plant organs at local, regional, national, and even global scales. Researchers have recently began focusing on the covariation of functional traits of different plant organs (Wang et al. 2014; Yan et al. 2016). Overall, most relevant research is limited to the site/local scale, with few studies demonstrating large-scale coevolution of different organs among different species.

3.1 Structure and Functions of Different Plant Organs

Plant organs can be divided into vegetative organs (i.e., leaves, branches, trunks, and roots) and reproductive organs (i.e., flowers, fruits, and seeds). These organs do not always coexist in all plants at the same time; for example, herbs lack trunks, and gymnosperms do not produce flowers and fruits. In addition to the variability in the type or number of organs, the internal structures of certain plant organs are diverse in the long term owing to the various external environments, which are commonly considered as the result of long-term adaptation.

Owing to the difference in development times of plant reproductive and vegetative organs, scientists should measure them separately at different stages, and most studies have focused on the functional traits of vegetative organs, particularly at large scales (Zhang et al. 2019). Therefore, we focused on the structure and functions of vegetative organs.

3.1.1 Structure and Functions of Leaf

The structure and functions of the leaves are closely related. Generally, leaves have similar structures, such as chloroplast and stomata, which serve diverse functions, such as photosynthesis, transpiration, even absorption, and reproduction in a limited number of plant species (Fig. 3.2). The relatively stable physical structure and chemical composition of leaves are essential to support the diverse functions with

Fig. 3.2 Complexity of leaf structure and multiple functional traits

complex physiological and biochemical reactions. In practice, scientists have attempted to explore these properties using leaf functional traits. For example, as a portal for the exchange of carbon dioxide and water between leaves and the external environment, stomatal morphological traits, excluding the open–close activity of stomata, have been widely used to study the long-term relationships between photosynthesis and transpiration.

Previous studies have demonstrated that plants can trade-off between photosynthesis and transpiration through long-term morphological evolution of stomata size and density and short-term opening and closing of stomata (Liu et al. 2018a, b, 2020). Palisade tissue and spongy tissue of leaves also play an important role in regulating plant growth and development (He et al. 2018a, b, c). Palisade tissue is found on the upper surface of leaves, because it is rich in chlorophyll, and is therefore closely linked to photosynthesis. Spongy tissue is located on the lower surface of leaves, which is sparsely arranged to cooperate with the stomata to regulate transpiration and gas exchange. Furthermore, scientists have revealed that various functional elements, such as nitrogen and phosphorus (Zhang et al. 2018a, b), chlorophyll (Li et al. 2018a, b; Zhang et al. 2020a, b), and nonstructural carbohydrates (Li et al. 2016), are essential participants in leaf structure and photosynthesis. Therefore, variations in leaf functional traits may reflect plant adaptive strategies and functional optimization from different perspectives (Sterner and Elser 2002; Wright et al. 2004a, b; Reich 2014). To date, the relationships between structure and functions of plant leaves and the underlying mechanism on a large scale still pose major challenges for us.

3.1.2 Structure and Functions of Trunks

The main structural components that support the entire aboveground portion of the plants are the trunks, including the branches in this book. They provide the necessary space for the growth and development of leaves, flowers, and fruits. Notably, the transport tissues consist of vessels and sieve tubes in trunks or stems and help transport water and nutrients from the roots upward to the leaves, while facilitating downward movement of photosynthates to the roots. Moreover, trunks, as important organs for storing energy and nutrients (Xu et al. 2020), play a crucial role in plant adaptation strategies, particularly nutrient uptake and absorption. Trunks can provide information about plant strategies by revealing variations in their morphological and chemical characteristics; however, this has not been emphasized in earlier research. For example, the thorns in trunks can protect plants from invasion by animals and pathogens, and succulent stems can retain more nutrients and water to adapt to extreme drought. Compared to those of herbaceous plants, the trunks or stems of woody plants are rich in lignin, which is conducive to supporting their huge aboveground parts. The various functional characteristics found in trunks or stems are the outcome of the long-term environmental adaptation of plants.

3.1.3 Structure and Functions of Roots

As one of the vegetative organs of seed plants, plant roots generally grow in natural soils. The development of roots marks a significant turning point in the ability of plants to anchor themselves and adapt to the generally arid environment. The thick roots, with a diameter of more than 2 mm, are primarily responsible for plant fixation and organic matter storage. Because the fine roots are less than 2 mm in diameter, they have high efficiency in absorbing water and nutrients from the soil to support the growth and development of plants and transporting them upward through vessels to the leaves (Bardgett et al. 2014). Therefore, in recent decades, there has been a significant focus on the development of roots and their spatial variation across various spatiotemporal scales. Researchers have developed a root-order-based method to classify the roots into five orders and have observed significant differences in the morphological traits, elemental content, and anatomical properties among these different root orders (Pregitzer et al. 2002). The root-order-based method is widely used to study the structure and function of fine roots at different scales (Guo et al. 2008). Accumulating evidence shows that morphological and chemical traits of roots are of great importance for the elucidation of belowground ecological processes and plant–soil interactions (Wang et al. 2018b). In general, morphological traits, such as root length, root diameter, and specific root length, as well as chemical traits of elemental content, root secretions, and mycorrhizal colonization, are important and need to be emphasized in the future.

3.2 Functional Trait Relationships Among Different Plant Organs

Functional traits of plants are relatively stable and measurable characteristics that reflect the long-term adaptation of plants to the environment and may affect the growth, development, and fitness of plants to some extent (Violle et al. 2007; He et al. 2019a, b). Therefore, plant functional traits help us to resolve some important issues in ecology, such as plant evolution, adaptation strategies, and the optimization of ecosystem productivity (Liu et al. 2021a, b, c). Theoretically, plant functional traits are indispensable as input parameters for most ecological process models in the future. In the global carbon cycle model, plant functional traits are the key parameters for exploring the dynamics of ecosystem productivity, as well as important references for classifying the global land surface as plant functional groups (Kucharik et al. 2000; Verburg and Johnson 2001). Plant functional traits can be divided into different groups to address specific ecological questions. Taking leaves as an example, functional traits can be defined as morphological traits (e.g., leaf size, leaf thickness, and specific leaf area), chemical traits (e.g., elemental content, chlorophyll content, and nonstructural carbohydrate content), and physiological traits (e.g., photosynthetic rate and respiration rate). Variations in plant functional traits may reflect plant evolution and ecosystem functions. It is common for leaf size to regulate leaf temperature, increase the photosynthetic rate, and determine the total leaf area of the community; therefore, leaf size serves as an effective indicator for predicting community productivity (Li et al. 2020). Chlorophyll, being a crucial component in photosynthesis, plays a key role in the capture of light. Therefore, chlorophyll content can serve as an indicator of the photosynthetic potential of plant communities (Croft et al. 2017). The long-term change in the symbiotic relationships between plant fine roots and mycorrhiza can largely explain the evolutionary trend of plants. As a global-scale meta-analysis, fine plant roots evolve to become thinner, more independent, and efficient, which is conducive to the spread of plants into historically more arid habitats (Ma et al. 2018).

In individual organs, such as leaves or roots, there are trade-offs between multiple functional traits. Wright et al. (2004a, b) proposed the concept of leaf economics spectrum (LES). Under LES, different functional traits are placed at different positions along the resource axes, leading to the classification of plants into resource acquisition and resource conservation groups; the former is characterized by rapid turnover and the later by slow turnover. Chave et al. (2009) further developed the wood economics spectrum (WES), which correlates wood density with plant growth and survival. Roumet et al. (2016) developed the root economics spectrum (RES) as a similar polarization of plant species along the resource axis. In general, the functional traits of different plant organs can reflect plant evolution and functional optimization from different perspectives. However, to date, most relevant studies have focused on a single organ and particularly investigated the intraspecific and interspecific variations and influencing factors to reveal plant response and adaptation mechanisms. Undeniably, these studies have significantly contributed to the long-term improvement of public understanding of plant evolution and their adaptation strategies to the environment. However, as previously explained, different plant organs are interconnected and cooperate, highlighting the importance of trade-offs between different functional traits of these organs to jointly determine optimal strategies during evolution and adaptation under long-term natural selection (Eviner and Chapin 2003; Tjoelker et al. 2005; He et al. 2020a, b, c). Overall, conducting research on the relationship between the structural characteristics of different plant organs and their functional traits is essential and promising.

3.3 Hypotheses on the Relationships of Functional Traits Among Different Plant Organs

Studies on the relationships of functional traits among different plant organs can be traced back to the Corner's rule proposed in 1949 (Corner 1949), which first demonstrated the quantitative relationships between leaves and branches in structure and quantity. This study has revealed that branch diameter is positively correlated with the total area of leaves growing on that branch. In the following 60 years, few studies have attempted to literally link plant functional traits among different organs. Therefore, it is still unclear whether plant functional traits are convergent or divergent between different organs and are allometric or isometric, particularly at large scales. Another limitation is the lack of a systematic and matched database of diverse functional traits, essential for any co-analysis. Additionally, the long-standing disregard of the significance of co-analyses among various plant organs compounds this issue, as these databases require significant material and human resources. Until recently, some researchers have established a database that includes consistent functional traits of leaves, branches, stems, and roots in over 100 terrestrial ecosystems in China, such as forests, shrublands, grasslands, and deserts. Using this database, scientists have analyzed the evolution of functional traits among different plant organs and proposed two seemingly paradoxical but complementary hypotheses: the convergent evolution hypothesis and divergent evolution hypothesis (Wang et al. 2018a; Zhang et al. 2020b).

3.3.1 Convergent Evolution Hypothesis

From a one-dimensional perspective, different plant organs are physically connected to each other. Vessels and sieve tubes are responsible for transferring water, mineral nutrients, and photosynthates among different plant organs. As the total amount of these transferable substances increases, the allocation to the various organs should increase allometrically. This biophysical property of plants may result in the covariation of different organs, supporting the convergent evolution hypothesis as shown in Fig. 3.3. Some previous studies have indirectly confirmed this hypothesis, and the Corner's rule serves as an illustrative example. The size of leaves is limited by the size of the branches to which they are attached (Olson et al. 2009), and larger leaves must be supported by larger branches (Corner 1949). Physiologically, the size ratio between leaves and branches is an inevitable result of water and nutrient requirements of plants (Gartner 1991; Niklas 1994; Brouat and McKey 2001). In addition, morphological traits of different plant organs, including leaf thickness, tissue density, root diameter, and seed mass, are correlated with each other (de la Riva et al. 2016), indicating the coupling, mutual influence, and convergent changes of different plant organs. Plant resource utilization strategies are also similar among different organs, as evidenced by the significant positive correlations between nutrient content, dry matter content, and lignin among different plant organs (Freschet et al. 2010). The examples provided above have largely confirmed the coupling relationships between leaves, branches, trunks, and roots in the cycle of substances and energy flow.

Recently, systematic analyses of many plant species at the regional scale have further demonstrated that the spatial pattern and covariation of plant functional elements among different organs, owing to the physical connectivity, are also consistent with the convergent evolution hypothesis (Zhang et al. 2018a, b, 2021a, b, c, d, e).

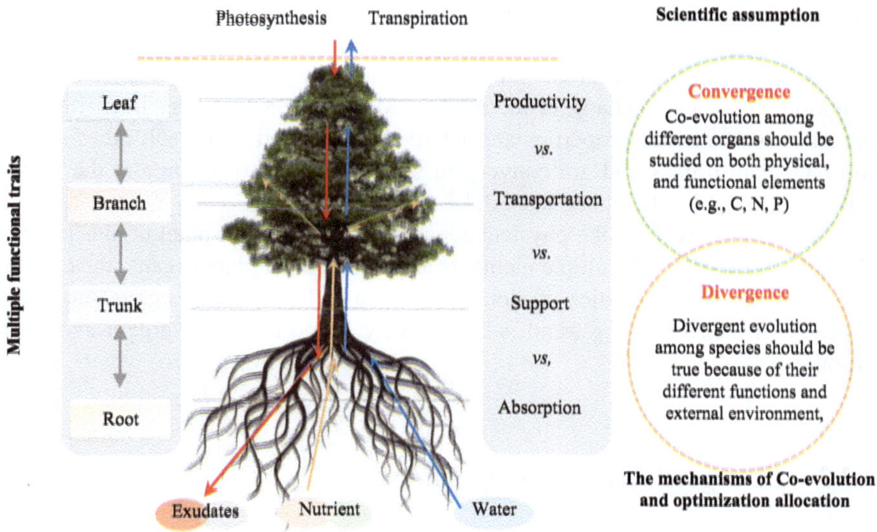

Fig. 3.3 Biological basis of coexistence for the coevolution hypothesis and optimization allocation hypothesis in terms of functional traits among different plant organs

Functional elements are the basic components of all living things on earth. They can be involved in various physiological and biochemical reactions in plants and support diverse physiological activities of plants. In general, there are 16 essential elements that regulate plant growth and development (Arnon and Stout 1939; Marschner 2011), such as carbon (C), hydrogen (H), oxygen (O), nitrogen (N), phosphorus (P), potassium (K), calcium (Ca), magnesium (Mg), sulfur (S), iron (Fe), manganese (Mn), zinc (Zn), copper (Cu), molybdenum (Mo), boron (B), and chlorine (Cl). These functional elements in plants, except C, are absorbed from the soil. Apart from these essential nutrients, plants passively absorb some harmful elements that are not conducive to plant growth and development, such as aluminum (Al), lead (Pb), and arsenic (As). In nature, these functional elements are transported through the vessels from soil to roots, stems, leaves, flowers, and fruits, thereby supporting physiological processes of plant species. Owing to the biophysical limitations, plant organs are assumed to change convergently with external environmental changes. The study on the variation and spatial patterns of elements between leaves and roots along the North–South Transect of Eastern China (NSTEC) has provided robust evidence (see Sect. 3.4.1 for details). In general, similar variations in various elements have been observed in both leaves and roots. Specifically, a negative correlation has been observed between the elemental content and the elemental coefficient of variation, and endogenous regulation considerably impacts macronutrients instead of the external environment (Zhao et al. 2016a, b). Furthermore, the latitudinal patterns of multiple elements in leaves and roots are mostly the same, which can also reflect the coevolution of different plant organs (Zhao et al.

2016a, b, 2018b). Another study on the stochiometric properties of C:N:P ratios in leaves, branches, trunks, and roots has established similar spatial patterns and influencing factors in different plant organs, thereby confirming the convergent evolution hypothesis (Zhang et al. 2018b). Moreover, researchers have revealed that the stoichiometric properties are positively correlated with each other among different plant organs and are convergent in the direction of change at the biome scale (Zhang et al. 2020b; see Sect. 3.4.2 for details).

In summary, owing to the physical connectivity of different plant organs and the continuous transport of multiple elements among different plant organs, the content and stoichiometric properties of elements among different plant organs act synergistically in their changing trend, which largely support the convergent evolution hypothesis.

3.3.2 Divergent Evolution Hypothesis

From a multidimensional perspective, different plant organs are responsible for distinct functions to support plant growth and development. Both the morphological structure and chemical composition of different plant organs are supposed to be closely correlated with their functions, allowing plants to effectively adapt to the unstable external environment. During long-term evolution, plants have developed the ability to regulate morphological traits and nutrient distribution among different organs to support their different functions in the physiological and ecological processes under different environments, particularly in these extreme conditions. For any plant species, prioritizing long-term survival over short-term survival is crucial, because survival of the population is essential in the long run. This is why functional elements in plants typically exhibit an "optimal need-based distribution." In addition, the microenvironments of different organs differ significantly, particularly in the aboveground parts (leaves) and underground parts (fine roots). The leaves are exposed to air and are directly affected by ambient temperature and precipitation, which is reflected in the significant degradation of leaf functional traits due to climate (Wright et al. 2017). The growth of fine roots in the soil is primarily influenced by the physical and chemical properties of soil. Moreover, these roots are necessary to compete with rhizosphere microbes for nutrients based on resource competition theory. However, as rhizosphere microbes are limited by carbon and energy supplies, there is a hypothesis of "carbon-to-nutrient exchange" between microbes and plants, which significantly enhances the complexity of root biological processes (Kuzyakov and Xu 2013). Some studies have shown that more than 80% of terrestrial plants are symbiotic with mycorrhiza, which could improve the ability of plants to absorb nutrients and compete (Tedersoo et al. 2020). Rhizosphere microbes and mycorrhizal symbiosis are two examples of unique factors that may increase the variations in root functional traits and, to some extent, weaken the relationship between root functional traits and soil physical and chemical properties. This could lead to a decoupling of the physiological restriction of "plant leaf–

branch–trunk–root on physical connectivity" that was previously mentioned. In summary, owing to their different activities and microenvironments, different plant organs must be relatively independent and divergent in their evolutionary strategies, thereby supporting the divergent evolution hypothesis (Fig. 3.3).

Herein, plant functional elements and their stoichiometric properties are used to demonstrate the divergent evolutionary rules of different plant organs. The elemental content of different plant organs is likely regulated by their different physiological activities. Some studies have revealed that more functional elements are allocated to more active organs (Zhang et al. 2018a, b). Compared to fine roots, leaves grow with a lower content of trace elements (Zhao et al. 2016a, b), which may be owing to the filtering effect of different plant organs on harmful elements (see Sect. 3.4.1 for details). These results confirmed, to some extent, the divergent evolution of different organs. Significant differences in stoichiometric element ratios between different plant organs have also been observed, including allometric growth rate, different developmental direction and evolutionary rate (Zhang et al. 2020b), as well as greater leaf homeostasis than other plant organs (Zhang et al. 2018b, 2020). The stoichiometric properties and varying elemental content in various plant organs can be considered as optimization strategies of plants to enhance adaptability throughout their long evolutionary history. Regarding the morphological structure of different plant organs, the differences of trait spectrum between leaves and fine roots also support the divergent evolution hypothesis (Fig. 3.3). Studies have reported that the trade-off between plant functional traits should be the result of plant adaptation to the environment and can be used as a proxy for plant adaptation strategy. Wright et al. (2004) demonstrated that there was an economic spectrum of leaf functional traits owing to differences in natural selection, environmental filtration, and ecological functions. The two ends of the leaf economic spectrum are the fast-growing species with high nutrient content, high specific leaf area, high photosynthetic rate, and short lifespan, and the slow-growing species with contrasting performance. Based on the leaf economic spectrum, global vegetation can be divided into various biogeographic communities according to leaf functional traits. This categorization can provide a theoretical basis to link biogeochemical cycle models with vegetation geography models to quantitatively assess the response of plant community structure and function to global climate change. Accordingly, fine roots play a crucial role as a vital functional organ in plants by absorbing water and nutrients from the soil. Fine roots and leaves are interrelated and play crucial roles in various physiological and ecological processes at the same time. Whether roots exhibit a similar economic spectrum is still unclear or controversial. Understanding the relationships among various functional traits of plant roots can not only promote the understanding of plant resource utilization strategies but also provide theoretical support for optimizing ecosystem process models (Norby and Jackson 2000; Reich et al. 2001).

Consistent with the convergent evolution hypothesis, roots are assumed to exhibit a similar economic spectrum to leaves owing to the interconnections between different plant organs. Species with high nutrient content, high specific root length, high respiration rate, and short lifespan and species with opposite functional traits are located on both end of the resource axis, respectively. Conversely, consistent with the divergent evolution hypothesis, roots exhibit multidimensional spectrum compared to leaves, as they grow in a complex environment, with high phylogenetic conservation and limitations imposed by rhizosphere microbes and mycorrhiza. Based on the coordinated measurement of various functional traits of leaves and roots from nine typical forest ecosystems along NSTEC, researchers have examined the relationships between the fine root trait spectrum and leaf trait spectrum (see Sect. 3.4.3 for details). The results showed that there were two principal axes for the functional traits of plant roots. One axis represented the characteristics of the functional element and reflected resource acquisition as a conservation axis. The other represented root morphological characteristics (Wang et al. 2018b). The independence of the functional element and morphological characteristics of fine roots suggests that the plant root functional traits are not one-dimensional, indirectly indicating the diverging spectrum pattern between leaves and roots. Based on this, researchers have further discussed the coupling and decoupling relationships between functional traits of leaves and fine roots based on their chemical and morphological traits. The significant positive correlation between elemental contents in leaves and fine roots confirms the convergent evolution hypothesis to some extent. However, the morphological traits of leaves and roots are not always coupled, confirming the divergent evolution hypothesis (Wang et al. 2017a, b, c). Similarly, using corresponding leaf traits (specific leaf weight, leaf density, maximum photosynthetic rate, and nitrogen content) and root functional traits (average root diameter, specific root length, specific root area, and root nitrogen content) of various plant species, Isaac et al. (2017) revealed that these root functional traits and leaf functional traits do not significantly correlate, and that the adaptive evolution of functional traits between fine roots and leaves is unaffected by environmental changes. Based on 13 temperate trees in Europe, Liese et al. (2017) also found that the leaf economic spectrum could not reflect the root economic spectrum. Root branching is closely related to the colonization of mycorrhiza, which determines the ability of roots to absorb nutrients to some extent, thereby affecting C:N ratio of roots, specific root length, and root life. These factors may contribute to the divergent development of functional traits between leaves and fine roots. However, further experimental data are required to validate this assumption across varying scales.

Many studies have found that the adaptation and optimization mechanisms of different plant organs are determined by the convergence and divergence of specific functional traits. The convergent evolution hypothesis and divergent evolution hypothesis coexist and are coupled with each other in different plant organs. Owing to physical connectivity, different plant organs can share nutrient pools and have the same energy source. For chemical functional traits, such as functional elements, their content would vary according to the different functions of organs; however, their trend of variation would be convergent owing to connectivity. However, morphological traits or the trade-offs between different functional traits are generally divergent and are determined by the different microenvironments and evolutionary strategies of different organs to help plants effectively adapt to the environment, optimize growth, and improve fitness.

3.4 Studies on the Relationships Between Functional Traits Among Different Plant Organs

In this section, we have compiled three research cases to study the evolution and optimization mechanisms of functional traits of various plant organs and elaborated the supporting evidence for the convergent evolution hypothesis and divergent evolution hypothesis at the regional scale. These cases focus on the variations in the content of several plant elements, variations in the stoichiometric properties of plant elements, and relationship between several functional traits of plants.

3.4.1 Variations and Influencing Factors of Functional Elements in Leaves and Roots

Chemical elements are fundamental components of living organisms. At the organ level, multiple elements jointly participate in various physiological and biochemical reactions within and between cells, supporting diverse physiological activities of organisms. At the community or ecosystem level, biogeochemical element cycles connect various living organisms, such as plants, animals, and soil microbes. As explained above, there are 16 essential elements for plant growth and development, such as carbon, nitrogen, phosphorus, sulfur, potassium, calcium, and magnesium. These elements play an important role in regulating plant growth and physiological metabolism, and their mutual relationship leads to enhanced homeostasis across various organisms, ranging from lower organisms to higher organisms (Sterner and Elser 2002; Yu et al. 2015). The essential elements should not be replaced by other elements, and the absence of any element can significantly impede plant growth and development, although some elements exhibit functional complementarity to some extent. Therefore, differences in the functions of different elements and plant requirements for specific elements can be reflected in terms of the elemental content in plant organs. In general, macroelements are involved in physiological and biochemical processes (Marschner 2011). Furthermore, the ratio of elements required to maintain maximum growth is similar across the majority of plant species (Ingestad and Agren 1988). Based on Liebing's law of minimum (Liebig 1840), any element below the optimal balance should become a limiting factor for plant growth. Therefore, plants should give priority to improve the absorption capacity of such limiting elements in the soils. However, if the content of a specific element exceeds the requirements of the plants, the absorption rate for that element should be inhibited and reduced to some extent (Chapin and Matson 2011). In addition to actively absorbing essential elements, plants should passively absorb some nonessential or even harmful elements. Practically, plants might adopt selective transport or active isolation strategies to avoid damage from some harmful elements or excessive essential elements. Therefore, it is theoretically possible that the variations

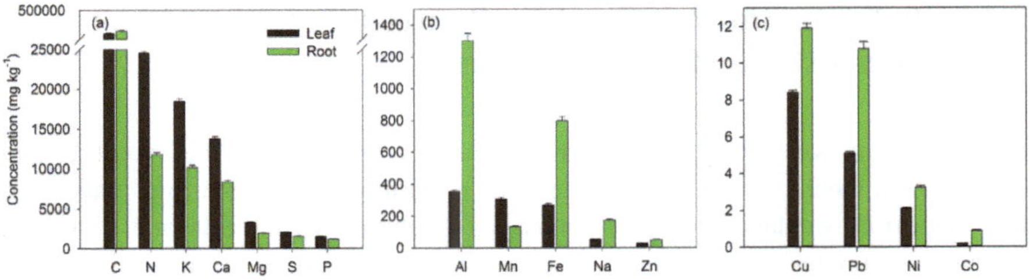

Fig. 3.4 Average concentrations of the 16 elements (C, N, K, Ca, Mg, S, P, Al, Mn, Fe, Na, Zn, Cu, Pb, Ni, and Co) in the leaves and roots of plants in China

in multiple functional elements in different plant organs can effectively reflect large-scale allocation and adaptation strategies of plant species.

Scientists selected nine typical forest ecosystems along the 3700-km-long NSTEC to investigate the spatial variations in multiple elements among different plant organs (Zhao et al. 2016a, b). Vegetation types varied from cold-temperate coniferous forests to temperate conifer broadleaf mixed forests, temperate deciduous broad-leaved forests, subtropical deciduous evergreen mixed forests, subtropical evergreen broad-leaved forests, and tropical monsoon rainforests, and were highly representative of the vegetation in the biome scale. The researchers first collected sun-exposed and fully expanded leaf samples and fine root samples with a diameter of less than 2 mm from 465 species during the peak of plant growth (July and August) and measured the contents of C, N, K, Ca, Mg, S, P, Al, Mn, Fe, Na, Zn, Cu, Pb, Ni, and Co.

The results showed significant differences between leaves and fine roots with seven orders of magnitude in the contents of 16 elements (Fig. 3.4). Leaves had higher content of macroelements than fine roots, while fine roots had higher trace element content (Zhao et al. 2016a, b), consistent the divergent evolution hypothesis to some extent.

The contents of 16 functional elements in leaves and fine roots varied significantly. From the most stable carbon to the most variable manganese in leaves, the coefficient of variation (CV) increased from 7.7% to 112%. Accordingly, the elemental CV in fine roots increased from 7.0% to 87.4%. Moreover, the elemental CV was significantly negatively correlated with elemental content in both leaves and roots (Fig. 3.5). Macroelements were more stable in plants than microelements. Therefore, given the variations in multiple elements, different plant organs follow the convergent evolution hypothesis to some extent.

Furthermore, the results showed that the plant taxon could effectively explain the spatial variations in elemental content in leaves and fine roots (Fig. 3.6). The explanatory power of plant taxon in relation to element variation increased with increasing elemental content, and similar trends were observed in both leaves and fine roots. A positive correlation was observed between elemental content and element variation at the species, family, and order levels (Zhao et al. 2016a, b), partially supporting the convergent evolution hypothesis.

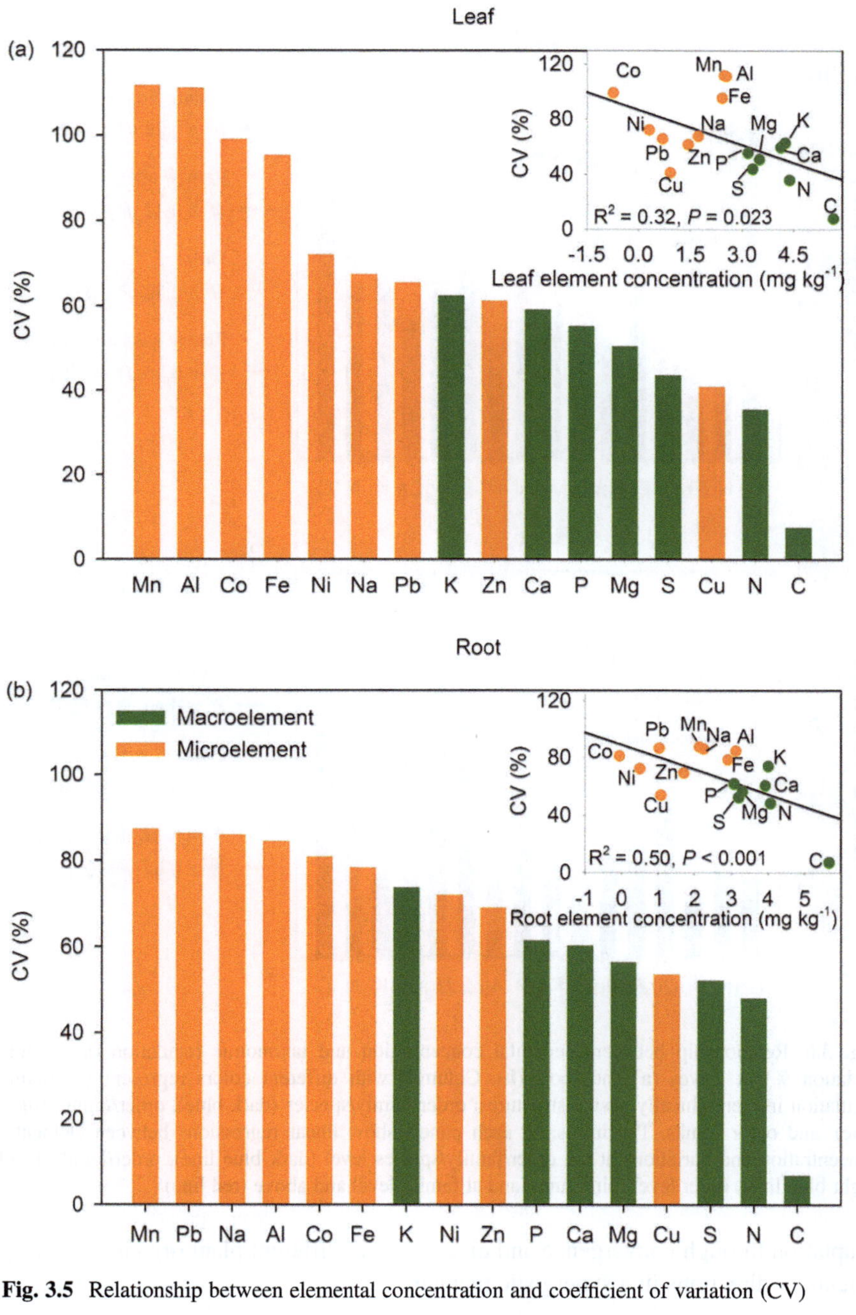

Fig. 3.5 Relationship between elemental concentration and coefficient of variation (CV)

Overall, the analysis of the contents and variation trends of multiple elements in leaves and fine roots confirmed the divergent evolution and convergent evolution hypotheses, respectively, suggesting that plants can optimize their growth and

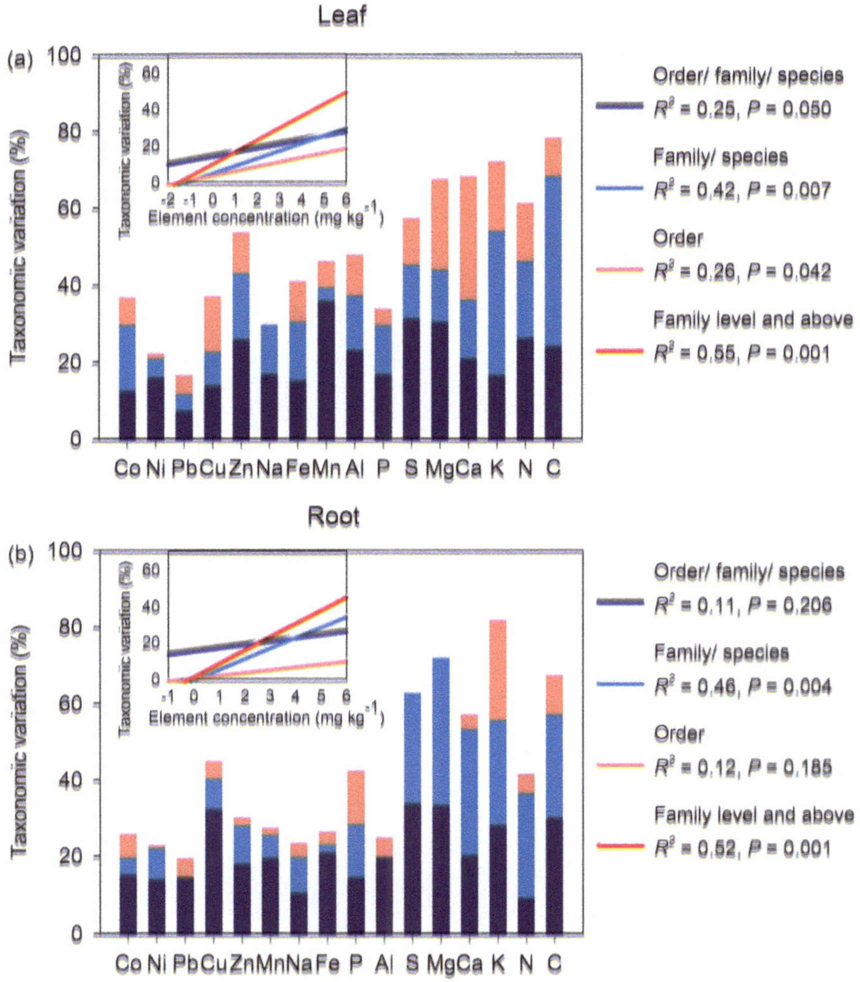

Fig. 3.6 Relationship between elemental concentration and taxonomic regulation (taxonomic variation %) in leaves (**a**) and roots (**b**). Columns with different colors represent taxonomic regulation in hierarchically nested structures: order/family/species (dark blue), order/family (light blue), and order (pink). The insets for each panels show linear regressions between elemental concentration and variations at the order/family/species level (dark blue line), order/family level (light blue line), order level (pink line), and at family level and above (red line)

adaptation through convergence and divergence in different plant organs to varying extents or directions in a synergistic manner.

3.4.2 *Variations and Evolutionary History of C:N Ratio in Leaves, Branches, Trunks, and Roots*

Carbon and nitrogen are crucial components of plant cell structure. They participate in most physiological and biochemical reactions to support growth and development of plants and regulate ecological processes, such as material cycling and energy flow at various scales. C:N ratio is typically used to reflect the balance of plant elements and the efficiency of nutrient utilization to some extent. Generally, high C:N ratio implies high nitrogen-use efficiency of plants. Plants with high C:N ratio are located at the slow-growing end of the LES (Agren 2004; Reich 2014). In addition, when decomposing, microbes prefer substrates with a low C:N ratio, providing more available nitrogen to terrestrial ecosystems (Cornwell et al 2008). Therefore, the C:N ratio has become a key parameter in many ecological process models (Kucharik et al. 2000; Zaehle et al. 2014). As it is an important plant functional trait, studying the variations and evolutionary history of C:N ratio among different organs can enhance our understanding of plant adaptation strategy and optimization mechanisms for these essential elements.

Twenty-eight typical forest communities (latitudinal gradient) were selected to collect leaf, branch, trunk, and root samples for trees and shrubs and leaf, stem, and root samples for herbs (Fig. 3.7). Carbon and nitrogen contents of plant organ samples from a total of 3739 species were consistently measured.

Scientists selected 28 typical natural forest ecosystems in China to study the variations in C:N ratio among different plant organs on a large scale (Zhang et al. 2020a). These sample sites were mostly in nature reserves or areas with little human disturbance. The study area spanned from tropical forests to cold-temperature forests. Researchers conducted a systematic field investigation of plant community composition, collecting samples of leaves, branches, trunks, and fine roots of all observed plant species during peak growing periods (July or August). For trees and

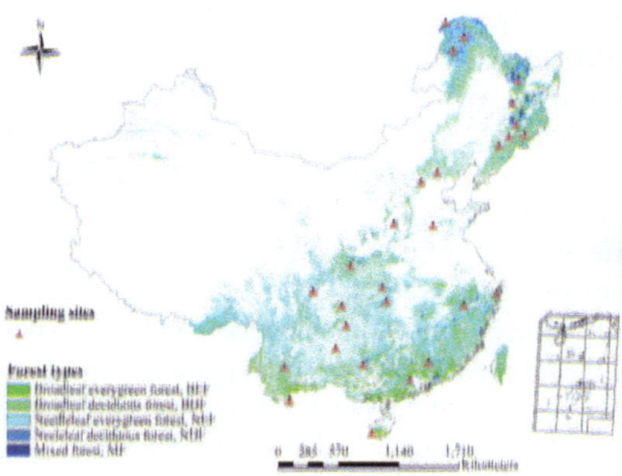

The number of sampled species

	Tree	Shrub	Herb	Total
BEF	895	903	622	2420
BDF	112	129	203	444
NEF	133	149	350	632
NDF	28	68	147	243
Total	1168	1249	1322	3739

Fig. 3.7 Locations of field investigation and number of plant species sampled

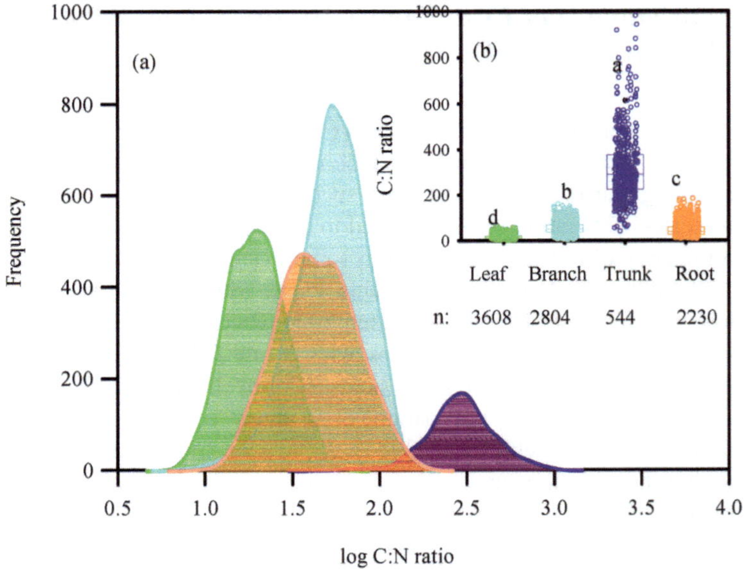

Fig. 3.8 Frequency distributions of C:N ratios of different plant organs fitted by kernel density curves (**a**). n = number of plant species. Different lowercase letters in (**b**) represent significant differences in C:N ratios among different plant organs at the $P = 0.05$ level

shrubs, leaves that were sun-exposed and fully expanded, branches with diameters less than 1 cm, trunks sampled using increment borer at breast height, and fine roots with diameters less than 2 mm were collected for analysis. For herbs, whole plants were collected in the field and brought to the laboratory, where they were divided into leaves, branches, trunks, and fine roots. In total, 2139 species were sampled using uniform procedure, and researchers measured the carbon and nitrogen contents of different plant organs for all plant species sampled.

The results showed that the differences in C:N ratio among different organs confirmed the divergent evolution hypothesis to some extent. The average C:N ratios were 21.05 ± 0.14, 57.85 ± 0.48, 319.04 ± 6.23, and 49.81 ± 0.58 in leaves, branches, trunks, and fine roots, respectively. These values were consistent with their functional activities (Fig. 3.8). Specifically, organs with high activity levels were found to have low C:N ratios owing to increased allocation of the limiting element N, as expected (Zhang et al. 2020a, 2021a, b, c, d, e).

Although the C:N ratios of different plant organs varied significantly, they all changed in the same direction, as evidenced by the positive correlations observed between any two plant organs in the allometric allocation models (Table 3.1). These results confirmed, to some extent, the convergent evolution hypothesis. However, the differences in functions and activities of different plant organs inevitably lead to the differentiation of C:N ratio among different organs, which was confirmed by the allometric variations in C:N ratios among different plant organs. Owing to its high

Table 3.1 Key parameters of C:N ratio allometric equations in different forest types

Forest type	y vs. x	PFG[a]	n	Slope	R^2	P	
BEF[b]	Branch vs. Leaf	Tree	767	1.158	0.347	< 0.001	***[c]
		Shrub	792	1.114	0.403	< 0.001	***
		Herb	295	1.857	0.447	< 0.001	***
	Trunk vs. Leaf	Tree	297	1.413	0.121	< 0.001	***
		Shrub	74	1.507	0.091	< 0.01	***
	Root vs. Leaf	Tree	386	1.307	0.234	< 0.001	***
		Shrub	455	1.324	0.279	< 0.001	***
		Herb	427	1.320	0.352	< 0.001	***
	Trunk vs. Branch	Tree	295	1.131	0.120	< 0.001	*
		Shrub	74	1.282	0.209	< 0.001	*
	Root vs. Branch	Tree	379	1.196	0.188	< 0.001	***
		Shrub	430	1.153	0.377	< 0.001	***
		Herb	219	0.802	0.379	< 0.001	***
	Trunk vs. Root	Tree	198	0.907	0.207	< 0.001	Ns
		Shrub	56	0.933	0.175	< 0.01	Ns
BDF	Branch vs. Leaf	Tree	84	1.116	0.363	< 0.001	Ns
		Shrub	98	1.352	0.305	< 0.001	**
		Herb	74	1.494	0.445	< 0.001	***
	Trunk vs. Leaf	Tree	36	1.262	0.018	0.438	
		Shrub	2				
	Root vs. Leaf	Tree	65	1.731	0.283	< 0.001	***
		Shrub	83	1.563	0.219	< 0.001	***
		Herb	141	1.645	0.245	< 0.001	***
	Trunk vs. Branch	Tree	33	−1.114	0.060	0.170	
		Shrub	2				
	Root vs. Branch	Tree	58	1.622	0.298	< 0.001	***
		Shrub	68	1.155	0.373	< 0.001	Ns
		Herb	60	1.285	0.206	< 0.001	*
	Trunk vs. Root	Tree	40	1.752	0.017	0.423	
		Shrub	2				
NEF	Branch vs. Leaf	Tree	117	0.968	0.362	< 0.001	Ns
		Shrub	138	1.217	0.316	< 0.001	**
		Herb	198	1.759	0.402	< 0.001	***
	Trunk vs. Leaf	Tree	77	1.236	0.103	< 0.01	Ns
		Shrub	12	1.651	0.138	0.234	
	Root vs. Leaf	Tree	90	1.109	0.254	< 0.001	Ns
		Shrub	105	1.218	0.304	< 0.001	*
		Herb	229	1.259	0.160	< 0.001	***
	Trunk vs. Branch	Tree	75	−1.100	0.004	0.574	
		Shrub	12	−1.544	0.028	0.603	
	Root vs. Branch	Tree	88	1.050	0.142	< 0.001	***
		Shrub	107	0.948	0.308	< 0.001	Ns

(continued)

Table 3.1 (continued)

Forest type	y vs. x	PFG[a]	n	Slope	R^2	P	
		Herb	155	0.721	0.242	< 0.001	Ns
	Trunk vs. Root	Tree	81	0.967	0.006	0.492	
		Shrub	10	0.844	0.050	0.534	
NDF	Branch vs. Leaf	Tree	27	0.876	0.129	0.066	
		Shrub	63	1.209	0.288	< 0.001	Ns
		Herb	82	1.499	0.519	< 0.001	***
	Trunk vs. Leaf	Tree	14	0.989	0.182	0.128	
		Shrub	2				
	Root vs. Leaf	Tree	20	1.021	0.160	0.081	
		Shrub	46	1.243	0.158	< 0.01	Ns
		Herb	85	1.446	0.111	< 0.01	***
	Trunk vs. Branch	Tree	14	−1.379	0.128	0.210	
		Shrub	2				
	Root vs. Branch	Tree	19	1.329	0.106	0.175	
		Shrub	47	1.014	0.330	< 0.001	Ns
		Herb	52	1.341	0.443	< 0.001	**
	Trunk vs. Root	Tree	14	0.738	0.012	0.705	
		Shrub	2				

[a]*PFG* plant functional group, *n* number, R^2 coefficient of determination, *P*, the significant level
[b]*BEF* broadleaf evergreen forest, *BDF* broadleaf deciduous forest, *NEF* needleleaf evergreen forest, *NDF* needleleaf deciduous forest
[e]***, **, and * denote significant difference between the slope of the equation and 1 at $P = 0.001$, $P = 0.01$, and $P = 0.05$, respectively; ns denotes no significant difference between the slope of the equation and 1 at $P = 0.05$. All C:N ratios were transformed to a log-log scale. The likelihood ratio test was used to examine differences between organs. The reduced major axis tests were not performed, if the number was less than 6

leaf activity, the leaf C:N ratio remained the most stable, with the velocity and rate of change being the slowest (Zhang et al. 2020a).

From an evolutionary perspective, leaves and branches developed a low C:N ratio, while the C:N ratio of trunks and roots remained consistent over time. Moreover, the C:N ratio of leaves and branches underwent varying changes during evolution, partially supporting the divergent evolution hypothesis (Fig. 3.9).

Based on the variation characteristics and evolutionary history of C:N ratio among different organs, we found that the convergent evolution and divergent evolution hypotheses were supported from different perspectives. Different organs, such as leaves, branches, trunks, and roots, have the same source of nutrients and energy owing to their structural connectivity. Therefore, if the environmental changes affect root uptake from the soil, the nutrient content or stoichiometric properties of other plant organs will be correspondingly affected. However, because different plant organs perform different functions and have different demands for element supply, plants can adopt allometric nutrient distribution in different organs to maintain the homeostasis of their active organs. This allows plant species to adapt

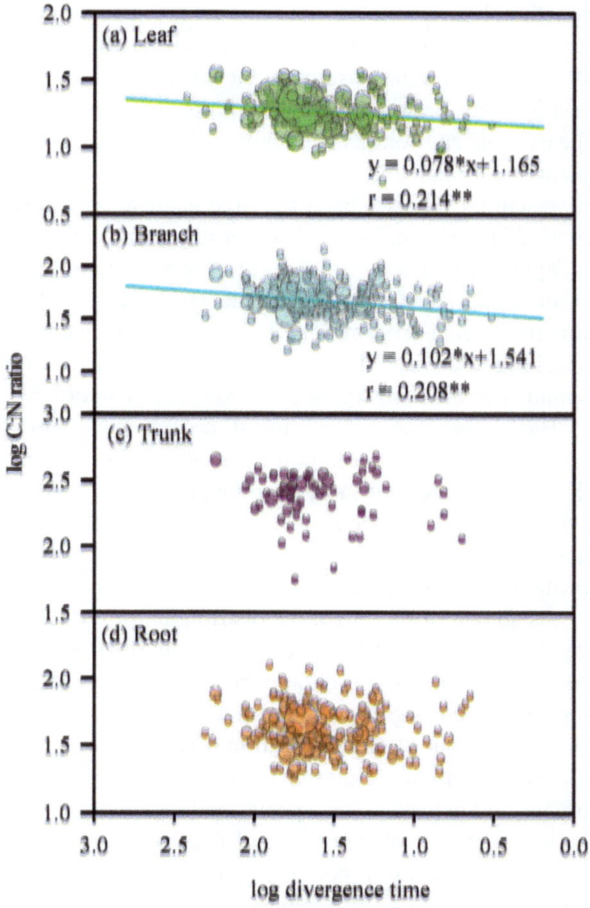

Fig. 3.9 Changes in C:N ratios of different organs over divergence time at the family level. *r* is the correlation coefficient; ** denotes $P < 0.01$. The family divergence time was defined as the time when the earliest genus within a family diverged. Linear regressions weighted by the number of species were used in (**a**–**d**). Large circle indicates that more species belonged to the family

to complex external environments, disturbances, and disaster events (Zhang et al. 2020a). In conclusion, passive convergent evolution and active divergent evolution work together to maintain the stable stoichiometric properties of active plant organs and ensure the maximum sustainable supply of limiting elements, which present important allocation mechanisms and approaches for optimizing the functions of plant species on a large scale.

3.4.3 Comparison of Trait Economic Spectrum Between Leaves and Fine Roots

Leaves and fine roots are the most metabolically active aboveground and belowground organs, respectively. Theoretically, there should be a strong correlation between their functional traits. Therefore, studying the relationship between functional traits of leaves and fine roots is expected to reveal plant resource utilization strategies and provide optimization parameters for ecosystem process models. However, the relationships are controversial to date. One argument is that the relationships between functional traits are one-dimensional. The selection effect of the external environment and biophysical constraints contribute to the aggregation of plant functional traits along the strategic axis from fast growth to slow growth, representing the axis from resource acquisition to resource conservation (Freschet et al. 2010; Reich 2014). Another argument is that different plant organs develop independently owing to the significant differences in microenvironmental constraints and self-functions (Craine et al. 2005). Therefore, relationships between functional traits are considered to be multidimensional, improving the coexistence of species and the stability of ecosystem structure and function. These concepts correspond with the convergent and divergent evolution hypotheses, respectively.

Based on nine typical forest ecosystems along the 3700-km-long NSTEC, researchers selected common trees, shrubs, and herbs and systematically collected the leaf and fine root samples. Five paired functional traits in leaves and roots were selected and consistently measured: leaf thickness vs. root diameter, specific leaf area vs. specific root length, leaf tissue density vs. root tissue density, leaf carbon content vs. root carbon content, and leaf nitrogen content vs. root nitrogen content.

As shown in Fig. 3.10, the distribution of ten leaf and absorptive root traits was significant for all 154 species. By comparing these paired functional traits, except carbon content, the morphological traits and elemental contents between leaves and roots showed significant positive correlations without considering phylogeny (Fig. 3.11). The strong correlation between leaf nitrogen and root nitrogen contents allowed nutritional characteristics to serve as a bridge to connect different plant organs and between plant species and the environment. The decoupling of morphological traits between leaves and fine roots should be determined by the different selection pressures of their microenvironments. Leaf functional traits are determined by maximizing photosynthesis rate and minimizing water loss, creating a one-dimensional trade-off relationship between leaf functional traits. Alternatively, roots are influenced by more complex biological (mycorrhizal colonization and rhizosphere microbes) and environmental (soil chemical and physical characteristics) factors, resulting in a very different evolutionary history for leaf and root morphological traits and, eventually, their decoupling (Wang et al. 2017a, b, c).

As plant leaves and fine roots are connected, the content of elements is generally significantly positively correlated, confirming the convergent evolution hypothesis. However, the decoupled leaf and root morphological traits and the different trait spectra in leaves and roots may reflect the functional differences among different plant organs, confirming the divergent evolution hypothesis to some extent.

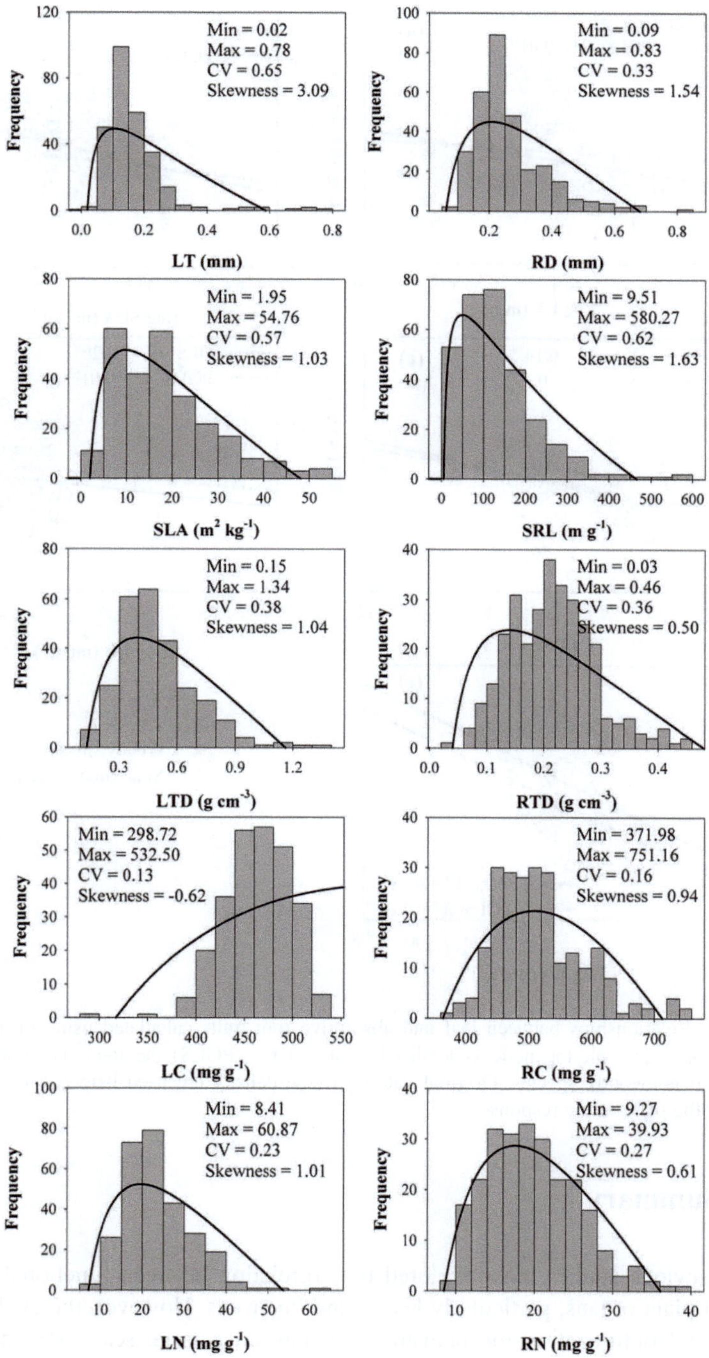

Fig. 3.10 Histograms showing the distribution of ten leaf and absorptive root traits for all 154 species. Bold curves indicate fitted log-normal curves. *Min* minimum value, *Max* maximum, *CV* coefficient of variation

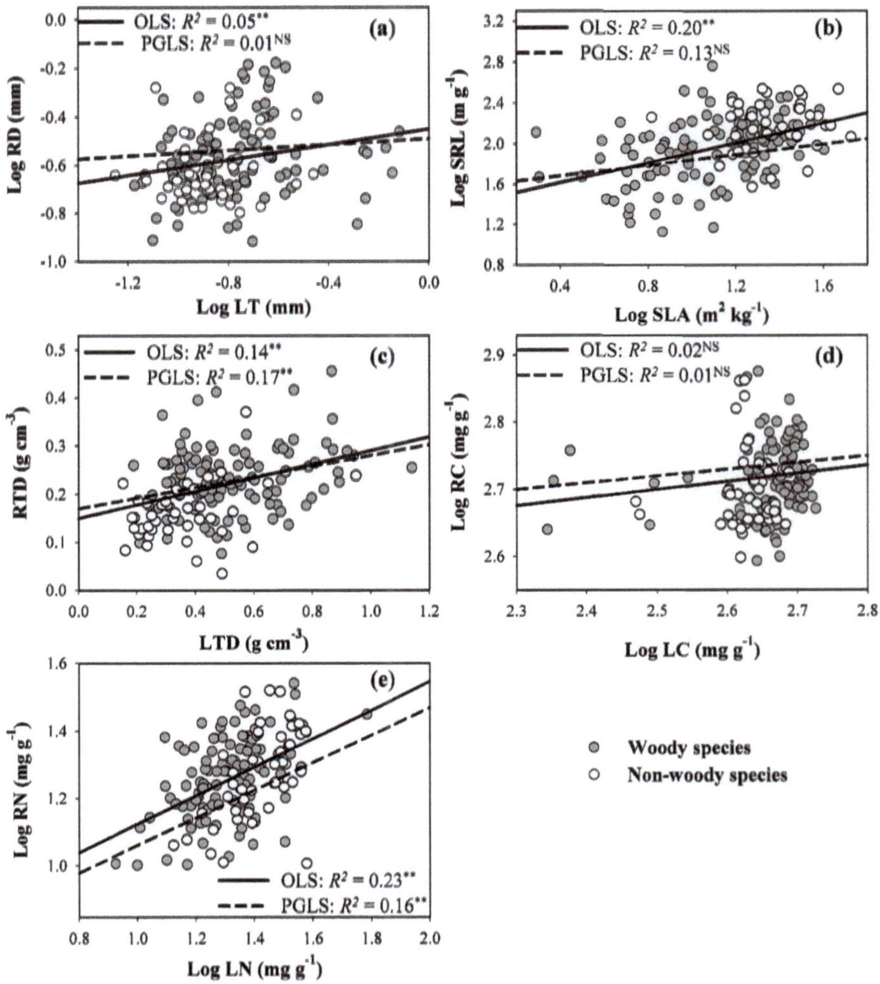

Fig. 3.11 Relationships between leaf and absorptive root traits calculated using ordinary least squares (OLS) and phylogenetic generalized least squares (PGLS) methods across all species, woody and nonwoody species. Original data on tissue density are used here because of issues regarding the scale of the response

3.5 Summary

Many previous studies have explored the correlation between functional traits of different plant organs, particularly leaves and fine roots. However, the evolutionary history and optimization mechanisms of plants on a large scale are unclear. In particular, the important role of branches and trunks in plant adaptation and nutrient distribution has been ignored.

Studying the coevolution of plant functional traits is just the beginning. More specifically, whether various functional traits of different plant organs fully support the coexistence of the convergent and divergent evolution hypotheses needs to be further verified, which will be a major challenge for a long time. In addition, existing studies are predominantly carried out at the species level. Although valuable for examining nutrient cycle and energy flow within individual plants, applying them on a larger scale is challenging. Furthermore, these studies may not offer a reliable theoretical basis or optimization parameters for ecological process models. To effectively advance research on the coevolution of multiple functional traits among different plant organs, the development of systematic and coordinated databases of plant functional traits at regional and even global scales is imperative.

The development of leaf and root trait spectrum has promoted the studies on coevolution of functional traits in different plant organs to some extent. However, comprehensive understanding of the trade-offs of functional traits has been somewhat constrained by current analytical techniques. Network analysis is a new approach to comprehensively capture and visualize the relationship between various functional traits and reveal the multidimensional adaptation mechanism of plant species (He et al. 2020a, b, c). The modular parameters should be helpful in determining whether multiple traits of different organs are convergent or divergent. Furthermore, plant communities are formed through the adaptation and evolution of different plant species to long-term environmental changes, and the composition of these communities is not random. Therefore, data on plant communities are fundamental to the study of ecosystem functions and regional ecological impacts. Furthermore, understanding the relationships between plant functional traits of different organs at the community level will be crucial in future research (He et al. 2019a, b). Research on functional traits at the community level is essential for developing new parameters and analysis methods to support ecologists in the future. Given the complex nature of functional trait coevolution, interdisciplinary collaboration in future research is required. We hope that plant physiologists, ecologists, evolutionary biologists, and computational scientists can deeply cooperate to overcome current limitations in both data collection and analysis.

References

Agren GI (2004) The C:N:P stoichiometry of autotrophs – theory and observations. Ecol Lett 7: 185–191

Arnon DI, Stout P (1939) The essentiality of certain elements in minute quantity for plants with special reference to copper. Plant Physiol 14:371–375

Bardgett RD, Mommer L, De Vries FT (2014) Going underground: root traits as drivers of ecosystem processes. Trends Ecol Evol 29:692–699

Brouat C, McKey D (2001) Leaf-stem allometry, hollow stems, and the evolution of *caulinary domatia* in myrmecophytes. New Phytol 151:391–406

Chapin FSIII, Matson PA (2011) Principles of terrestrial ecosystem ecology, 2rd edn. Springer, New York

Chave J, Coomes DA, Jansen S et al (2009) Towards a worldwide wood economics spectrum. Ecol Lett 12:351–366
Chen WL, Zeng H, Eissenstat DM et al (2013) Variation of first-order root traits across climatic gradients and evolutionary trends in geological time. Glob Ecol Biogeogr 22:846–856
Cheng S, Xian W, Fu Y et al (2019) Genomes of subaerial zygnematophyceae provide insights into land plant evolution. Cell 179:1057–1067
Corner EJH (1949) The durian theory of the origin tree. Ann Bot 13:367–414
Cornwell WK, Cornelissen JHC, Amatangelo K et al (2008) Plant species traits are the predominant control on litter decomposition rates within biomes worldwide. Ecol Lett 11:1065–1071
Craine JM, Lee WG, Bond WJ et al (2005) Environmental constraints on a global relationship among leaf and root traits of grasses. Ecology 86:12–19
Croft H, Chen JM, Luo XZ et al (2017) Leaf chlorophyll content as a proxy for leaf photosynthetic capacity. Glob Chang Biol 23:3513–3524
Evert RF, Eichhorn SE (2013) Raven biology of plants. W. H. Freeman and Company, New York
Eviner VT, Chapin FS (2003) Functional matrix: a conceptual framework for predicting multiple plant effects on ecosystem processes. Annu Rev Ecol Evol Syst 34:455–485
Freschet GT, Cornelissen JHC, van Logtestijn RSP et al (2010) Evidence of the 'plant economics spectrum' in a subarctic flora. J Ecol 98:362–373
Gartner BL (1991) Stem hydraulic-properties of vines vs shrubs of western poison oak, Toxicodendron-diversilobum. Oecologia 87:180–189
Guo D, Xia M, Wei X et al (2008) Anatomical traits associated with absorption and mycorrhizal colonization are linked to root branch order in twenty-three Chinese temperate tree species. New Phytol 180:673–683
He NP, Liu CC, Tian M et al (2018a) Variation in leaf anatomical traits from tropical to cold-temperate forests and linkage to ecosystem functions. Funct Ecol 32:10–19
He NP, Liu CC, Zhang JH et al (2018b) Opportunities and challenges for plant trait research: from organs to communities. Acta Ecologic Sinica 38:6787–6796
He NP, Zhang JH, Liu CC et al (2018c) Patterns and influencing factors of traits in forest ecosystems: synthesis and perspectives on the synthetic investigation from the north-east transect of eastern China (NETEC). Acta Ecol Sin 38(18):6359–6382
He NP, Liu CC, Piao SL et al (2019a) Ecosystem traits linking functional traits to macroecology. Trends Ecol Evol 34:200–210
He L, Chen JM, Liu J et al (2019b) Diverse photosynthetic capacity of global ecosystems mapped by satellite chlorophyll fluorescence measurements. Remote Sens Environ 232:111344
He NP, Li Y, Liu CC et al (2020a) Plant trait networks: improved resolution of the dimensionality of adaptation. Trends Ecol Evol 35:908–918
He NP, Liu CC, Xu L et al (2020b) Ecosystem traits open new insights into macroecology: opportunity and challenge. Acta Ecol Sin 40(8):2507–2522
He P, Gleason SM, Wright IJ et al (2020c) Growing-season temperature and precipitation are independent drivers of global variation in xylem hydraulic conductivity. Glob Chang Biol 26: 1833–1841
Ingestad T, Agren GI (1988) Nutrient-upteke and allocation at strady-state nutrition. Physiol Plant 72:450–459
Isaac ME, Martin AR, Virginio ED et al (2017) Intraspecific trait variation and coordination: root and leaf economics spectra in coffee across environmental gradients. Front Plant Sci 8:1196
Kenrick P, Crane PR (1997) The origin and early evolution of plants on land. Nature 389:33–39
Kucharik CJ, Foley JA, Delire C et al (2000) Testing the performance of a dynamic global ecosystem model: water balance, carbon balance, and vegetation structure. Glob Biogeochem Cycles 14:795–825
Kuzyakov Y, Xu XL (2013) Competition between roots and microorganisms for nitrogen: mechanisms and ecological relevance. New Phytol 198:656–669

La Riva EGD, Olmo M, Poorter H et al (2016) Leaf mass per area (LMA) and its relationship with leaf structure and anatomy in 34 mediterranean woody species along a water availability gradient. PLoS One 11:e0148788

Li NN, He NP, Yu GR et al (2016) Leaf non-structural carbohydrates regulated by plant functional groups and climate: evidences from a tropical to cold-temperate forest transect. Ecol Indic 62: 22–31

Li Y, He NP, Hou JH et al (2018a) Factors influencing leaf chlorophyll content in natural forests at the biome scale. Front Ecol Evol 6:64

Li Y, Liu CC, Zhang JH et al (2018b) Variation in leaf chlorophyll concentration from tropical to cold-temperate forests: association with gross primary productivity. Ecol Indic 85:383–389

Li YQ, Reich PB, Schmid B et al (2020) Leaf size of woody dicots predicts ecosystem primary productivity. Ecol Lett 23:1003–1013

Liebig JV (1840) Organic chemistry in its application to vegetable physiology and agriculture. Prentice Hall, New York

Liese R, Alings K, Meier IC (2017) Root branching is a leading root trait of the plant economics spectrum in temperate trees. Front Plant Sci 8:315

Liu CC, He NP, Zhang JH et al (2018a) Variation of stomatal traits from cold-temperate to tropical forests and association with water use efficiency. Funct Ecol 32:20–28

Liu J, Milne RI, Cadotte MW et al (2018b) Protect third pole's fragile ecosystem. Science 362: 1368–1368

Liu CC, Li Y, Zhang JH et al (2020) Optimal community assembly related to leaf economic-hydraulic-anatomical traits. Front Plant Sci 11:2222341

Liu CC, Li Y, Yan P et al (2021a) How to improve the predictions of plant functional traits on ecosystem functioning? Front Plant Sci 12:622260

Liu XR, Chen HX, Sun TY et al (2021b) Variation in woody leaf anatomical traits along the altitudinal gradient in Taibai Mountain, China. Glob Ecol Conserv 26:e01523

Liu C, Li Y, Xu L et al (2021c) Stomatal arrangement pattern: a new direction to explore plant adaptation and evolution. Front Plant Sci 12:655255

Liu CC, Li Y, He NP et al (2022a) Differential adaptation of lianas and trees in wet and dry forests revealed by trait correlation networks. Ecol Indic 135:108564

Liu CC, Sack L, Li Y et al (2022b) Contrasting adaptation and optimization of stomatal traits across communities at continental-scale. J Exp Bot 73:6405–6416

Liu H, Ye Q, Simpson KJ et al (2022c) Can evolutionary history predict plant plastic responses to climate change? New Phytol 235:1260–1271

Ma ZQ, Guo DL, Xu XL et al (2018) Evolutionary history resolves global organization of root functional traits. Nature 555:94–97

Marschner H (2011) Marschner's mineral nutrition of higher plants. Academic Press

Morris JL, Puttick MN, Clark JW et al (2018) The timescale of early land plant evolution. Proc Natl Acad Sci USA 115:E2274–E2283

Niklas KJ (1994) The allometry of safety-factors for plant height. Am J Bot 81:345–351

Norby RJ, Jackson RB (2000) Root dynamics and global change: seeking an ecosystem perspective. New Phytol 147:3–12

Olson ME, Aguirre-Hernandez R, Rosell JA (2009) Universal foliage-stem scaling across environments and species in dicot trees: plasticity, biomechanics and Corner's rules. Ecol Lett 12:210–219

Pech M, Karim Z, Yamamoto H et al (2011) Elongation factor 4 (EF4/LepA) accelerates protein synthesis at increased Mg^{2+} concentrations. Proc Natl Acad Sci USA 108:3199–3203

Pregitzer KS, DeForest JL, Burton AJ et al (2002) Fine root architecture of nine north American trees. Ecol Monogr 72:293–309

Reich PB (2014) The world-wide 'fast-slow' plant economics spectrum: a traits manifesto. J Ecol 102:275–301

Reich PB, Tilman D, Craine J et al (2001) Do species and functional groups differ in acquisition and use of C, N and water under varying atmospheric CO_2 and N availability regimes? A field test with 16 grassland species. New Phytol 150:435–448
Roumet C, Birouste M, Picon-Cochard C et al (2016) Root structure-function relationships in 74 species: evidence of a root economics spectrum related to carbon economy. New Phytol 210: 815–826
Sterner RW, Elser JJ (2002) Ecological stoichiometry: the biology of elements from molecules to the biosphere. Princeton University Press, Princeton
Tedersoo L, Bahram M, Zobel M (2020) How mycorrhizal associations drive plant population and community biology. Science 367:867–871
Tjoelker MG, Craine JM, Wedin D et al (2005) Linking leaf and root trait syndromes among 39 grassland and savannah species. New Phytol 167:493–508
Verburg PSJ, Johnson DW (2001) A spreadsheet-based biogeochemical model to simulate nutrient cycling processes in forest ecosystems. Ecol Model 141:185–200
Violle C, Navas ML, Vile D et al (2007) Let the concept of trait be functional! Oikos 116:882–892
Wang N, Gao J, Zhang SQ et al (2014) Variations in leaf and root stoichiometry of Nitraria tangutorum along aridity gradients in the Hexi corridor, Northwest China. Contemp Probl Ecol 7:308–314
Wang RL, Wang QF, Zhao N et al (2017a) Complex trait relationships between leaves and absorptive roots: coordination in tissue N concentration but divergence in morphology. Ecol Evol 7:2697–2705
Wang R, Gibson CD, Berry TD et al (2017b) Photooxidation of pyrogenic organic matter reduces its reactive, labile C pool and the apparent soil oxidative microbial enzyme response. Geoderma 293:10–18
Wang H, Prentice IC, Davis TW et al (2017c) Photosynthetic responses to altitude: an explanation based on optimality principles. New Phytol 213:976–982
Wang H, Harrison SP, Prentice IC et al (2018a) The China plant trait database: toward a comprehensive regional compilation of functional traits for land plants. Ecology 99:500–500
Wang S, Wang X, Han X et al (2018b) Higher precipitation strengthens the microbial interactions in semi-arid grassland soils. Glob Ecol Biogeogr 27:570–580
Willis K, McElwain J (2014) The evolution of plants. Oxford University Press
Wright IJ, Reich PB, Westoby M et al (2004a) The worldwide leaf economics spectrum. Nature 428:821–827
Wright IJ, Groom PK, Lamont BB et al (2004b) Leaf trait relationships in Australian plant species. Funct Plant Biol 31:551–558
Wright IJ, Dong N, Maire V et al (2017) Global climatic drivers of leaf size. Science 357:917–921
Xu L, He NP, Yu GR (2020) Nitrogen storage in China's terrestrial ecosystems. Sci Total Environ 709:136201
Yan Z, Li P, Chen Y et al (2016) Nutrient allocation strategies of woody plants: an approach from the scaling of nitrogen and phosphorus between twig stems and leaves. Sci Rep 6:20099
Yu Q, Wilcox K, La Pierre K et al (2015) Stoichiometric homeostasis predicts plant species dominance, temporal stability, and responses to global change. Ecology 96:2328–2335
Zaehle S, Medlyn BE, De Kauwe MG et al (2014) Evaluation of 11 terrestrial carbon-nitrogen cycle models against observations from two temperate free-air CO_2 enrichment studies. New Phytol 202:803–822
Zhang JH, Zhao N, Liu CC et al (2018a) C:N:P stoichiometry in China's forests: From organs to ecosystems. Funct Ecol 32:50–60
Zhang JH, He NP, Liu CC et al (2018b) Allocation strategies for nitrogen and phosphorus in forest plants. Oikos 127:1506–1514
Zhang H, Xiang Y, Irving LJ et al (2019) Nitrogen addition can improve seedling establishment of N-sensitive species in degraded saline soils. Land Degrad Dev 30:119–127
Zhang JH, He NP, Liu CC et al (2020a) Variation and evolution of C:N ratio among different organs enable plants to adapt to N-limited environments. Glob Chang Biol 26:2534–2543

Zhang Y, Li Y, Wang RM et al (2020b) Spatial variation of leaf chlorophyll in northern hemisphere grasslands. Front Plant Sci 11:1244
Zhang JH, Li MX, Xu L et al (2021a) C:N:P stoichiometry in terrestrial ecosystems in China. Sci Total Environ 795:148849
Zhang JH, Ren TT, Yang JJ et al (2021b) Leaf multi-element network reveals the change of species dominance under nitrogen deposition. Front Plant Sci 12:580340
Zhang Y, He NP, Li MX et al (2021c) Community chlorophyll quantity determines the spatial variation of grassland productivity. Sci Total Environ 801:149567
Zhang ZH, Hou JH, He NP (2021d) Predictability of functional diversity depends on the number of traits. J Resour Ecol 12:332–345
Zhang Y, Migliavacca M, Penuelas J et al (2021e) Advances in hyperspectral remote sensing of vegetation traits and functions. Remote Sens Environ 252:112121
Zhao N, Yu GR, He NP et al (2016a) Coordinated pattern of multi-element variability in leaves and roots across Chinese forest biomes. Glob Ecol Biogeogr 25:359–367
Zhao WL, Chen YJ, Brodribb TJ et al (2016b) Weak co-ordination between vein and stomatal densities in 105 angiosperm tree species along altitudinal gradients in Southwest China. Funct Plant Biol 43:1126–1133

Chapter 4
Functional Trait Diversity and Its Relationship to Plant Adaptation and Functional Optimization

Plants have many indispensable and important functional traits that help them respond and adapt to changing environmental conditions, optimize their functions, and even reproduce across generations. Plant functional trait diversity, also known as plant functional diversity, is a comprehensive index for characterizing the distribution range and dispersion of plant functional traits in a community. It reflects the overall differences, adaptation, and optimization mechanisms of functional traits in plant communities. Because they account for redundant species and interspecific complementary effects within communities, these functional diversity parameters are uniquely advantageous in predicting community mechanisms and the functional evolution of ecosystems. Since 2010, these concepts, their theory, and methodology have been progressively developed, and the application of functional diversity parameters is increasingly integrated into several important fields of ecological studies. However, the selection of suitable parameters for the different functional diversity is a major challenge for researchers to date. Furthermore, it is highly debatable whether the number of functional traits and their selection have an impact on functional diversity parameters. In this chapter, we review the definition of plant functional diversity, introduce its origin and significance, and systematically summarize several common functional diversity parameters and their algorithms, which include functional richness, functional evenness, and functional dispersion. We then present an analysis of the influence of the number of functional traits on functional diversity parameters at the regional scale, the changing patterns of plant functional trait diversity at the large scale, and their relationship with productivity. Finally, we explore the development direction of plant functional diversity to provide novel insights to upcoming researchers.

Plant functional traits are relatively stable and measurable parameters that are ecologically important for long-term response and adaptation to the external environment (Violle et al. 2007; He et al. 2018a, b, c, 2019a, b). Specifically, they can be divided into morphological traits, such as plant height, seed size, individual size, and leaf lifespan (Shipley et al. 2006a, b); physiological traits, such as chlorophyll, nitrogen, phosphorus, and water contents (Li et al. 2018a, b, 2022a, b, c, d); and

N. He et al., *Plant Functional Ecology*,
https://doi.org/10.1007/978-981-96-8264-5_4

life history traits, such as life type, flowering period, and seed longevity (Verheyen et al. 2003). Theoretically, functional plant traits can reflect the response and adaptation of plants to the growth environments and are the important basis for plants to achieve their functional stability (Cornelissen et al. 2003; Dawei Zhou 2009). Therefore, most studies on plant functional traits focus on how plants adapt to environmental changes, optimize resource utilization, or stabilize productivity (Lei et al. 2016; Xu et al. 2019). Furthermore, the question of how specific parameters of plant functional traits can be used to predict the functions of individual plants, communities, and ecosystems has recently become a hot scientific topic.

In natural communities, all plants possess multiple functional traits that work together through multiple mechanisms to aid in their adaptation to complex and changing external environments and achieve relative functional stability (He et al. 2020a, b, c). To scientifically represent the diversity, distribution, and range of functional traits within plant communities, researchers have developed functional trait diversity (Petchey and Gaston 2006), simply referred to as functional diversity. Over the past decade, related studies on functional trait diversity have steadily increased, and the relationships between functional trait diversity and ecosystem function have gradually become the core of ecological research (Villéger et al. 2008; Schleuter et al. 2010).

With the establishment of conceptual systems for functional trait diversity in plants, the parameters and corresponding calculation methods have been continuously updated and refined (de Bello et al. 2011; Legras et al. 2018). Currently, the quantitative assessment of functional trait diversity relies on various theoretical frameworks and the resulting development of multiple computational methods. These parameters can be used to delineate the spatial patterns of functional traits within a particular species or community. Each parameter is derived from specific mathematical algorithms, such as raw data, distance matrices, and convex packet volumes.

Based on the spatial distribution of functional traits, researchers have developed an increasing number of parameters to describe functional diversity (Schleuter et al. 2010; Pavoine and Bonsall 2011). These diverse parameters have contributed to our understanding of the environmental adaptation strategies of natural plant communities to some extent; however, the selection dilemma and even inappropriate selection based on too many parameters have been a major challenge for scientists so far. Therefore, it is crucial to comprehensively compare the ecological significance of different parameters and select the appropriate functional diversity parameter for specific scientific problems.

However, most practically relevant studies have focused on constructing functional diversity parameters and verifying the importance of functional diversity, although there is a lack of studies on the underlying mechanisms of functional trait diversity in maintaining community structure and functional stability. Notably, functional trait diversity is an integrated parameter of multiple functional traits in plants. Furthermore, whether the number and selection of functional traits affect functional diversity parameters and their ecological significance remain unclear.

4.1 Definition of Functional Trait Diversity

The extent and number of functional traits and their intrinsic cooperation mechanisms are essential for plants to adapt to the environment and to maintain their relative functional stability. Integrating multiple functional traits, exploring the interactions of different plant species in a community, and predicting functional traits with ecosystem functions are of great importance in functional ecology (Lei et al. 2016). Therefore, functional trait diversity in plants has been recognized by most ecologists since their emergence (Tilman 2001). Although many ecologists have attempted to uncover the ecological significance of functional trait diversity, it has not been well conceptualized and defined in the early stages of research (Hooper et al. 2005; Mason et al. 2005; Mouchet et al. 2010). Consequently, functional trait diversity was initially characterized through exploratory and qualitative descriptions.

There are currently multiple definitions of functional trait diversity. Initially, Tilman (2001) defined functional diversity as the range of functional trait values and variability exhibited by all plant species within a community, thereby influencing ecosystem function to some extent. In this concept, Tilman emphasized the importance of variability of functional traits present within a given community. Subsequently, other researchers have provided different definitions of functional diversity from different perspectives, such as the diversity of functional traits within a community (Tesfaye et al. 2003), the number and types of functional traits within an ecosystem and their distribution (Díaz and Cabido 2001), and types and distribution of functional traits of organisms in an ecosystem. Functional trait diversity (or functional diversity) represents the spatial distribution and extent of multiple functional traits within a plant community. To date, the concept of functional trait diversity has gained broad acceptance among researchers, although there is some controversy within scientific communities.

Functional trait diversity is used to describe the overall variation or diversity of plant functional traits within a community (Petchey and Gaston 2006). They are more holistic and abstract parameters that are believed to play a crucial role in driving ecosystem functions. This is because plant functional trait diversity considers redundant species and interspecific complementarity in plant communities (Díaz and Cabido 2001). Therefore, theoretically, it is a matter of linking multiple plant functional traits with ecosystem functions and quantitatively revealing ecosystem functions and even multifunctionality (Schleuter et al. 2010; Yan et al. 2023a, b, c). However, there are some dilemmas for scientists to resolve, for example, whether functional trait diversity values are influenced by these selected traits as well as the number of traits and how functional diversity is related to ecosystem functions in natural communities at different scales.

4.2 Classification of Functional Trait Diversity

Specifically, quantifying functional trait diversity in a multidimensional resource space is useful for understanding community composition, which is further linked to ecosystem functions. With the development of functional trait diversity, researchers have gradually realized that a single index of functional diversity is insufficient to accurately and comprehensively represent multiple functional traits. Therefore, numerous functional diversity indices have been continuously developed and refined. To facilitate the differentiation (or ecological significance) of different parameters, researchers have established a classification of functional diversity indices with three dimensions, as shown in Fig. 4.1, thereby comprehensively studying their role in regulating plant growth and development in natural communities (Mason et al. 2005; Ricotta 2007; Villéger et al. 2008; Mouchet et al. 2010). Overall, the relatively commonly accepted classification is to divide functional trait diversity into three levels: functional richness, functional homogeneity, and functional dispersion (Walker et al. 1999; Mason et al. 2005; Schleuter et al. 2010; Pavoine and Bonsall 2011). Practically, functional richness, functional evenness, and functional dispersion are independent of each other but can quantify different aspects of functional trait diversity from different perspectives, which can synergistically reveal the response of plant communities to the external environment and the maintenance mechanisms of ecosystem functions.

Herein, functional richness describes the ecological niche space that plant species occupy in a community through multiple functional traits. These multiple functional traits are tied to individual species and may reflect the species adaptation to the environment (Ansquer et al. 2009; Reich 2014). However, the adaptation strategies of the same species to different environments inevitably vary, indicating that there is intraspecific variation within a specific community. Moreover, within different plant communities with the same species or species richness, functional traits exhibit different levels of adaptation to different environments. Thus, both functional richness and species richness are somewhat related but fundamentally different

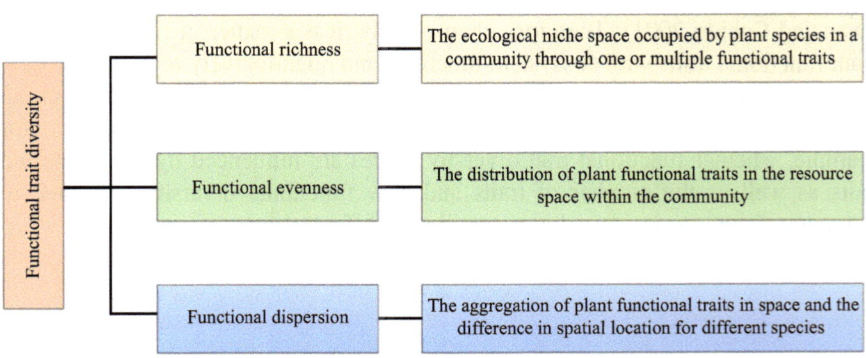

Fig. 4.1 Three dimensions of plant functional trait diversity and their connotations

(Kunstler et al. 2016). High functional trait richness can reflect higher efficiency of resource-environmental space use by plant species in the environment and potentially higher productivity of plant communities than low functional trait richness (Yuan et al. 2019). In contrast, low functional richness implies that the community has more unused resource space and is more vulnerable to invasion by exotic species (Pyles et al. 2020). In conclusion, these parameters of functional richness are essential for productivity optimization, productivity stability, and environmental resilience to disturbances (Macdougall et al. 2013; Correia et al. 2018).

The parameters associated with functional evenness are primarily used to characterize the distribution of plant functional traits in resource space within the community, with emphasis on characterizing whether the functional trait values are uniform or not. High functional trait uniformity implies that species and their functional traits are evenly distributed in ecological niche space within the community, and there is no better resource use preference. Conversely, communities with a greater preference for resource use should have greater assemblages of plant species (Kearsley et al. 2019).

Functional dispersion is primarily used to quantify the aggregation of plant functional traits in space and the difference in spatial location of different species, that is, the difference in the inherent variation of species and functional traits. High species abundance and low functional traits can lead to increased functional dispersion, and vice versa. Furthermore, high species abundance can lead to low variation in functional dispersion for equally high functional traits (Mason et al. 2005). Overall, these functional dispersion parameters can be a good measure to describe the degree of variation and competition between plant communities at different scales (Tordoni et al. 2019).

4.3 Calculation of Functional Trait Diversity

To calculate the indices of plant functional diversity, researchers must first conduct a systematic field survey to sample and measure multiple functional trait values for each species within the communities, then construct a matrix of multiple functional traits related to these plant species, and finally quantify functional diversity in terms of functional richness, functional evenness, and functional dispersion (Fig. 4.2). As too many parameters (or calculation methods) are developed for functional trait diversity, we only introduce a few common parameters and the corresponding calculation methods (Laliberté and Legendre 2010; Mouchet et al. 2010).

The method of calculating functional diversity indices is shown in Fig. 4.3, and it involves three main steps: convex packet volume, the center of gravity method, and shortest distance/tree (Song et al. 2011; Lei et al. 2016).

There are three methods for visualizing functional diversity: the convex packet volume method, multivariate mean difference, and homogeneity. A species (spi) with two functional trait eigenvalues (functional trait I and functional trait II) is

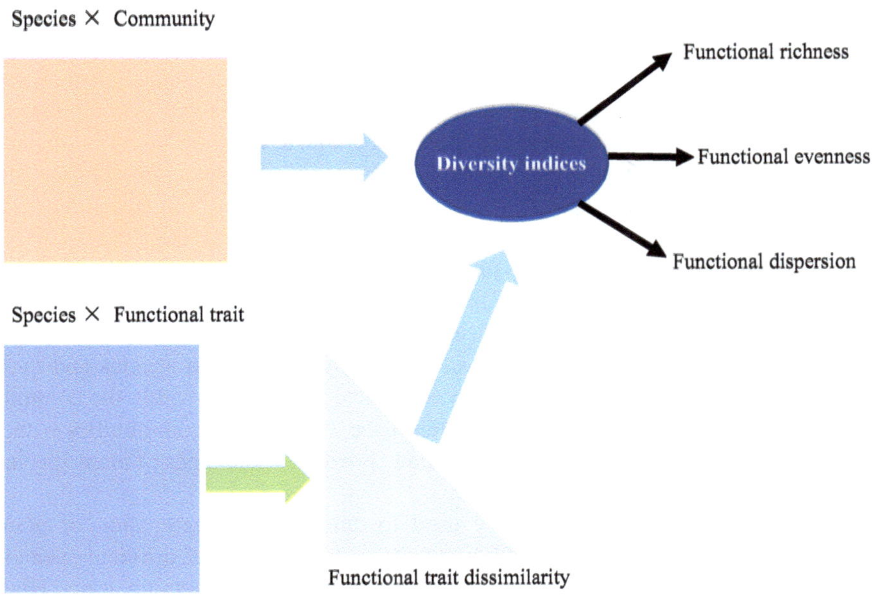

Fig. 4.2 Basic idea of calculating functional trait diversity in plants

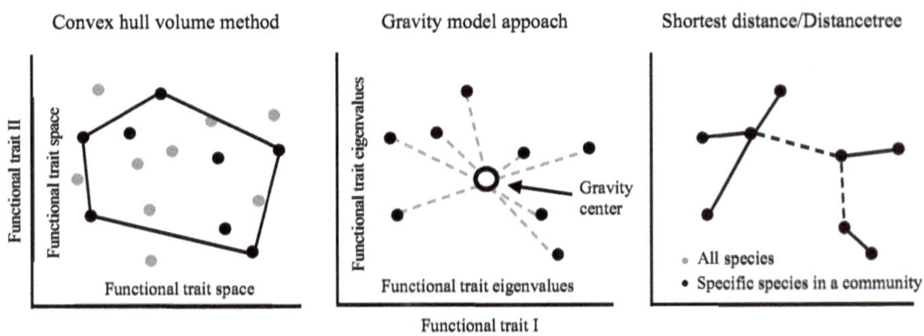

Fig. 4.3 Calculation method of functional diversity index

represented by each point in the functional trait space. The position of a given point on the coordinate axis reflects the position of the species in two-dimensional trait space. Black points indicate species in the target community, and gray points indicate species in other communities. In specific calculations, researchers can perform calculations on more than two functional trait matrices, and two (or more) functional traits in the graph can be replaced by multiple axes, where different matrices can be processed using PCoA (Villéger et al. 2008).

4.3.1 Calculation of Functional Richness

Functional richness represents the extent of the functional trait space in a specific community. For a single functional trait, the value of functional richness is the difference between its maximum and minimum in the community.

1. Multidimensional functional richness (FR_{lm})

FR_{lm} is the total volume of functional trait space for all plant species in a community, which can be simultaneously calculated for multiple functional traits as follows (Schleuter et al. 2010):

$$FR_{lm} = \int_{i}^{max} [f_i(x)]dx \tag{4.1}$$

$$f_i(x) = \exp\left[-\frac{1}{2}\left(x-\mu_i\right)^T {\sum}_i^{-1} (x-\mu_i)\right] \tag{4.2}$$

where x is the value of the functional trait, i is the species, μ_i is the mean of functional trait for species i, Σ_i is the variance/covariance matrix of functional traits for each species, and $f_i\ (x)$ is the dependent function of the functional trait space of all plant species in the community.

FR_{lm} represents the variations in individual functional traits and gaps in the spatial extent of multidimensional functional traits. Its disadvantage is the high data requirement, which should include data on the intraspecific variations in functional traits.

2. Functional richness index (FRic)

Functional richness index (FRic), which corresponds to the volume of a multidimensional convex package for multiple functional traits of species in a community (Villéger et al. 2008), is extremely complex to calculate. This requires the use of computer software (http://www.Pricklysoft.org/software/raithull.html) and open-source programs in the R Language Toolkit (http://www.ecolag.univmontp2.fr/index.php?option=com_content&%20task=view%20&%20id=219&Itemid=125).

3. Functional tree distance (functional attribute diversity, FAD)

FAD is a multidimensional functional richness index that uses the Euclidean distance rule to calculate the sum of distances of species in pairs in a data array assembled in functional trait space (Walker et al. 1999).

$$ED_{ij} = \sqrt{\sum_{t=1}^{T} \left(x_{tj} - x_{ti}\right)^2} \tag{4.3}$$

$$\mathrm{FAD} = \sum\nolimits_{i=1}^{S} \sum\nolimits_{j>1}^{S} \mathrm{ED}_{ij} \quad (4.4)$$

where T is the number of functional traits; x_{ti} and x_{tj} are the values of population i and population j for functional trait T, respectively; ED_{ij} is the Euclidean distance between species i and species j; and S is the number of species.

4. Functional trait average distance (modified FAD, MFAD)

MFAD is a functional richness index obtained by improving the FAD calculation considering species abundance (Schmera et al. 2009). The calculation formula is as follows:

$$\mathrm{MFAD} = \frac{\mathrm{FAD}}{S} \quad (4.5)$$

In the formula, FAD is the multidimensional functional richness index, and S is the number of species in a plant community.

4.3.2 Calculation of Functional Evenness

Mason et al. (2005) proposed the functional evenness index to reveal the distribution pattern of functional traits in the space occupied by species. A high functional homogeneity index implies a very regular distribution of multiple functional traits; conversely, a low functional homogeneity index indicates the presence of gaps in species distributions in space. Mason et al. (2005) have developed the specifics of the calculation technique. Practically, it is necessary to consider the functional evenness of multiple functional traits and to further calculate the multidimensional functional evenness index (FEve). In theory, the bell-shaped curves show the variety of species in the ecotone space, the vertical dashed lines indicate the range of species occupying the ecotone space in the community, and the vertical axis represents the multiplicity.

Functional evenness is influenced by abundance and functional traits of both species and is calculated using the ratio of the tree distance between the paired species to their branch lengths, thereby linking all species and functional traits together. The minimum spanning trees were constructed as a multidimensional space using species and multiple functional traits, and the uniformity of branch lengths for the minimum spanning tree was calculated (Villéger et al. 2008). The specific equations are as follows:

$$\mathrm{FEve} = \frac{\sum_{b=1}^{s-1} \min\left(\mathrm{PEW}_b * \frac{1}{S-1}\right) - \frac{1}{S-1}}{1 - \frac{1}{S-1}} \tag{4.6}$$

$$\mathrm{PEW}_b = \frac{\mathrm{EW}_b}{\sum_{b=1}^{S-1} \mathrm{EW}_b} \tag{4.7}$$

$$\mathrm{EW}_b = \frac{d_{ij}}{w_i + w_j} \tag{4.8}$$

where S is the number of species, PEW_b is the locally weighted average uniformity, EW_b is the uniformity obtained from the weighted average, w_i is the relative multiplicity of species i, and d_{ij} is the Euclidean distance between species i and species j.

4.3.3 Calculation of Functional Dispersion

Functional dispersion is commonly expressed as a functional dispersion index, which can be primarily used to characterize the maximum dispersion of the spatial distribution for the multiplicity of functional traits in communities (Mason et al. 2005). Among them, high aggregation or multiplicity of edge species in the functional trait space may lead to high functional dispersion. Herein, the formulae for calculating functional dispersion for single functional traits have been described according to the priority order. In practice, the vertical axis represents the multiplicity, the bell-shaped curve indicates the multiplicity of species in the ecological niche space, and the vertical dashed lines represent the range of species filling the ecological niche space in the community.

1. Functional divergence index (FDiv)

Functional divergence index (FDiv) is a multidimensional index used to quantify the aggregation of various functional traits in the distribution of resource space. The closer the dominant species in a community are to the center of gravity, the lower the functional dispersion. Like the calculation methods for the multidimensional functional richness index, FDiv is calculated from the multidimensional spatial convex packet volume by calculating the average distance from the center of gravity between different functional traits in the multidimensional convex packet volume (Villéger et al. 2008) and then determine the dispersion index based on the multidimensional weights. The specific calculation formula is presented as follows:

$$g_k = \frac{1}{S}\sum\nolimits_{i=1}^{S} x_{ik} \tag{4.9}$$

$$\mathrm{dG}_i = \sqrt{\sum_{k=1}^{\mathrm{T}} (x_{ik} - g_k)^2} \tag{4.10}$$

$$\overline{\mathrm{dG}} = \frac{1}{S}\sum_{i=1}^{S} \mathrm{dG}_i \tag{4.11}$$

$$\Delta \mathrm{d} = \sum_{i=1}^{S} w_i * \left(\mathrm{dG}_i - \overline{\mathrm{dG}}\right) \tag{4.12}$$

$$\Delta|\mathrm{d}| = \sum_{i=1}^{S} w_i * \left|\mathrm{dG}_i - \overline{\mathrm{dG}}\right| \tag{4.13}$$

$$\mathrm{FDiv} = \frac{\Delta \mathrm{d} + \overline{\mathrm{dG}}}{\Delta|\mathrm{d}| + \overline{\mathrm{dG}}} \tag{4.14}$$

2. Functional dispersion index (FDis)

FDis is a new index developed as an improvement of FDiv. FDis indices consider the relative multiplicity and evenness of plant species and therefore represent a more intuitive index of functional dispersion in the community. The weighted center of gravity is first calculated, the distance between paired species is then calculated, and finally the weighting methods ensure that these parameters themselves are no longer influenced by species richness. The calculation equations are as follows:

$$\mathrm{c} = [c_i] = \frac{\sum a_j x_{ij}}{\sum a_j} \tag{4.15}$$

$$\mathrm{FDis} = \frac{\sum a_j z_j}{\sum a_j} \tag{4.16}$$

where a_j is the relative multiplicity of species j, x_{ij} is the i-functional trait of species j, and z_j is the distance of the weighted center of gravity c of species j.

3. Rao quadratic entropy index (quadratic entropy, RaoQ)

Rao's parameters have defined a quadratic entropy equation that is primarily used to quantify the diversity and variability of functional traits for plant species within the community (Pavoine et al. 2005). The specific equations are described as follows:

Table 4.1 Methods to calculate the diversity parameters of plant functional traits in natural communities

Classification	Parameters	Name and abbreviation	Features and significance
Functional richness	Feature richness index	Functional richness index, FRic	The total volume of functional trait space for all plant species in a community, considering multiple functional traits simultaneously
	Functional tree distance	Functional attribute diversity, FAD	By calculating the sum of the distances of paired species in the data array consisting of functional trait space
	Average distance of functional traits	Modified FAD, MFAD	Considering species abundance, the indices are calculated as improved FAD
Functional evenness	Functional evenness index	Functional evenness index, FEve	The degree of uniformity in the distribution of plant functional traits within the community across ecological space
Functional dispersion	Functional divergence index	Functional divergence, FDiv	Quantifying the aggregation of plant functional traits in a community's resource space
	Functional dispersion index	Functional dispersion, FDis	Considering the relative multiplicity and evenness of species, it is a more intuitive index to describe functional dispersion in a community
	Secondary entropy index	Quadratic entropy, RaoQ	Quantifying the diversity and variability of plant functional traits within a community

$$d_{ij} = \frac{1}{T}\sum\nolimits_{1}^{T}\left(x_{tj} - x_{ti}\right)^2 \tag{4.17}$$

$$\text{RaoQ} = \sum\nolimits_{i=1}^{S-1}\sum\nolimits_{j=i+1}^{S} d_{ij}p_i p_j \tag{4.18}$$

where S is the number of species; T is the number of functional traits; x_{tj} and x_{ti} are the t-functional traits of species j and i, respectively; p_i is the relative abundance of species I; and d_{ij} is the distance between species i and j in the space of functional traits.

In summary, we have summarized the commonly used methods for calculating plant functional trait diversity parameters in natural communities, as shown in Table 4.1 below:

4.4 Major Advances in Studying the Diversity of Functional Trait

4.4.1 Advances in Functional Trait Diversity in Plant Communities

Functional trait diversity aims to represent the diversity, range, and space distribution of multiple plant functional traits within a community. In contrast to traditional species taxonomic diversity, these functional trait diversity parameters consider redundant species and interspecific complementarity in plant communities and can therefore organically link plant functional traits to ecosystem functions to a large extent (Díaz and Cabido 2001; Schleuter et al. 2010). Increasing evidence has shown that the functional trait diversity of plants has the potential to accurately predict ecosystem functioning and even multifunctionality at local scales (Hulot et al. 2000; Heemsbergen et al. 2004; Li et al. 2022a, b, c, d; Yan et al. 2023a, b, c).

Since the 1990s, researchers have conducted many studies on the relationships between functional trait diversity (or functional diversity) and ecosystem functions and have generally considered it as one of the main mechanisms through which diversity can influence ecosystem functions to some extent. Therefore, functional trait diversity is generally considered a specific parameter of diversity (Loreau 2000; Loreau et al. 2001. Tilman 2001). Theoretically, overlapping ecological niches are common in natural plant communities, where plant functional traits more directly reflect species' responses and their adaptations to the environment and can more specifically represent the spatial distribution and diversity of available resources. Therefore, to study the relationships between diversity and ecosystem functions, more attention should be paid to functional traits and not just species diversity itself.

In the early phase of functional trait diversity, most studies have assumed that species richness corresponds to high functional richness. Therefore, species richness has also been used to represent functional richness in some studies. However, Díaz and Cabido (2001) opined that species diversity can only replace functional diversity if the ecological niche space is linearly related to species richness. However, high interspecific ecotone overlap and large intraspecific variation are observed in most natural plant communities. Therefore, in these studies of diversity-ecosystem functions, species diversity cannot simply replace functional trait diversity. Moreover, experimental results have shown that changes in the biodiversity of plant communities are typically not accompanied by changes in the diversity of functional traits. However, there is a lack of uniform and convenient methods for measuring functional trait diversity in practice. Therefore, there are still many studies that use species richness as a simple and easily measurable indicator, instead of functional trait diversity to investigate diversity-ecosystem functions.

Regarding the methods to calculate plant functional trait diversity, previous studies have typically classified plant species in these communities into different functional groups based on a specific functional trait, and the abundance of functional groups was then used as an indicator of functional trait diversity. The richness of plant functional group may play an important role in regulating primary productivity of the ecosystem (Hooper et al. 2005); however, they ignored the importance of species richness, leading to increased uncertainty (Díaz and Cabido 2001). Over the last decade, researchers have developed several indices for calculating functional trait diversity in plant communities (Villéger et al. 2008; Schleuter et al. 2010), which have largely compensated for the shortcomings of the abovementioned classification methods. First, Mason et al. (2005) classified functional trait diversity into three components, such as functional richness, functional evenness, and functional divergence, based on the components of species diversity. Subsequently, studies have accurately quantified the abstract concepts, which is crucial for the development of functional trait diversity (Petchey and Gaston 2006). Concurrently, some researchers have created or enhanced these functional diversity index techniques, which greatly aided later development (Petchey and Gaston 2002; Cornwell et al. 2006).

With the development of functional trait diversity indices, they were gradually applied in ecological studies. Dong et al. (2017a, b) found that species diversity, functional diversity, and their relationships with primary productivity differ significantly in alpine meadow and alpine grassland communities, and that functional diversity can explain greater variation in ecosystem productivity than species diversity. Researchers have demonstrated that the impact of diversity on aboveground carbon stocks is mediated by functional diversity and functional dominance, suggesting that ecological niche complementarity and selection effects are not the only factors affecting carbon stocks (Mensah et al. 2016). Furthermore, the effect of functional diversity is larger than the effect of functional dominance, depending on maximum plant height, which explains the effect of vertical stratification in forests (Ottaviani et al. 2020a, b).

With the rapid evolution of functional traits in recent decades, researchers have begun focusing on them and made great progress, particularly regarding how the diversity of functional traits can be used to explain the spatiotemporal variations in ecosystem productivity (Xu et al. 2018), which is an effective way to explore functional trait diversity and ecosystem functions in the future (Mcelroy et al. 2012). The relationships between productivity and functional trait diversity are important for predicting ecosystem response to different global change scenarios (Griffin et al. 2009; Meira-Junior et al. 2016). However, most of these relationships have been observed in grasslands or some field-controlled experiments at the local scale (Niu et al. 2014; Pakeman 2014; Liu et al. 2015a, b; Milcu et al. 2016). However, establishing the relationships between functional trait diversity and productivity in natural ecosystems is a challenge, particularly at large scales. Hence, we have reported here, in detail, recent research in large natural communities, aiming to investigate the relationships between functional trait diversity and ecosystem functions in forest ecosystems at the biome scale (Zhang et al. 2021a, b, c, d, e; Li et al. 2022a, b, c, d; Yan et al. 2023a, b, c).

4.4.2 *Spatial Variations in Functional Trait Diversity and Linking to Productivity*

Because of the complexity of the relationships between diversity and productivity, some studies have attempted to examine the effects of diversity on ecosystem productivity in field-controlled plots (Hooper et al. 2005), and some studies have established the links between functionality diversity and ecosystem functions (Díaz and Cabido 2001; Schumacher and Roscher 2009; Rawat et al. 2020). However, determining the relationship between functional trait diversity and productivity in natural ecosystems, particularly at large scales, remains a major challenge (Li et al. 2022a, b, c, d).

Large-scale forest ecosystems are characterized by long growth histories, large plant species, complex local environments, and overlapping effects of biotic and abiotic factors (He et al. 2018a, b, c; Li et al. 2018a, b). Therefore, the mechanism of how functional trait diversity in large forest communities regulates ecosystem productivity is still unclear. Li et al. (2022a, b, c, d) selected nine natural forests along the North–South Transect of Eastern China to survey and collect all observed species in the sample plots (Fig. 4.4). In total, 366 tree species and their leaf samples are collected. Researchers then measured 10 functional plant traits that are closely related to productivity and assigned leaf size, leaf specific area, stomata size, stomata density, leaf chlorophyll content, leaf nitrogen content, and leaf phosphorus content. Based on the trait data, three functional diversity indices [functional richness index (Fric), functional evenness index (FEve), and functional dispersion index (FDiv)] were calculated for each community to determine the relationships between functional diversity and ecosystem productivity in forest communities at the biome scale (Li et al. 2022a, b, c, d).

Using a "systematic and consistent" protocol, Dr. He′s team measured functional trait data. The study focused on spatial variations in plant functional diversity and how it relates to gross primary productivity (GPP) in large-scale natural forests (He et al. 2019a, b, 2020a, b, c). The functional diversity indices of richness (FRic), evenness (FEve), and dispersion (FDiv) varied continuously, but in different directions, owing to climate (primarily temperature and precipitation). In particular, FRic increased with increasing temperature and precipitation, while FDiv decreased (Fig. 4.5). This variation in functional trait diversity can explain the mechanisms of environmental adaptation and functional optimization of forest communities.

Among the three parameters of functional trait diversity, GPP increased with FRic and decreased with FDiv, and FEve did not correlate with GPP. Therefore, FRic and FDiv, rather than FEve, were identified as important indicators to influence large-scale GPP. Notably, functional trait diversity, total soil nitrogen (STN), annual precipitation (MAP), and mean annual temperature (MAT) together explained 90.42% of the total variation in GPP. The combination of functional trait diversity and abiotic factors can more effectively explain the spatial variations in ecosystem productivity, broadening the potential application of functional trait diversity (Li et al. 2022a, b, c, d). The results of this study provide novel insights into

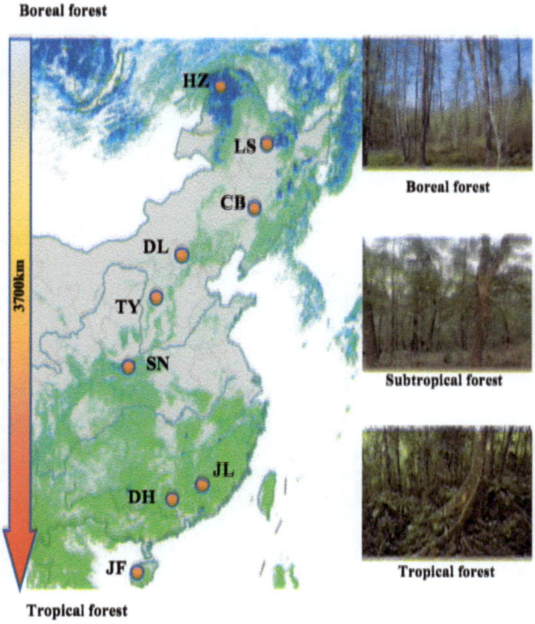

Category	Traits
Morphological traits	Leaf area (LA)
	Leaf dry weight (LDW)
	Leaf water content (LWC)
	Specific leaf area (SLA)
	Leaf thickness (LT)
Stoichiometric	Leaf carbon content (C)
	Leaf nitrogen content (N)
	Ratio of C:N
	Leaf Ferrum content (Fe)
	Leaf Kalium content (K)
	Leaf Magnesium content (Mg)
	Leaf Phosphorus content (P)
	The ratio C:P
Chlorophyll traits	Leaf Chlorophyll a content (*Chl a*)
	Leaf Chlorophyll b content (*Chl b*)
	Total chlorophyll content (*Chl*)
	Ratio of *Chl a*: *Chla*
Stomatal traits	Stomatal pore length (SPL)
	Stomatal pore width (SPW)
	Stomatal area (SA)
	Stomatal length (SL)
	Stomatal density (SD)
	Stomatal area fraction (P)
Anatomical traits	upper epidermal cells width (UCW)
	Palisade tissue thickness(PT)
	Sponge tissue thickness(ST)
	Ratio of PT:ST
	lower epidermal cells width (LCW)
	vessels width (VW)
Other traits	Soluble sugar content (SSC)
	Starch content (SC)
	Non-structure carbohydrate (NSC)
	Ratio of SSC:SC
	Leaf calorific value (LCV)

366 species and 34 traits

Fig. 4.4 Spatial distribution of nine typical forest communities from the boreal to the tropical zones. There were 34 plant functional traits, and 366 trees were systematically measured

evaluating the adaptive mechanisms of plant communities and their GPP response to global change scenarios (Fig. 4.6). Overall, by integrating different components of functional trait diversity in response to ecosystem functioning, it is possible to improve our understanding of functional trait diversity and the underlying mechanisms of ecosystem productivity optimization and stability in the future.

4.4.3 Functional Trait Diversity at the Regional Scale and Its Association with Interannual Variability in Productivity

The GPP of the ecosystem is one of the important components of the carbon cycle, and its annual variability can affect the carbon balance of the ecosystem in the long term (Ballantyne et al. 2012; Fernández-Martínez et al. 2019; Piao et al. 2020). Both climate change and biodiversity loss have significant effects on the interannual variability of the GPP, such as altered rainfall frequency or increased temperatures that can lead to interannual fluctuations in the productivity of terrestrial ecosystems.

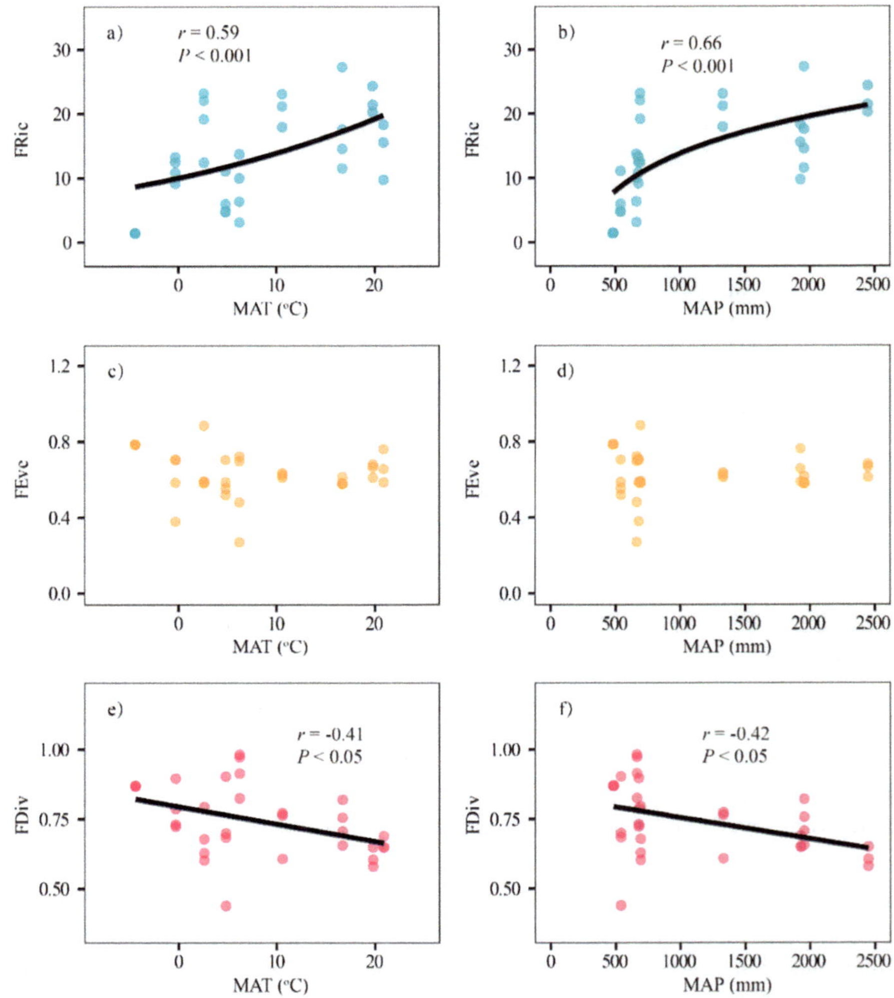

Fig. 4.5 Relationship between functional trait diversity of forest communities and climatic factors. (*FRic* functional richness index, *FEve* functional evenness index, *FDiv* functional dispersion index, *MAT* mean annual temperature, *MAP* annual rainfall)

Although there is ample evidence that biodiversity may stabilize ecosystem productivity and reduce interannual variability in terrestrial ecosystem productivity (Isbell et al. 2015; Oehri et al. 2017; Craven et al. 2018; García-Palacios et al. 2018), the role of multiple diversity (e.g., taxon diversity, functional diversity, and phylogenetic diversity) have not been fully understood, particularly across multiple ecosystem types (Yan et al. 2023a, b, c). Moreover, the availability of soil nutrients, such as nitrogen and phosphorus, is believed to influence the interannual variability of ecosystem productivity (Fernández-Martínez et al. 2019). Previous studies have shown that high ecosystem productivity typically results in low interannual

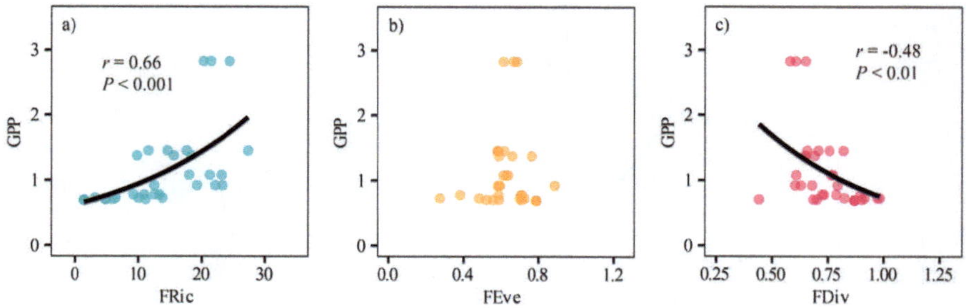

Fig. 4.6 Functional diversity in relation to the total primary productivity (GPP) of ecosystems. (*FRic* functional richness index, *FEve* functional evenness index, *FDiv* functional dispersion index)

variability (Roy et al. 2001). As soil nutrient levels increase, the number of plant species in ecosystems is expected to increase at the local scale, thereby indirectly reducing interannual variability in ecosystem productivity (Oehri et al. 2017). However, the direct and indirect effects of soil nutrients on the interannual variability of GPP have not been well demonstrated on a large scale.

Considering the abovementioned gaps, researchers have recently assessed the relative effects of climate factors, soil nutrients, and diversity on the role of interannual variability in the GPP across multiple ecosystem types, including forests, grasslands, and deserts. These plant functional trait data were derived from the China_Traits database using a systematic and consistent measurement protocol. Specifically, these functional trait data included (1) the data of 454 community composition collected in China, (2) plant functional trait data and phylogenetic data for more than 2500 plant species, and (3) accompanying measured soil nutrient data. Using these data from 72 typical natural ecosystems, researchers have systematically examined the effects of environmental factors (temperature, precipitation, and soil nutrients) and multifaceted diversity (species richness, functional diversity, and phylogenetic diversity) on the interannual variability of the PP and their pathways (e.g., direct and indirect).

The results of the study showed that soil nutrients and functional diversity are the two most important factors influencing the interannual variability of GPP (Fig. 4.7). Across all ecosystem types, ecosystems with rich soil nutrients and high functional diversity had low interannual variability in GPP (Fig. 4.8). The findings provide strong evidence that ecosystems with high biodiversity have low interannual variability in productivity. Ecosystems rich in soil nutrients are better protected from climate change than those poor in soil nutrients, because they reduce interannual variability in ecosystem productivity. Furthermore, the results suggest that multidimensional biodiversity, not only species richness, but also functional and phylogenetic diversity, plays a critical role in mitigating the impacts of climate change on the productivity of terrestrial ecosystems.

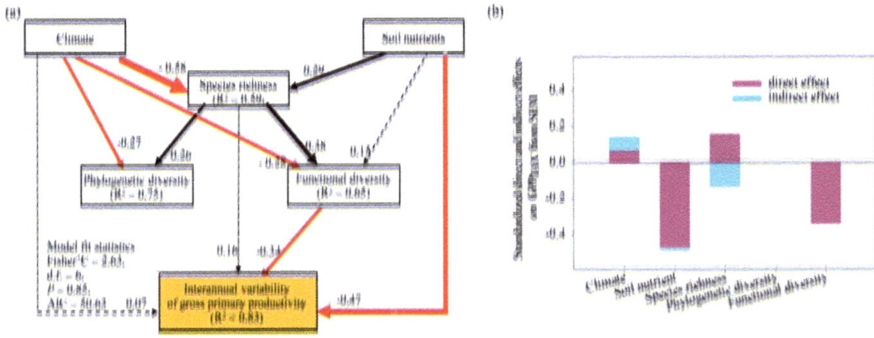

Fig. 4.7 Piecewise structural equation model revealing the direct and indirect effects of multiple abiotic and biotic factors on the interannual variability in gross primary productivity (GPP_{IAV}) at 72 sites. Black arrows indicate positive effects, while red arrows indicate negative effects; solid lines indicate statistically significant effects, while dashed lines indicate nonsignificant effects; the width of each significant path is proportional to its standardized path coefficient

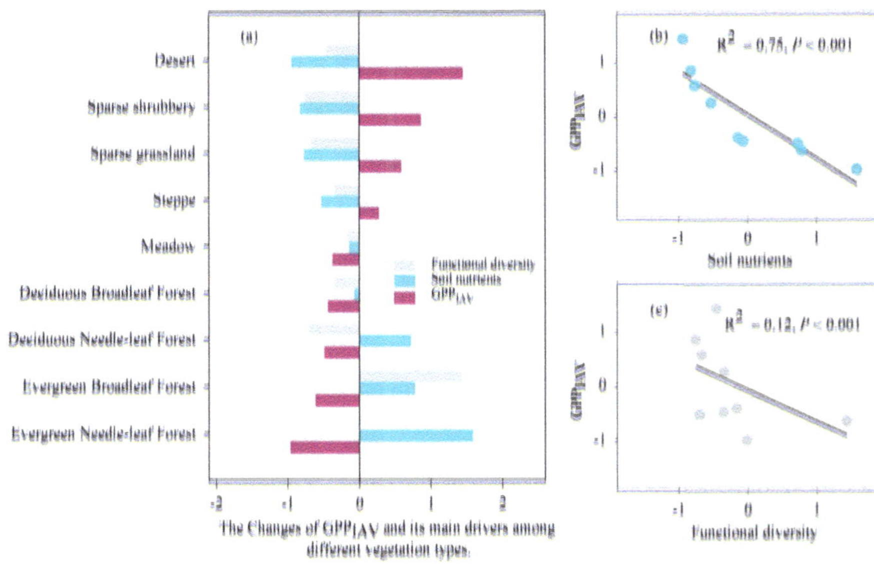

Fig. 4.8 Pattern of interannual variability in gross primary productivity (GPP_{IAV}), soil nutrients, and functional diversity among different vegetation types (**a**), and the bivariate relationship between GPP_{IAV} and soil nutrients and functional diversity (**b–c**)

4.4.4 Effect of the Number of Functional Traits on the Stability of Functional Trait Diversity

In nature, multiple functional traits work together to enable plants to perform multiple functions; therefore, functional diversity can accurately assess and predict

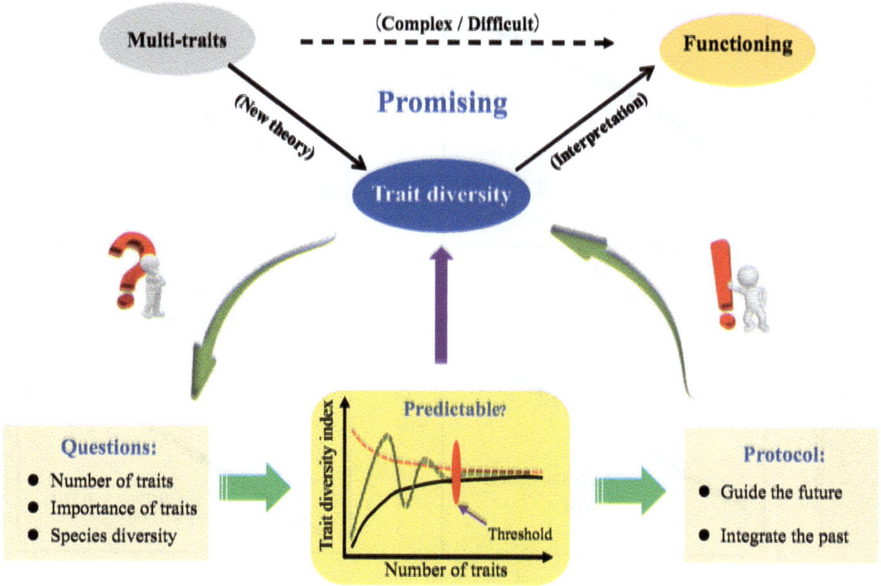

Fig. 4.9 Opportunity and challenge of functional diversity. The trait diversity framework holds promise for dealing with the complex relationships in natural communities. Specifically, these parameters of functional diversity are predictable (such as invariable, regular, or irregular) in different scenarios of different numbers of traits, diverse importance of different traits, and different species diversity

ecosystem functions (Petchey and Gaston 2006). Similarly, functional homogeneity and functional dispersion can provide information about the stabilization and adaptation mechanisms of community composition (Wang et al. 2011a, b, c; Komac et al. 2015). In the last decade, functional diversity has received considerable attention, and research on the quantitative relationship between functional diversity and ecosystem functions is rapidly increasing. However, it is controversial how these parameters of functional trait diversity can be scientifically calculated and used. In specific studies, researchers tend to select some functional trait parameters according to their scientific goals or data availability, and the obvious difference in data sources may easily lead to inconsistent or poorly comparable results (Zhang et al. 2021a, b, c, d, e). No researcher has attempted to examine the effect of the number of functional traits on the stability of functional trait diversity, although it is crucial (de Bello et al. 2009; Spasojevic and Suding 2012; Scheiner et al. 2017). In other words, are these functional diversity parameters relatively stable or comparable under changing scenarios of selected functional traits in terms of importance, number of functional traits, and species diversity in the community? Therefore, validation based on a large amount of functional trait data is urgently needed (Fig. 4.9).

The effect of "functional trait number or functional trait selection" on functional trait diversity must be scientifically determined. What is the spatial pattern of

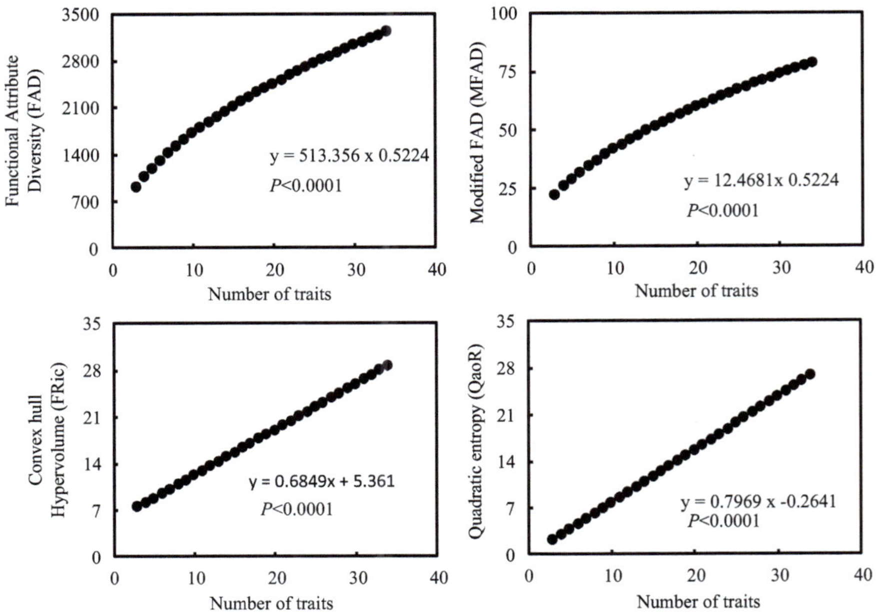

Fig. 4.10 Relationships between functional diversity and trait number. Four out of seven parameters of functional traits were predictable, namely functional attribute diversity (FAD), modified FAD (MFAD), convex hull hypervolume (FRic), and quadratic entropy (QaoR). When the number of traits was increased from 3 to 34, an average of 100 random combinations were calculated

variation for this effect? Recently, researchers measured 34 leaf functional traits in 366 plant species from nine typical forest communities in the North–South Transect of Eastern China to explore them (Zhang et al. 2021a, b, c, d, e; Fig. 4.9). The main objectives of this study were (1) to demonstrate the predictability of these functional diversity indicators by varying the number of selected functional traits and (2) to investigate how to scientifically calculate and predict functional diversity at the site-to-region scale to provide a theoretical basis for future studies on functional diversity.

The results showed that altering the number of traits had different effects on the parameters of trait diversity. FAD, MFAD, FRic, and RaoQ were predictable using specific equations (Fig. 4.10 and Table 4.2). Theoretically, the measurement of functional richness depends on understanding spatial niche distribution by measuring multiple traits of each species. The more functional traits involved, the larger the niche space that can be quantified (Mason et al. 2005). Therefore, these parameters of trait richness can be measured and are expected to be comparable to the total resource or niche space of different communities (de la Riva et al. 2018).

Three parameters of trait richness increased with the number of traits selected. Specifically, FRic showed a linear increasing trend ($R^2 = 0.9$), and FAD and MFAD were well fitted as power functions (Table 4.2). Furthermore, RaoQ was linearly correlated with the number of selected traits, like FRic. The plausible explanation is

Table 4.2 Protocol of trait diversity indices and their predictability

Trait diversity	Specific indexes	Predictable	Fitted equation
Trait richness	Functional attribute diversity (FAD)	Yes	Y = 513.356x^0.5224
	Modified FAD (MFAD)	Yes	Y = 12.4681x^0.5224
	Convex hull hypervolume (FRic)	Yes	Y = 0.6849x + 5.3610
Trait evenness	Functional evenness (FEve)	No	
Trait divergence	Functional divergence (FDiv)	No	
	Functional dispersion (FDis)	No	
	Quadratic entropy (RaoQ)	Yes	Y = 0.7968x-0.2641

that the calculated equation contains information on both functional richness and trait dispersion; however, functional richness is dominant. Furthermore, the general equations of FAD, MFAD, FRic, and RaoQ were the same in different forest communities from tropical to boreal zones, although some detailed information differed between these functional richness parameters.

Some parameters of trait diversity were unpredictable, such as FEve, FDiv, and FDis. These parameters changed irregularly with altering traits, and therefore, it was difficult to capture their changing trends. FEve, FDiv, and FDis were commonly summarized as indicators for quantifying the range of traits and the degree of distribution rules (Kraft and Ackerly 2010). The three parameters of trait diversity in terms of prevalence and distribution were more flexible than the parameters of functional richness. In other words, the selection of traits or trait number were more important for the parameters.

Trait diversity offers a promising perspective to explain variations in community or ecosystem functions, such as productivity, nutrient uptake, flexibility, and resilience (Cornwell et al. 2006). Naturally, they may also show how susceptible the community is to anthropogenic disruptions (Biswas et al. 2019). Therefore, trait diversity parameters are widely used in numerous fields of ecology (Petchey et al. 2004). However, our results showed that only part of the trait diversity (primarily functional richness) was predictable, suggesting that the theoretical basis for the application of trait diversity need to be further investigated. Scientists are generally unable to evaluate the repeatability or comparability of different studies owing to differences in the traits and numbers selected. Theoretically, it is speculated that the basis of the calculated trait diversity is like species diversity (Petchey and Gaston 2002). The theoretical basis of the latter is the theory of niche complementarity. The differentiation of different species is the differentiation of traits. Plant traits may more directly reflect resource acquisition and niche complementation of a species (Mason et al. 2005). Therefore, using functional diversity to directly measure all aspects of the spatial distribution of traits in a niche may also help test these theoretical mechanisms. However, the calculation methods of trait diversity cannot be ignored, because of the difference in the relative importance of specific traits. These functions should serve as a protocol guide for future studies or be integrated into other studies.

The number of traits is a non-negligible factor in the study of trait diversity; however, there is no implementation standard to refer to. Therefore, in previous studies, very few traits (three or four) were used to calculate trait diversity (Staples et al. 2019). Researchers should carefully design experimental protocols for their own experimental purposes. Some researchers may consider functional redundancy and avoid using more functional traits in their analyses (Kearsley et al. 2019). This is a potential question to determine which functional trait plays a more important role in the community, and it is also a question of trait importance. For natural communities, functional redundancy must exist, and functional redundancy can ensure the stability and anti-interference ability of the community (Pillar et al. 2013). Herein, the prediction equations for FRic and RaoQ increased nonlinearly with the number of traits, likely owing to functional redundancy between traits. Furthermore, the selection of additional traits is encouraged to further explore ecosystem processes and functions, reduce possible errors, and promote comparison among future studies.

4.5 Perspectives of Plant Functional Trait Diversity

1. Do not be confused by multiple functional diversity parameters

Various calculation methods and evaluation parameters have been developed to assess the functional diversity of traits (Villéger et al. 2008; Laliberté and Legendre 2010). Therefore, selecting appropriate parameters from these calculation methods and parameters is crucial. Selecting a uniform method for describing the functional trait space, such as functional dendrogram or multivariate space, is recommended, whenever possible; and this method should be used to examine its properties in multiple dimensions such as richness, homogeneity, and dispersion. Otherwise, the results obtained using different calculation methods may be problematic. Similarly, special care must be taken when comparing results from different studies, as they may be calculated using different methods and metrics.

2. A high degree of uncertainty in the calculation process itself should be fully understood

Ideally, the same (or convergent) results would be achieved if different methods were used to compute similar metrics, such as functional richness calculated using tree diagrams or convex hulls. However, practically, the results obtained using different methods typically differ significantly (Mammola and Cardoso 2020; Wong and Carmona 2021). Therefore, it is necessary to fully acknowledge the uncertainty of various parameters resulting from the calculation methods used.

3. Emphasize the influence of the selected functional traits and the number of functional traits

The parameter values of functional trait diversity are typically influenced by the selection of functional traits and the number of functional traits (Zhang et al.

2021a, b, c, d, e). When the selected functional traits or the number of traits change, the functional trait diversity parameters should also change. This should be emphasized when comparing different studies. Practically, it is important not to simply compare functional diversity with different trait selection and number, as they are not interchangeable, and being deceptive in this regard is incorrect.

4. Rational choice of sensible parameters and calculation methods

Various methods and metrics have been developed to assess functional trait diversity, and additional new parameters may be developed in the future. Therefore, selecting the appropriate methods and parameters is not only confusing but also a major challenge for researchers conducting studies on functional trait diversity.

5. More metrics should be developed in the future

The existing parameters of functional trait diversity are insufficient to comprehensively examine the distribution properties within each functional trait space and address questions related to ecological adaptation and evolution. One possible approach is the disorder framework (Guillerme et al. 2020), which offers a structured method for developing and evaluating new metrics suited to specific datasets and problems.

4.6 Summary

All plants in nature have many indispensable and important functional traits, which synergistically help them perform multiple functions. Traditionally, it is difficult to scientifically uncover the response and adaptation mechanisms of plants in nature based on a single or few functional traits. Therefore, researchers have developed a new conceptual system of functional trait diversity, which can reflect the overall differences, adaptation mechanisms, and function optimization mechanisms of plant communities through comprehensive parameters that characterize the distribution range and the degree of aggregation and dispersion of multiple functional traits. In this chapter, we reviewed the theoretical foundation and construction methods of different plant functional diversity and presented recent advances. Specifically, we reported a series of studies, with a large amount of supporting data, to investigate the spatial variability of large-scale plant functional diversity and its relationship to productivity. Furthermore, we discussed the dependence of plant functional diversity parameters on the number of selected functional traits and discussed their limitations and development directions in the future.

References

Ansquer P, Duru M, Theau JP et al (2009) Functional traits as indicators of fodder provision over a short time scale in species-rich grasslands. Ann Bot 103:117–126

Ballantyne A, Alden C, Miller J et al (2012) Increase in observed net carbon dioxide uptake by land and oceans during the past 50 years. Nature 488:70–72

Biswas SR, Mallik AU, Braithwaite NT et al (2019) Effects of disturbance type and microhabitat on species and functional diversity relationship in stream-bank plant communities. For Ecol Manag 432:812–822

Cornelissen JHC, Lavorel S, Garnier E et al (2003) A handbook of protocols for standardised and easy measurement of plant functional traits worldwide. Aust J Bot 51:335–380

Cornwell WK, Schwilk DW, Ackerly DD (2006) A trait-based test for habitat filtering: convex hull volume. Ecology 87:1465–1471

Correia DLP, Raulier F, Bouchard M et al (2018) Response diversity, functional redundancy, and post-logging productivity in northern temperate and boreal forests. Ecol Appl 28:1282–1291

Craven D, Eisenhauer N, Pearse WD et al (2018) Multiple facets of biodiversity drive the diversity-stability relationship. Nat Ecol Evol 2:1579–1587

de Bello F, Thuiller W, Lepš J et al (2009) Partitioning of functional diversity reveals the scale and extent of trait convergence and divergence. J Veg Sci 20:475–486

de Bello F, Lavorel S, Albert CH et al (2011) Quantifying the relevance of intraspecific trait variability for functional diversity. Methods Ecol Evol 2:163–174

de la Riva EG, Violle C, Perez-Ramos IM et al (2018) A multidimensional functional trait approach reveals the imprint of environmental stress in mediterranean woody communities. Ecosystems 21:248–262

Díaz S, Cabido M (2001) Vive la différence: plant functional diversity matters to ecosystem processes. Trends Ecol Evol 16:646–655

Dong SK, Tang L, Zhang SF et al (2017a) Relationship between plant species diversity and functional diversity in alpine grasslands. J Ecol 37:1472–1483

Dong N, Prentice IC, Evans BJ et al (2017b) Leaf nitrogen from first principles: field evidence for adaptive variation with climate. Biogeosciences 14:481–495

Fernández-Martínez M, Sardans J, Chevallier F et al (2019) Global trends in carbon sinks and their relationships with CO_2 and temperature. Nat Clim Chang 9:73–79

García-Palacios P, Gross N, Gaitán J et al (2018) Climate mediates the biodiversity-ecosystem stability relationship globally. Proc Natl Acad Sci 115:8400–8405

Griffin JN, Mendez V, Johnson AF et al (2009) Functional diversity predicts overyielding effect of species combination on primary productivity. Oikos 118:37–44

Guillerme T, Puttick MN, Marcy AE et al (2020) Shifting spaces: which disparity or dissimilarity measurement best summarize occupancy in multidimensional spaces? Ecol Evol 10:7261–7275

He NP, Liu CC, Tian M et al (2018a) Variation in leaf anatomical traits from tropical to cold-temperate forests and linkage to ecosystem functions. Funct Ecol 32:10–19

He NP, Liu CC, Zhang JH et al (2018b) Opportunities and challenges for plant trait research: from organs to communities. Acta Ecologic Sinica 38:6787–6796

He NP, Zhang JH, Liu CC et al (2018c) Patterns and influencing factors of traits in forest ecosystems: synthesis and perspectives on the synthetic investigation from the north-east transect of eastern China (NETEC). Acta Ecol Sin 38(18):6359–6382

He NP, Liu CC, Piao SL et al (2019a) Ecosystem traits linking functional traits to macroecology. Trends Ecol Evol 34:200–210

He L, Chen JM, Liu J et al (2019b) Diverse photosynthetic capacity of global ecosystems mapped by satellite chlorophyll fluorescence measurements. Remote Sens Environ 232:111344

He NP, Li Y, Liu CC et al (2020a) Plant trait networks: improved resolution of the dimensionality of adaptation. Trends Ecol Evol 35:908–918

He NP, Liu CC, Xu L et al (2020b) Ecosystem traits open new insights into macroecology: opportunity and challenge. Acta Ecol Sin 40(8):2507–2522

He P, Gleason SM, Wright IJ et al (2020c) Growing-season temperature and precipitation are independent drivers of global variation in xylem hydraulic conductivity. Glob Chang Biol 26: 1833–1841

Heemsbergen DA, Berg MP, Loreau M et al (2004) Biodiversity effects on soil processes explained by interspecific functional dissimilarity. Science 306:1019–1020

Hooper DU, Chapin FS III, Ewel JJ et al (2005) Effects of biodiversity on ecosystem functioning: a consensus of current knowledge. Ecol Monogr 75:3–35

Hulot FD, Lacroix G, Lescher-Moutoué F et al (2000) Functional diversity governs ecosystem response to nutrient enrichment. Nature 405:340–344

Isbell F, Craven D, Connolly J et al (2015) Biodiversity increases the resistance of ecosystem productivity to climate extremes. Nature 526:574–577

Kearsley E, Hufkens K, Verbeeck H et al (2019) Large-sized rare tree species contribute disproportionately to functional diversity in resource acquisition in African tropical forest. Ecol Evol 9:4349–4361

Komac B, Pladevall C, Domènech M et al (2015) Functional diversity and grazing intensity in sub-alpine and alpine grasslands in Andorra. Appl Veg Sci 18:75–85

Kraft NJB, Ackerly DD (2010) Functional trait and phylogenetic tests of community assembly across spatial scales in an Amazonian forest. Ecol Monogr 80:401–422

Kunstler G, Falster DS, Coomes DA et al (2016) Plant functional traits have globally consistent effects on competition. Nature 529:204–207

Laliberté E, Legendre P (2010) A distance-based framework for measuring functional diversity from multiple traits. Ecology 91:299–305

Legras G, Loiseau N, Gaertner JC (2018) Functional richness: overview of indices and underlying concepts. Acta Oecol 87:34–44

Lei AJ, Kong DL, Li XM et al (2016) Plant functional traits, functional diversity, and ecosystem function: progress and perspectives. Biodiversity 24:922–931

Li NN, He NP, Yu GR et al (2016) Leaf non-structural carbohydrates regulated by plant functional groups and climate: evidences from a tropical to cold-temperate forest transect. Ecol Indic 62: 22–31

Li Y, He NP, Hou JH et al (2018a) Factors influencing leaf chlorophyll content in natural forests at the biome scale. Front Ecol Evol 6:64

Li Y, Liu CC, Zhang JH et al (2018b) Variation in leaf chlorophyll concentration from tropical to cold-temperate forests: association with gross primary productivity. Ecol Indic 85:383–389

Li X, Li MX, Xu L et al (2022a) Allometry and distribution of nitrogen in natural plant communities of the Tibetan plateau. Front Plant Sci 13:845813

Li Y, Hou J, Xu L et al (2022b) Variation in functional trait diversity from tropical to cold-temperate forests and linkage to productivity. Ecol Indic 138:108864

Li Y, Liu CC, Sack L et al (2022c) Leaf trait network architecture shifts with species-richness and climate across forests at continental scale. Ecol Lett 25:1442–1457

Li J, Chen X, Niklas KJ et al (2022d) A whole-plant economics spectrum including bark functional traits for 59 subtropical woody plant species. J Ecol 110:248–261

Liu B, Li H, Zhu B et al (2015a) Complementarity in nutrient foraging strategies of absorptive fine roots and arbuscular mycorrhizal fungi across 14 coexisting subtropical tree species. New Phytol 208:125–136

Liu J, Zhang X, Song F et al (2015b) Explaining maximum variation in productivity requires phylogenetic diversity and single functional traits. Ecology 96:176–183

Loreau M (2000) Biodiversity and ecosystem functioning: recent theoretical advances. Oikos 91:3–17

Loreau M, Naeem S, Inchausti P et al (2001) Biodiversity and ecosystem functioning: current knowledge and future challenges. Science 294:804–808

Macdougall AS, Mccann KS, Gellner G et al (2013) Diversity loss with persistent human disturbance increases vulnerability to ecosystem collapse. Nature 494:86–89

Mammola S, Cardoso P (2020) Functional diversity metrics using kernel density n-dimensional hypervolumes. Methods Ecol Evol 11:986–995
Mason NW, Mouillot D, Lee WG et al (2005) Functional richness, functional evenness, and functional divergence: the primary components of functional. Oikos 111:112–118
Mcelroy MS, Papadopoulos YA, Adl MS (2012) Complexity and composition of pasture swards affect plant productivity and soil organisms. J Plant Sci 92:687–697
Meira-Junior MS, Imaña-Encinas J, Pinto JR et al (2016) Functional diversity influence in forest wood stock: a study of the brazilian savanna. Biosci J 32:1619–1631
Mensah S, Veldtman R, Assogbadjo AE et al (2016) Tree species diversity promotes aboveground carbon storage through functional diversity and functional dominance. Ecol Evol 6:7546–7557
Milcu A, Eugster W, Bachmann D et al (2016) Plant functional diversity increases grassland productivity-related water vapor fluxes: an Ecotron and modeling approach. Ecology 97: 2044–2054
Mouchet MA, Villéger S, Mason NW et al (2010) Functional diversity measures: an overview of their redundancy and their ability to discriminate. Funct Ecol 24:867–876
Niu K, Choler P, De Bello F et al (2014) Fertilization decreases species diversity but increases functional diversity: a three-year experiment in a Tibetan alpine meadow. Agric Ecosyst Environ 182:106–112
Oehri J, Schmid B, Schaepman-Strub G et al (2017) Biodiversity promotes primary productivity and growing season lengthening at the landscape scale. Proc Natl Acad Sci USA 114:10160–10165
Ottaviani G, Keppel G, Gotzenberger L et al (2020a) Linking plant functional ecology to Island biogeography. Trends Plant Sci 25:329–339
Ottaviani G, Molina-Venegas R, Charles-Dominique T et al (2020b) The neglected belowground dimension of plant dominance. Trends Ecol Evol 35:763–766
Pakeman RJ (2014) Leaf dry matter content predicts herbivore productivity, but its functional diversity is positively related to resilience in grasslands. PLoS One 9(7):e101876
Pavoine S, Bonsall MB (2011) Measuring biodiversity to explain community assembly: a unified approach. Biol Rev 86:792–812
Pavoine S, Ollier S, Pontier D (2005) Measuring diversity from dissimilarities with Rao's quadratic entropy: are any dissimilarities suitable? Theor Popul Biol 67:231–239
Petchey OL, Gaston KJ (2002) Functional diversity (FD), species richness and community composition. Ecol Lett 5:402–411
Petchey OL, Gaston KJ (2006) Functional diversity: back to basics and looking forward. Ecol Lett 9:741–758
Petchey OL, Hector A, Gaston KJ (2004) How do different measures of functional diversity perform? Ecology 85:847–857
Piao S, Wang X, Wang K et al (2020) Interannual variation of terrestrial carbon cycle: issues and perspectives. Glob Chang Biol 26:300–318
Pillar VD, Blanco CC, Muller SC et al (2013) Functional redundancy and stability in plant communities. J Veg Sci 24:963–974
Pyles MV, Magnago LF, Borges ER et al (2020) Land use history drives differences in functional composition and losses in functional diversity and stability of Neotropical urban forests. Urban For Urban Green 49:126608
Rawat M, Arunachalam K, Arunachalam A et al (2020) Relative contribution of plant traits and soil properties to the functioning of a temperate forest. Catena 194:104671
Reich PB (2014) The world-wide 'fast-slow' plant economics spectrum: a traits manifesto. J Ecol 102:275–301
Ricotta C (2007) A semantic taxonomy for diversity measures. Acta Biotheor 55:23–33
Roy J, Mooney HA, Saugier B (2001) Terrestrial global productivity. Elsevier
Scheiner SM, Kosman E, Presley SJ et al (2017) Decomposing functional diversity. Methods Ecol Evol 8:809–820

Schleuter D, Daufresne M, Massol F et al (2010) A user's guide to functional diversity indices. Ecol Monogr 80:469–484
Schmera D, Erős T, Podani J (2009) A measure for assessing functional diversity in ecological communities. Aquat Ecol 43:157–167
Schumacher J, Roscher C (2009) Differential effects of functional traits on aboveground biomass in semi-natural grasslands. Oikos 118:1659–1668
Shipley B, Lechowicz MJ, Wright I et al (2006a) Fundamental trade-offs generating the worldwide leaf economics spectrum. Ecology 87:535–541
Shipley B, Vile D, Garnier E (2006b) From plant traits to plant communities: a statistical mechanistic approach to biodiversity. Science 314:812–814
Song YT, Wang P, Zhou DW (2011) Methods for calculating functional diversity in plant communities. J Ecol 30:2053–2059
Spasojevic MJ, Suding KN (2012) Inferring community assembly mechanisms from functional diversity patterns: the importance of multiple assembly processes. J Ecol 100:652–661
Staples TL, Dwyer JM, England JR et al (2019) Productivity does not correlate with species and functional diversity in Australian reforestation plantings across a wide climate gradient. Glob Ecol Biogeogr 28:1417–1429
Tesfaye M, Dufault NS, Dornbusch MR et al (2003) Influence of enhanced malate dehydrogenase expression by alfalfa on diversity of rhizobacteria and soil nutrient availability. Soil Biol Biochem 35:1103–1113
Tilman D (2001) Functional diversity. In: Levin SA (ed) Encyclopedia of biodiversity. Elsevier, New York
Tordoni E, Petruzzellis F, Nardini A et al (2019) Make it simpler: alien species decrease functional diversity of coastal plant communities. J Veg Sci 30:498–509
Verheyen K, Honnay O, Motzkin G et al (2003) Response of forest plant species to land-use change: a life-history trait -based approach. J Ecol 91:563–577
Villéger S, Mason NW, Mouillot D (2008) New multidimensional functional diversity indices for a multifaceted framework in functional ecology. Ecology 89:2290–2301
Violle C, Navas ML, Vile D et al (2007) Let the concept of trait be functional! Oikos 116:882–892
Walker B, Kinzig AP, Langridg J (1999) Plant attribute diversity, resilience, and ecosystem function: the nature and significance of dominant and minor species. Ecosystems 2:95–113
Wang J, Mo H, Wang F et al (2011a) Exploring the network structure and nodal centrality of China's air transport network: a complex network approach. J Transp Geogr 19:712–721
Wang L, Wang N, You X et al (2011b) Analysis of chain saw selective felling operations on damage rate of residual trees during winter time in a mixed conifer-broad-leaved forest in China. For Prod J 61:283–289
Wang RZ, Huang WW, Chen L et al (2011c) Anatomical and physiological plasticity in Leymus chinensis (Poaceae) along large-scale longitudinal gradient in Northeast China. PLoS One 6: e26209
Wong MK, Carmona CP (2021) Including intraspecific trait variability to avoid distortion of functional diversity and ecological inference. Methods Ecol Evol 12:946–957
Xu Z, Li M, Zimmermann NE et al (2018) Plant functional diversity modulates global environmental change effects on grassland productivity. J Ecol 106:1941–1951
Xu L, Yu GR, He NP (2019) Increased soil organic carbon storage in Chinese terrestrial ecosystems from the 1980s to the 2010s. J Geogr Sci 29:49–66
Yan P, Fernández-Martínez M, Van Meerbeek K et al (2023a) The essential role of biodiversity in the key axes of ecosystem function. Glob Chang Biol 29:4569–4585
Yan P, Zhang JH, He NP et al (2023b) Functional diversity and soil nutrients regulate the interannual variability in gross primary productivity. J Ecol 111:1094–1106
Yan P, He NP, Yu KL et al (2023c) Integrating multiple plant functional traits to predict ecosystem productivity. Commun Biol 6:239
Yuan ZQ, Ali A, Wang SP et al (2019) Temporal stability of aboveground biomass is governed by species asynchrony in temperate forests. Ecol Indic 107:105661

Zhang JH, Li MX, Xu L et al (2021a) C:N:P stoichiometry in terrestrial ecosystems in China. Sci Total Environ 795:148849

Zhang JH, Ren TT, Yang JJ et al (2021b) Leaf multi-element network reveals the change of species dominance under nitrogen deposition. Front Plant Sci 12:580340

Zhang Y, He NP, Li MX et al (2021c) Community chlorophyll quantity determines the spatial variation of grassland productivity. Sci Total Environ 801:149567

Zhang ZH, Hou JH, He NP (2021d) Predictability of functional diversity depends on the number of traits. J Resour Ecol 12:332–345

Zhang Y, Migliavacca M, Penuelas J et al (2021e) Advances in hyperspectral remote sensing of vegetation traits and functions. Remote Sens Environ 252:112121

Zhou DW (2009) Advances in plant functional ecology. Acta Ecol Sin 29:5644–5655

Chapter 5
Plant Trait Networks and Multidimensional Adaptive Mechanisms

In nature, plants have various functional traits, that cooperate or complement each other to aid in their adaptation to complex environments. Therefore, the complex relationships between various functional traits can reflect the adaptive strategies of plant survival, growth, and reproduction and their response to the environment to a large extent. At present, most research is limited to simple relationships between single or a few specific traits, and multidimensional approaches and quantitative methods are still very rare, limiting our comprehensive understanding of plant adaptation mechanisms. Traditional analysis methods, such as correlation analysis, principal component analysis, path analysis, and structural equation model, may not always fully elucidate the complex relationships between various functional traits, although they have their advantages. In bridging this gap, researchers have introduced innovative concepts like multidimensional network analysis and plant trait networks (PTNs) or plant functional trait networks, which contribute to a new research framework for studying complex biological systems. PTNs are defined as multidimensional networks of various functional traits, each having its overall characteristics (or node characteristics) that depict the complex relationships among diverse traits, as well as the response and adaptation strategies of individual plants, functional groups, and communities to environmental change and disturbance. Based on this, five important network parameters and four core node parameters of PTNs are defined in terms of their physiological and ecological significance. Theoretically, PTNs have the potential to capture and visualize the relationships among plant traits in multiple dimensions, providing a novel perspective on understanding plant responses and adaptation strategies to environmental or resource changes. By introducing the concept, theory, parameters, and methods of PTNs, the scientific nature and adaptability of PTNs are discussed herein based on examples from forest transects in eastern China and global-scale leaf functional traits data. Overall, the theoretical framework of PTNs provides a new approach for exploring the multidimensional adaptation mechanisms of plants at different scales and their responses to future global changes.

N. He et al., *Plant Functional Ecology*,
https://doi.org/10.1007/978-981-96-8264-5_5

Plant functional traits are measurable, relatively stable, and closely related to various plant functions, which are manifested through long-term evolution and adaptation processes (Lavorel and Garnier 2002; Violle et al. 2007; He et al. 2018a, b, c, 2019a, b). In nature, all plants possess various functional traits, such as leaf size, leaf thickness, leaf functional elemental content, height, root diameter and length, seed size, and quality (Cornelissen et al. 2003). These functional traits are both independent and interrelated, and their relationships, including synergies, trade-offs, coupling, and decoupling, are highly complex (Chave et al. 2009; Reich 2014; Diaz et al. 2016; He et al. 2020b).

Since the 1990s, research on plant functional traits has made significant advancements, such as stoichiometric theory, leaf economics spectrum theory, and metabolic theory (Enquist et al. 1998; Elser et al. 2000; Wright et al. 2004a, b). However, most early studies primarily focused on individual functional traits or a few functional traits, examining their spatial variation, synergistic relationships, and connections with the environment. In complex nature, changes in any single functional trait alone are insufficient to fully elucidate plant responses and adaptations to environmental changes. To meet long-term survival and reproductive requirements, plants must optimize individual functions or balance functions through the interaction of multiple functional traits, forming various adaptation mechanisms and growth strategies. Numerous studies have reported these close correlations between multiple functional traits; however, previous studies have mostly focused on changes in correlations between single or pairwise traits, neglecting the complex relationships among multiple traits (Moles et al. 2006; Chave et al. 2009; Moles et al. 2009; Wright et al. 2017). Theoretically, scientists can reveal and fully explain plant adaptation strategies to environmental changes only by comprehensively considering these complex relationships (Fig. 5.1).

Reductionism focuses on the microscopic properties of objects, emphasizing the transition from macro to micro; however, system theory focuses on the macroscopic properties and complexity of objects, emphasizing the transition from micro to macro. Complexity science aims to reveal the universality behind complex systems through interdisciplinary approaches (Anderson 1999). In comparison, reductionism has been the primary approach dominating the scientific world to date, emphasizing the analysis of scientific theories rather than synthesis. Generally, complexity theory is still in its early stages and lacks a universally recognized conceptual framework and disciplinary structure. However, its interdisciplinary paradigm, multi-scale perspectives, and holistic worldview provide new scientific paradigms that allow us to explore various complex natural systems and address complex issues. In 1977, the renowned scientist Ilya Prigogine was awarded the Nobel Prize in Chemistry for his pioneering theory of dissipative structures, which is considered a major achievement in complexity science. In 2021, complexity science once again received recognition from the Nobel Prize, with the award given to Japanese–American scientist Syukuro Manabe for his work on "physical modeling of the Earth's climate, quantifying its variability, and reliably predicting global warming." Within complexity science research, complex networks are a relatively new idea and concept, as well as a new approach to studying complex systems.

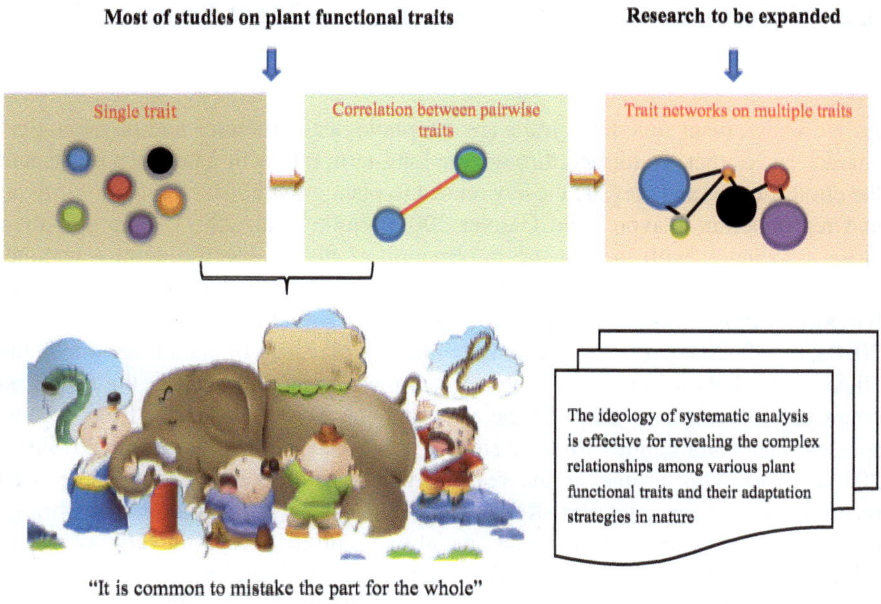

Fig. 5.1 Systematic analysis of multiple functional traits is required to fully elucidate the complex adaptation mechanisms of plants in a changing environment

Currently, complex networks have been widely applied in various fields, such as molecular biology, social sciences, and transportation studies. By studying complex networks, scientists can quantify and predict various complex relationships in nature. Inspired by complexity science and based on the understanding of plants as complex systems, He et al. (2020b) proposed that complex networks can provide an effective way to explore the complex relationships between multiple plant traits and help reveal the multidimensional adaptation strategies of plants in response to a changing environment. Although network analysis has been widely applied in various fields, its application in the study of plant functional traits is rare. This is largely explained by the lack of a well-defined concept and methodological approach for research on PTNs, as well as the need for the systematic and coordinated collection of data on plant functional traits.

To overcome these two major challenges, scientists recently incorporated the concept of complex networks into the study of plant functional traits and developed a new theoretical framework and methodology for PTNs, which provides a new approach for exploring the multidimensional adaptation strategies of plants at different scales (He et al. 2020a, b, c). In this chapter, we present a systematic review of the theoretical frameworks, key parameters, and computational methods of PTNs and introduce several recent applications in natural ecosystems. We hope that this overview will promote further advancements in the field.

5.1 Germination of Network Ideology on Plant Functional Traits

As we know, plant functional traits are morphological structures and physiological characteristics that are formed through the long-term interaction between plants and the environment, and they are closely related to plant growth, development, survival, and reproduction (Lavorel and Garnier 2002; Violle et al. 2007). Owing to their close association with the genetic information of plants, specific plant functional traits are characterized by relative stability and exhibit a certain degree of spatio-temporal variation or plasticity in response to environmental changes (He et al. 2018a, b, c, 2020a, b, c). The relative stability and plasticity of plant functional traits allow researchers to observe the changes in these traits and analyze the strategies employed by plants to adapt to the environment at various levels, from organs, species, and communities to ecosystems (Reich et al. 1997; He et al. 2019a, b, 2020a, b, c). This is one of the reasons why plant functional traits have been receiving increasing attention. Researchers aim to use functional traits to bridge the gap between studies at different spatial and temporal scales in ecology and to drive the comprehensive and rapid development of the discipline.

Theoretically, plant functional traits show multifunctionality. On the one hand, each functional trait of a plant species has multiple functions and mechanisms. Take stomata as an example—they are typically composed of two guard cells and are important channels for gas exchange between plants and the external environment, and their form and behavior are closely related to plant photosynthesis and transpiration (Hetherington and Woodward 2003). Stomata are typically similar to gates that control the entry and exit of CO_2 and water, playing a crucial role in various functions, such as light energy utilization efficiency, water-use efficiency, and heat balance. The presence of trichomes on leaves not only reflects sunlight to reduce radiation damage to leaves but also effectively prevents the loss of leaf moisture (Sack and Buckley 2020). On the other hand, no single function of plant species can be accomplished by a single functional trait, tissue, or organ. For example, photosynthesis is primarily the result of the optimization of both mesophyll and stomatal tissues. Numerous studies have revealed close relationships between various functional traits in plants, and even multiple functional traits jointly regulate the same function. For example, leaf vein traits and stomatal traits are closely related and jointly control plant water-use efficiency (Ackerly 2004; Wei et al. 2023). During the process of plants adapting to environmental changes, change in a single functional trait alone is not sufficient to meet their requirement of environmental adaptation in the long run, and the synergy among multiple functional traits is necessary to achieve the balance and optimization of multiple functions.

Currently, studies on plant functional traits have gradually shifted from focusing on single functional traits to multiple functional traits (La Riva et al. 2016; Bruelheide et al. 2018; Yin et al. 2018). The most representative achievement in this area is the leaf economics spectrum (LES) theory proposed by Wright et al. (2004a, b, 2017). Subsequently, researchers have developed economic spectra for

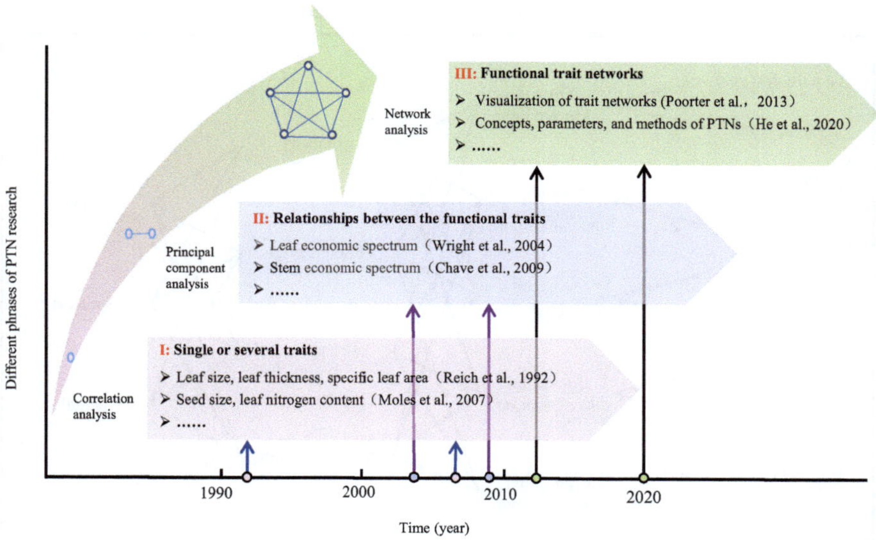

Fig. 5.2 Development of a comprehensive approach for the analysis of multiple traits, from single functional traits to functional trait networks. *PTN* plant trait network

other plant organs, such as stems, roots, and other physiological trait spectra (such as hydraulic traits), as well as the overall plant economic spectrum (Sack et al. 2003; Chave et al. 2009; Freschet et al. 2010; Sack et al. 2013; Reich 2014; Roumet et al. 2016). However, some studies have found that the complex relationships between these functional traits can change with the functional group and the environment (Wright and Suttongrier 2012), exhibiting distinct multidimensional characteristics (He et al. 2020b).

Guided by reductionism, and despite the current focus on the "constellation" of multiple functional traits, research still tends to simplify the approach through dimensional reduction (Fig. 5.2). For example, when multiple functional traits are closely related, they can be collectively reduced to a single dimension. Methods such as principal component analysis provide theoretical and analytical support for dimensional reduction. Therefore, leaf economics spectra are considered to vary along a single axis, ranging from fast-growing, short-lived tissues to slow-growing, long-lived tissues (Wright et al. 2004a, b). Researchers have recently conducted extensive research on the spatiotemporal variations of individual functional traits in plants, the correlations between pairwise functional traits, and PTNs involving multiple functional traits, yielding fruitful results. The development processes for a comprehensive approach to the analysis of multiple traits are broadly summarized in Fig. 5.2.

This simplification along a single dimension can mask the overall adaptation patterns of plants in the natural world. As mentioned earlier, plant functional traits are not only related to individual functions but also jointly contribute to multiple functions, such as plant survival, stress resistance, and reproductive capacity

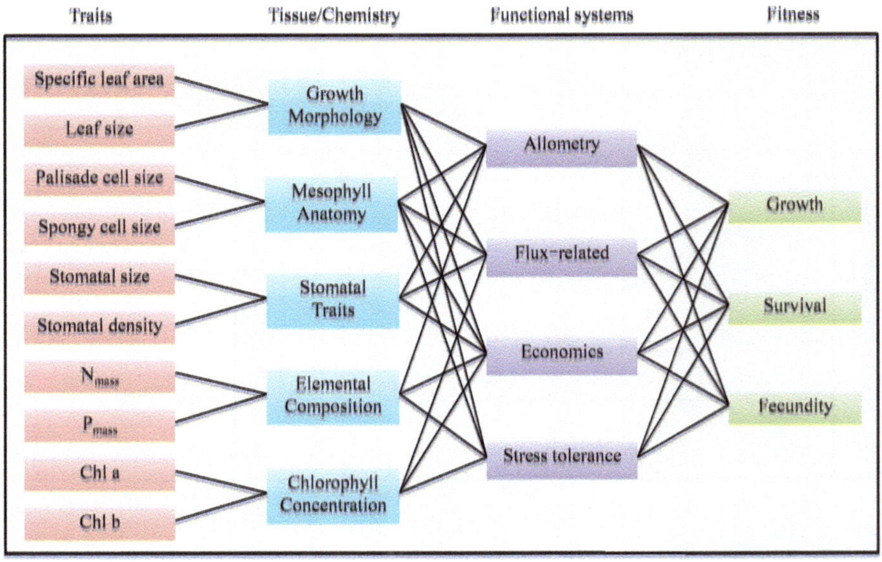

Fig. 5.3 Complexity of correlations among multiple traits, within and among organs and functional systems, and their contribution to fitness

(Bruelheide et al. 2018; Yin et al. 2018; Sack and Buckley 2020). Elucidating the complex relationships among multiple functional traits is essential for interpreting the adaptive mechanisms of plant phenotypes (Figs. 5.2 and 5.3). However, explaining such complex relationships among multiple functional traits and how these vary with species, habitats, and environments remains a significant research challenge.

5.2 Development of Network Technologies in Plant Functional Trait Research

A network is a collection $N = \{V, E\}$, composed of a set of nodes $V = \{v1, v2, \ldots, vn, \ldots$and a set of edges $E = \{e1, e2, \ldots, em, \ldots$Nodes are the basic elements of the network, and the edges represent the connections between the nodes. In different applications, nodes and edges can represent different entities and their relationships. Networks have evolved from graph theory and developed from random networks to complex networks. The introduction of the small-world network model and the scale-free network model represents a new era in network research (Watts and Strogatz 1998; Barabasi and Albert 1999; Strogatz 2001). From the perspective of scientific development, since the emergence of Newtonian mechanics, reductionism has remained the dominant approach in many fields of modern science. It aims to break systems down into numerous basic units, analyze and decompose systems, and explore the details of systems by revealing the compositional elements or components of systems, but neglects how these elements combine to form systems. Therefore, while reductionism is applicable to simple systems, it is limited when dealing with complex systems. Complex networks have emerged in the twenty-first century as a new and important approach that is being applied to the study of various complex systems. All complex system structures are complex networks, and various complex systems in nature can be described using different types of networks.

Theoretically, a network can be a collection of possible interacting units in a system, representing a set of nodes connected by edges. The complex system of plant functional traits in nature can also be described as a complex network. Currently, the ideas and concepts of complex networks have been gradually applied to many fields. In social sciences, the networks formed by the interactions among individuals in society are called social networks (Liljeros et al. 2001); and in biology, researchers successfully constructed a metabolic model of the bacterium *Haemophilus influenzae* in 1995, which included 488 reactions and 343 metabolites corresponding to 296 metabolic genes, forming a huge network (Fleischmann et al. 1995). Jeong et al. (2000) used various biological substrates (such as ATP and ADP) as nodes and chemical reactions involving these substrates as edges to construct a complex network of metabolism and then explored the metabolic processes in 43 different biological tissues. In addition to metabolic networks (Guimera and Amaral 2005), complex networks are also widely applied in transportation networks (Wang and Wang 2011; Wang et al. 2011a, b, c), gene networks (Monaco et al. 2018), and microbial networks (Wang et al. 2018a, b, c, d; Feng et al. 2022) (Table 5.1).

Proulx et al. (2005) were the first to propose applying the theory and methods of complex networks to ecology and evolution. They reviewed the latest applications of network ideas in the fields of genetics and proteins and opined that these studies, by applying existing network analysis techniques, provided novel insights into the organization and function of biological systems. They also observed that recognizing the commonalities of network ideology and concepts in evolutionary and ecological applications to create a predictive science of biological networks is a major challenge of the future. However, they only suggested to integrate the ideas of complex networks into ecological researchers, without specific definitions, parameters, and implementation plans.

In the case of plant functional traits, network analysis helps visualize the multidimensional relationships and interdependencies among multiple functional traits, rather than solely focusing on the two-dimensional relationships between single or pairwise traits, which was the traditional approach. Inspired by previous research, recent studies have gradually elucidated the basic structure of plant functional trait networks (Osnas et al. 2018) and attempted to visualize the interdependencies among multiple traits in a network format (Poorter et al. 2013; Sack et al. 2013; Poorter et al. 2014; Schneider et al. 2017; Kleyer et al. 2018). Researchers have also started to quantify these relationships using network parameters (Kleyer et al. 2018; Flores-Moreno et al. 2019). However, PTNs are in their infancy; the theoretical framework of plant functional trait networks is not clear, and further research is needed to develop methods for constructing plant functional trait

Table 5.1 Applications of complex networks in different fields

Name	Nodes	Edges	Definition	Examples
Social networks	People and social positions	Kinship, business relationships, and friendships	A social network refers to a relatively stable relationship system formed by the interaction between individual members of society. The social network focuses on the interaction and connection between people, and social interaction will affect people's social behavior	Liljeros et al. (2001)
Traffic networks	Traffic stations	Subway lines and airplane routes	The transportation network refers to the overall transportation network composed of various transportation networks and post-power grids	Wang and Wang (2011), Wang et al. (2011a, b, c)
Protein interaction networks	Protein molecules	Molecule interactions	Protein interaction networks are composed of individual proteins and their interactions with each other, which participate in biological signal transmission, gene expression regulation, energy and material metabolism, and cell-cycle regulation	Mongiovì and Sharan (2013)
Microbial networks	Microbial species	Microbial interactions	Microbial networks refer to the interactions between different microbial species or the coexistence modes of species in different environments	Feng et al. (2022)
Metabolic networks	Molecules	Biochemical reactions	Metabolic networks refer to the networks formed by all biochemical reactions in the cell, reflecting the interactions between all compounds involved in metabolic processes and all catalytic enzymes	Fleischmann et al. (1995), Guimera and Amaral (2005)
Information networks	Website pages and author's work	Network link address and literature citations	Information networks refer to the channels of electronic information transmission and is a general term for the lines and equipment that constitute this channel	El Gamal and Kim (2011)
Plant functional trait networks	Functional traits	Synergies and trade-offs between functional traits	According to the multidimensional networks of various functional traits in plants, the overall characteristics can reflect the response and adaptation mechanisms of plant individuals and communities to the environment, as well as the mechanisms of plant community structure maintenance and productivity optimization	He et al. (2020b)

networks, identifying ecologically significant network parameters and collecting extensive accompanying data. Moreover, the strategies employed by different plant life forms and functional groups to adapt to their environment through changes in relationships between plant functional traits are not yet clear, particularly in the context of large-scale natural communities.

Ecology is the scientific study of relationships between organisms and their environment. Plants are the source of material and energy for ecosystems (including humans); therefore, understanding plant adaptation mechanisms to the environment and the stability of community productivity have always been the core scientific questions. As mentioned above, functional traits are the manifestations of plant adaptation to the environment, and there are close relationships among functional traits. Changes in a single functional trait alone cannot achieve a plant's response and adaptation to the environment. Although research on the changes in individual or multiple functional trait indicators have specific scientific value, it is difficult to fully elucidate the overall relationships among functional traits and the multidimensional response mechanisms of functional traits. Therefore, applying the ideas of complex networks and constructing new plant functional trait networks offer a promising means of understanding the complex relationships among multiple functional traits and their comprehensive adaptation strategies.

5.3 Definition and Implications of PTNs

Based on previous research, He et al. (2020b) applied the concept of complex networks to the study of functional traits and introduced the concept of PTNs. Herein, PTNs are defined as multidimensional networks composed of interrelationships among multiple functional traits. The overall characteristics or node properties of PTNs represent the complex relationships among plant traits and reveal the responses and adaptation strategies of individuals, functional groups, and communities to environmental changes or disturbances (Fig. 5.4). In PTNs, nodes represent traits, and edges represent the relationships between pairs of functional traits.

Using spatial topology techniques, PTNs have the potential to comprehensively capture and visualize the connections between plant traits, providing a novel perspective for understanding plant responses and adaptation strategies to environmental or resource changes from a multidimensional perspective (Fig. 5.4). In the specific operational process, PTNs can be further subdivided into more operational leaf trait networks (LTNs), root trait networks (RTNs), flower trait networks, multiple element networks (MENs), and other trait networks, although the approach remains the same as outlined in the following sections. Crucially, PTNs can be combined with the classic ecological approaches, such as evolutionary history, environmental filtering, and competitive differentiation, to develop and test related mechanisms and hypotheses and to explore plant response and adaptation strategies (Fig. 5.5).

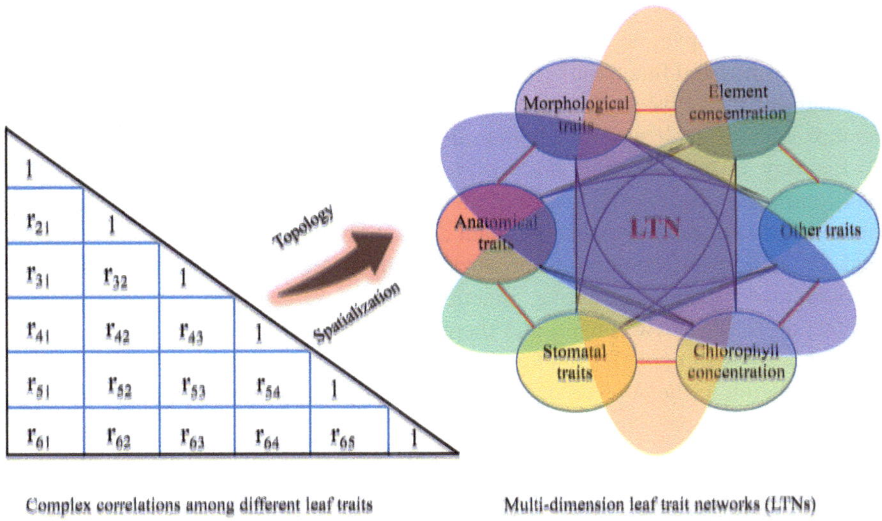

Fig. 5.4 Establishing multiple dimensions of leaf trait networks (LTNs) based on the correlations among leaf functional traits. r_{21}: the correlation coefficient between trait 1 and trait 2

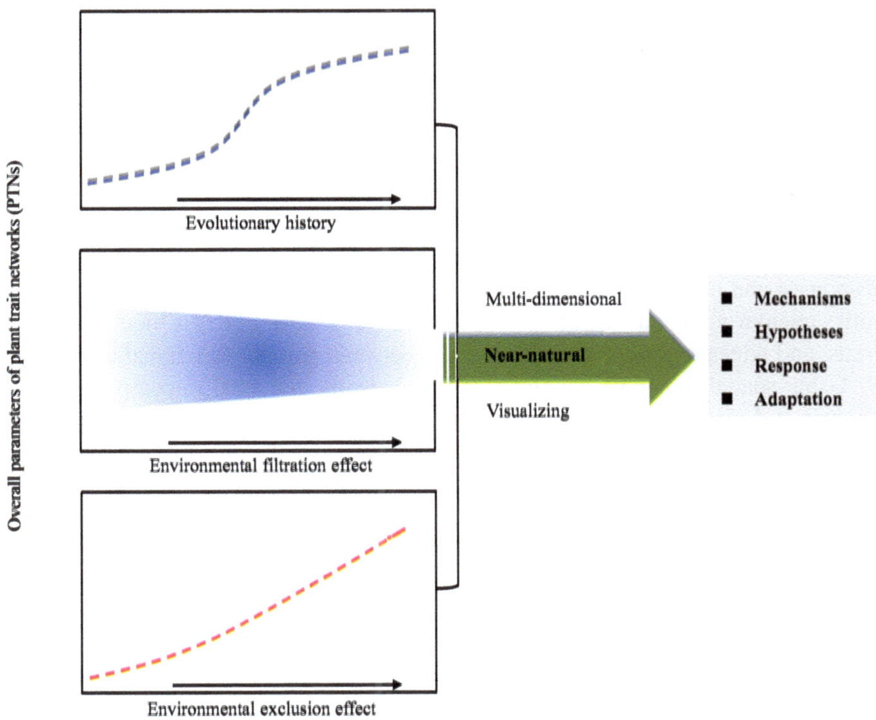

Fig. 5.5 Theoretical system of plant functional trait networks provides a new way of multidimensional quantification and visualization in studying the adaptation of plants to changing environments

PTNs typically exhibit the following basic characteristics: (1) in a specific environment, any functional trait of a plant may become a central trait within the network, with irreplaceable or indispensable significance; (2) the interactions between functional traits should hold ecological significance; (3) through the analysis of overall and key node parameters of PTNs, adaptive mechanisms that cannot be revealed by the correlations between single traits or pairs of traits can be identified. According to the classification of plant functional traits, PTNs can be classified based on different plant organs, such as leaf trait networks and RTNs, or based on functional systems, such as economic trait networks or hydraulic trait networks, as well as based on other criteria, such as MENs. They can be flexibly applied in the research process based on scientific purposes or data availability.

The theoretical framework of PTNs also encompasses many classical theories in ecology. For example, leaf functional element networks (Yan et al. 2018; Oliveras et al. 2020) can, to some extent, reflect Liebig's law of the minimum and the theory of ecological stoichiometry, particularly hypotheses related to limiting elements (Elser et al. 2000; Sterner and Elser 2002). When external resources or environmental conditions change significantly, plants may adjust their functional trait networks to adapt. Therefore, the functional trait network is not fixed but changes with the environment. Assuming that in an ideal situation, plants possess a convergent or relatively stable spherical element network, plants can adapt to the environment through adjusting the complex relationships within the network, changing the overall shape or core functional elements, thereby improving their fitness (Fig. 5.6). When resources or environmental conditions change, the functional element network may deform, become concave, or stretch in response to these changes; however, the degree of shift is limited. If the environment becomes extremely harsh or resources are extremely scarce, exceeding the threshold of shift, the entire functional element network will collapse, resulting in the death and disappearance of plants from the community. Generally, under conditions of environmental change or disturbance, plants can adapt by moving key nodes and altering the overall structure of the functional trait network to meet various functional requirements for survival, growth, and long-term reproduction.

The application of PTNs will enhance our understanding of plant multidimensional adaptation strategies in the following ways:

1. Visualizing the complex relationships among multiple functional traits: Derived from graph theory, PTNs have the capacity to describe the relationships between various functional traits using edges and nodes, visualizing the complex relationships among multiple functional traits and making them easier to understand (Poorter et al. 2013, 2014; Osnas et al. 2018; Sack and Buckley 2020).
2. Elucidating the "spectra" or "dimensions" formed by multiple functional traits: PTNs consist of multiple functional modules, and highly correlated functional traits are aggregated into one "functional module." Each functional module may correspond to one or more "functional trait spectra." For example, the LES focuses on the relationships among key economic traits (Wright et al. 2004a, b; Shipley et al. 2006a, b), which may correspond to the productivity module

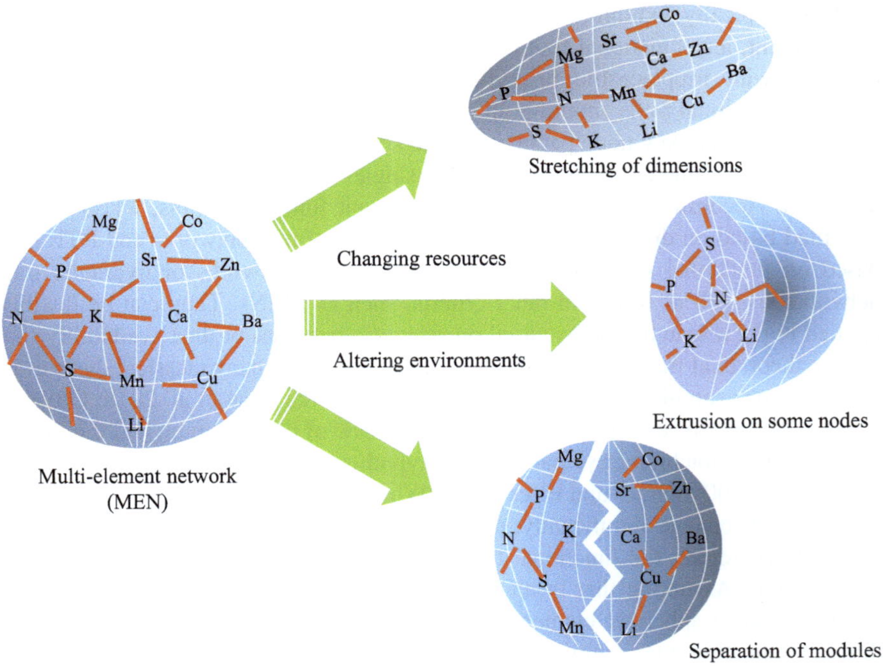

Fig. 5.6 Multielement networks of plant leaves and their potential responses to changing resources and environments

(Fig. 5.4). In relatively humid environments with sufficient temperature and precipitation, the development of resource acquisition functional traits is beneficial for increasing productivity. In PTNs, the LES is assembled into a separate module. PTNs can highlight the adaptive strategies of plants in multiple dimensions, such as productivity, temperature, and water, and can unify or accommodate various current theories and viewpoints (Fig. 5.4).

3. Quantifying the overall dependencies among multiple functional traits: The relationships among functional traits may originate from biological mechanisms or mathematical statistical correlations. However, not all functional traits are directly connected. Indirect connections between two functional traits may require one or more intermediate functional traits as bridges. Therefore, PTNs with high edge density, short average path length, and short diameter indicate strong coordination among multiple functional traits (Flores-Moreno et al. 2019).
4. Identifying key functional traits: In functional trait networks, functional traits with high connectivity are called “hub traits,” and those with high intermediacy are called “intermediate traits” (Kleyer et al. 2018). For example, the biomass allocation characteristics and stem specific length in perennial herbaceous plants can be considered central functional traits, and Li et al. (2022a, b, c, d) found that leaf economics traits related to leaf thickness were central in the leaf functional trait network of forests in eastern China.

5. Exploring contrasting changes in functional traits and their relationships in different environments: Relationships between functional traits can change with environmental changes. For example, economic traits and hydraulic traits can decouple in relatively humid areas but couple in relatively dry areas (Sack et al. 2003; Li et al. 2015; Yin et al. 2018). For example, Li et al. (2022a, b, c, d) found that functional trait networks in temperate forests with favorable temperature and water conditions were closely related and complex.
6. Quantifying plant phenotypic integration: Phenotypic integration research primarily focuses on determining whether there is a strong correlation between functional traits within a single functional unit (or functional module) and whether there is linkage between modules (Murren 2002). The modular nature of PTNs can be used to measure the connectivity among functional trait modules, thereby aiding in interpreting the issues of phenotypic integration (Murren 2002). Furthermore, PTNs have many potential applications that warrant further exploration and expansion in the future. In summary, PTNs provide an efficient method for understanding plant multidimensional adaptability through the relationships between functional traits.

5.4 Construction Methods and Key Parameters of PTNs

5.4.1 Methods for Constructing Functional Trait Networks in Plants

Networks are typically composed of a series of nodes and edges. Therefore, researchers have designated plant functional traits as nodes and the relationships among functional traits as edges, thereby constructing PTNs. To build PTNs, the first step is to compute a matrix of relationships between functional traits. For bivariate relationships between functional traits, quantification metrics such as the Pearson correlation coefficient, Spearman correlation coefficient, phylogenetic correlation coefficient, and other parametric tests can be used (Kleyer et al. 2018; Flores-Moreno et al. 2019). Notably, there are close relationships among plant functional traits for three main reasons (Sack and Scoffoni 2013; Sack et al. 2013). First, the association between functional traits is a result of structural optimization, such as the relationship between structure and physiology, where the number and size of structures directly determine physiological activity or growth processes. Second, the association between plant functional traits is a result of functional balance (or redundancy), where two independent functional traits may contribute to high-level functions. For example, stomatal morphological traits and leaf vein traits both play crucial roles in water conduction and optimization. Third, the association between functional traits is a result of selection for ecological niche optimization, in which environmental selection shapes adaptive choices. For example, in conditions with limited light, monosporic plants with more branching veins are more likely to produce fleshy fruits.

To avoid erroneous correlations between functional traits, researchers must set a threshold (e.g., $|r| > 0.2$, $p < 0.05$) to determine if there is a correlation between functional traits (Kleyer et al. 2018). Subsequently, relationships above the threshold are set to 1, while relationships below the threshold are set to 0, thereby forming an adjacency matrix $A = [a_{i,j}]$ where $ai, j \in [0,1]$; therefore, the adjacency matrix A only shows whether there is a connection between pairs of plant functional traits. Finally, by running code using the "igraph" package in R software, PTNs can be visualized, and the relevant parameters can be calculated (Fig. 5.7).

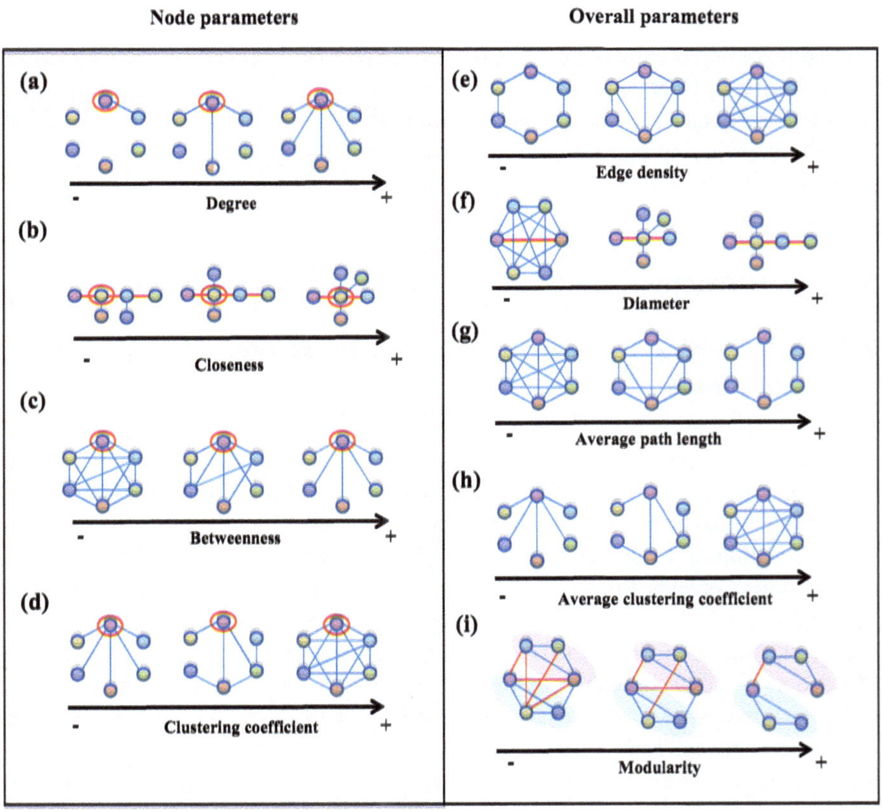

Fig. 5.7 Key parameters of plant trait networks (PTNs): The node parameters of PTNs are degree (**a**), closeness (**b**), betweenness (**c**), and clustering coefficient (**d**). The general parameters of PTNs are edge density (**e**), diameter (**f**), average path length (**g**), average clustering coefficient (**h**), and modularity (**i**). Each circle represents a specific plant functional trait, and the edges represent the relationships between functional traits. For each parameter, the three networks from left to right represent increasing parameter values

5.4.2 Important Parameters of PTNs

Typically, researchers utilize node parameters to further identify key nodes in PTNs as well as general parameters of the network to characterize the tight relationships between functional traits. In the rest of this chapter, we focus on the definitions, calculation methods, and ecological significance of four node parameters and five general parameters of PTNs (Table 5.2).

1. Node parameters of PTNs

Degree, closeness, betweenness, and clustering coefficient are commonly used to identify the topological roles of different functional traits in PTNs (Fig. 5.7); these parameters are described as follows:

① Degree

The degree is typically used to describe the number of edges connected by nodes; that is, the degree value of node v_i is denoted as k_i (Deng et al. 2012). Degree is a parameter that describes the centrality of the nodes in a network. Taking the leaf functional trait network as an example, the largest leaf functional trait in the network indicates the functional trait that is most connected with other functional traits; therefore, this is identified as a "central functional trait" of plants, which may have a core regulatory effect on the entire phenotype:

$$k_i = \sum_{j \neq i} a_{ij} \tag{5.1}$$

where a_{ij} represents the strength of the relationship between nodes v_i and v_j.

② Closeness

Closeness is another important index to represent the functional properties of network centers. Closeness, c_i, of node v_i refers to the reciprocal of the average distance between node v_i and other nodes. Functional traits with high tightness are closely related to other functional traits:

$$C_i = \frac{n-1}{\sum_{j=1}^{n-1} d_{ij}} (i \neq j) \tag{5.2}$$

where c_i refers to the closeness of node I, d_{ij} is the shortest distance between traits v_i and v_j, and n is the number of LTN traits.

③ Betweenness

Betweenness is an important parameter for describing the connection between nodes in a network. The betweenness of a node is the number of all the shortest paths through it. In PTNs, the large functional traits are the "bridges" or "intermediaries" that connect the functional modules. The environment screening of these functional

Table 5.2 Common parameters of plant trait networks (PTNs) and their ecological significance

Sort	Argument	Definition	Ecological significance
General parameters of the network	Edge density	The ratio of the number of observed edges to the maximum potential number of edges in a trait network	A high edge density suggests good synergy between the traits and high resource utilization efficiency or production efficiency of plants
	Average path length	Among the paths connecting two traits in the trait network, the shortest is called the "shortest path," and the average value of all the shortest paths is the "average path length"	A short average path length suggests good synergy between the traits and high resource use efficiency or productivity of the plants
	Diameter	The longest of all shortest paths in a trait network	A short diameter suggests good synergy between the traits and enhanced efficiency of plant in resource use or production
	Average clustering coefficient	The average clustering coefficient of all nodes in the network	A low average clustering coefficient suggests good synergy between specific traits; however, they are more inclined to form functional groups to achieve specific functions
	Modularity	Trait networks can be divided into subnetworks or modules, and modularity is used to describe the degree of separation between subnetworks (or modules)	A high degree of modularity suggests clear boundaries between functional modules, close internal connections within functional modules, and weak connection between functional modules, implying that the functional modules possess specific characteristics to perform specific functions
Node parameters of the network	Degree	The number of all edges connected to this node	The trait that has the most connections with other traits has the highest degree and is referred to as the "central trait" of the network, which may play a central regulatory role that affects the whole phenotype
	Closeness	The inverse of the mean value of the shortest path that connects the trait to other traits	The high tightness of a trait suggests that the trait is closely related to many other traits in the network and serves as a central trait in the network
	Betweenness	The number of all shortest paths through this node	A trait with a high intermediate number suggests that the trait is a "bridge" or "intermediary" connecting functional modules, and the environmental screening of the trait can greatly affect

(continued)

Table 5.2 (continued)

Sort	Argument	Definition	Ecological significance
			the coordination relationship between multiple functional modules
	Clustering coefficient	The trait that is associated with a specific trait is the adjacent trait of that specific trait, and the clustering coefficient is the probability of the association between the adjacent traits	The high clustering coefficient of a trait suggests that it is the central trait of the network function module and plays an important role in performing a specific function

traits can greatly affect the coordination relationship between multiple functional modules. The betweenness of nodes, v_i, is represented as follows:

$$B_i = \sum\nolimits_{jk} \sigma(j, i, k) \tag{5.3}$$

where $\sigma(j, i, k)$ is the number of shortest paths between traits v_j and v_k across node v_i.

④ Clustering coefficient

The clustering coefficient is a parameter used to measure the clustering of network nodes. A trait connected to node v_i is the adjacent functional trait of v_i. The clustering coefficient of node v_i is the ratio between the true number of connected edges and the maximum number of theoretically connected edges among all its adjacent functional traits. Functional traits with high clustering coefficients are the center of a functional subgroup in PTNs. Thus,

$$CC_i = \frac{2l_i}{t_i(t_i - 1)} \tag{5.4}$$

where l_i is the number of links between traits connected to focal trait v_i, and t_i is the number of traits connected to focal trait v_i.

The greater the clustering coefficient of a functional trait, the closer the correlation between the other traits directly related to the functional trait (i.e., the adjacent traits of the functional trait). In the whole network, the clustering coefficient of all nodes is equal to 1.

2. General parameters of PTNs

① Edge density (*ED*)

The *ED* describes the density of the connected edges between nodes in a network; that is, the proportion of actual connections among traits out of all possible connections. High *ED* values indicate great synergy between functional traits and high resource utilization efficiency or production efficiency:

$$ED = \frac{2L}{n \cdot (n-1)} \tag{5.5}$$

where L is the number of edges in the network, and n is the number of node traits.

② Diameter (D)

D refers to the maximum shortest path distance between any two traits, and AL is the mean shortest path between all traits. PTNs with high D and AL indicate high overall independence between functional traits. In other words, low D and AL indicate better synergy between functional traits and high resource utilization or production efficiency of plants:

$$D = \max \{d_{ij}\} \ (i \neq j) \tag{5.6}$$

$$AL = \frac{1}{n \cdot (n-1)} \sum_{i \neq j} d_{ij} \tag{5.7}$$

where d_{ij} is the shortest distance between focal node trait v_i and node trait v_j; and n is the number of traits in the PTNs.

③ Average clustering coefficient (AC)

The AC is defined as the average of the clustering coefficients (CCs) of all traits in a PTN. Thus, the AC quantifies the average probability across each trait in the network; a high AC indicates that specific functional traits, but not all traits, in the network have good coordination, and they are inclined to form functional groups or functional modules to achieve specific functions:

$$AC = \frac{1}{n} \sum_{i=1}^{n} \frac{2l_i}{t_i(t_i - 1)} \tag{5.8}$$

where l_i is the number of links between traits that are related to trait v_i, and t_i is the number of traits that are related to trait v_i.

④ Modularity

Functional trait networks can be divided into multiple subnetworks or functional modules, and modularity is used to describe the degree of separation between these subnetworks or functional modules (Armbruster et al. 2014). In PTNs, modularity is used to measure the degree of good covariation of functional traits and the degree of separation between different clustering modules. The higher the degree of network modularity, the stronger the internal connections within the module and the weaker the external connection. A highly modular network indicates that functional traits are divided into different functional modules that perform specific functions in different "dimensions":

$$Q = \frac{\sum\left[\left(A_{ij} - \frac{k_i \cdot k_j}{2m}\right) \cdot \tau\right]}{2m} \tag{5.9}$$

where m is the number of edges, A_{ij} is the element of the A-adjacency matrix of row i and column j, k_i is the order of i, and k_j is the order of j; it is 1 if j and i are the same, and, otherwise, 0 (Newman and Girvan 2004).

5.5 Application of Studying Plant Functional Traits

5.5.1 *Spatial Variation and Mechanism of LTNs in Forest Plants in Eastern China*

Plants adapt to their environment through the synergy or optimization of multiple functional traits, and the complex relationships among these traits can be represented as a multidimensional network based on correlations. Currently, limited research has been conducted on the functional trait networks between communities along different climate gradients, and their spatial variation patterns, influencing factors, and regulatory mechanisms are still unclear (He et al. 2020b). Based on previous theories in ecological physiology, several hypotheses have been proposed. For example, the modular complexity of LTNs increases with species richness, climate warmth, and humidity, reflecting reduced limitations on leaf economics traits that disproportionately contribute to multiple plant functions. These functional traits may be central to LTNs, and their variations may cause significant phenotypic changes.

To explore these hypotheses, Li et al. (2022a, b, c, d) established LTNs using consistently measured data for 35 leaf functional traits from nine typical forests and 394 tree species along a north–south transect of eastern China (NSTEC), spanning 3700 km (Fig. 5.8). They constructed LTNs for each plant community and calculated network properties such as ED, diameter, average path length, AC, and modularity to reveal the closeness and complexity of the relationships among leaf functional traits. They also verified the above-stated scientific hypotheses by combining community species richness and climate data. In addition, they calculated the degree, closeness centrality, betweenness centrality, and clustering coefficient of each node in the network to identify key functional traits in the network.

As shown in Fig. 5.8, the different forest communities along the transect exhibited significantly different LTNs. Specifically, LTNs showed a trend of transitioning from simple networks with low connectivity in cold-temperate needle leaf forests with low species diversity at high latitudes to complex networks with high connectivity in tropical rainforests (Fig. 5.9). The diameter and average path length of the LTNs were correlated with latitude. Therefore, forest communities in cold climates have simple and loose LTNs, while tropical forests with high species richness have compact and complex networks (Fig. 5.9). Overall, the phenotypic structure of these LTNs demonstrate a more complex trend in climatically resource-rich areas, reflecting high species richness and functional diversity.

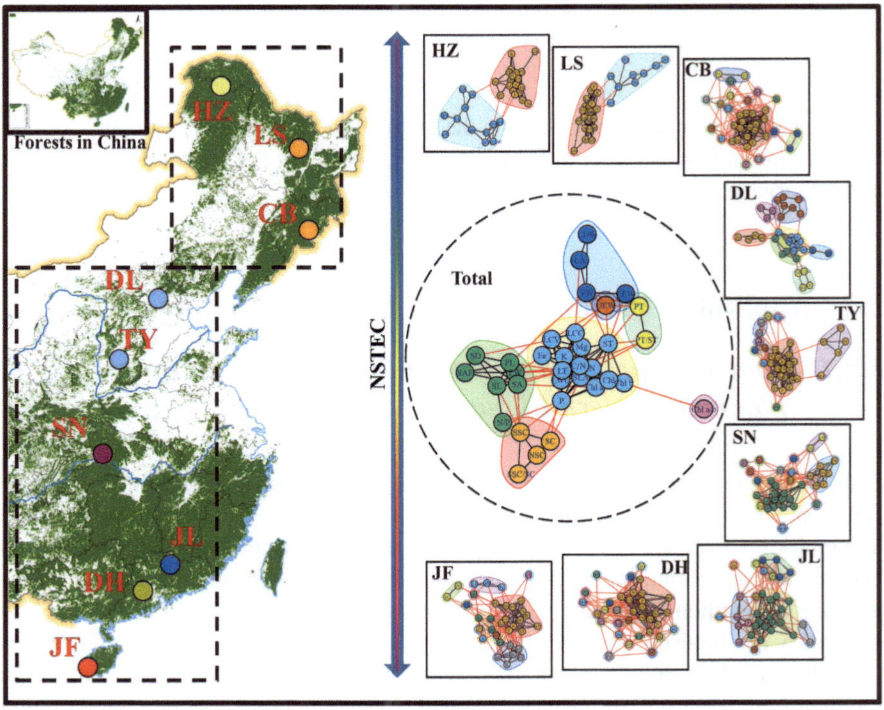

Fig. 5.8 Leaf functional trait networks in nine forest communities along a north–south transect of eastern China (NSTEC). Huzhong (HZ), Liangshui (LS), Changbai (CB), Dongling (DL), Taiyue (TY), Shennongjia (SN), Jiulian (JL), Dinghu (DH), Jianfengling (JF)

The degree and closeness centrality of leaf thickness and its related leaf economics traits were high, indicating their closest correlations with other traits (Fig. 5.10). In other words, in all the forests studied, leaf economics traits serve as hub traits in LTNs with the highest connectivity. Variations in these functional traits can cause changes in other functional traits, thereby affecting overall plant phenotype. This study provides case and methodological support for researching PTNs by systematically analyzing and interpreting variations and adaptive strategies of plant LTNs (Li et al. 2022a, b, c, d).

5.5.2 Spatial Variation of LTNs and Adaptation Mechanisms

Plants optimize various functions by adjusting the relationships among multiple functional traits, enabling them to develop multiple strategies to cope with changing environments, efficiently compete for resources, and promote growth (Fig. 5.11a). Therefore, uncovering the interdependencies and spatial variation patterns among various plant functional traits contributes to a better understanding of plant

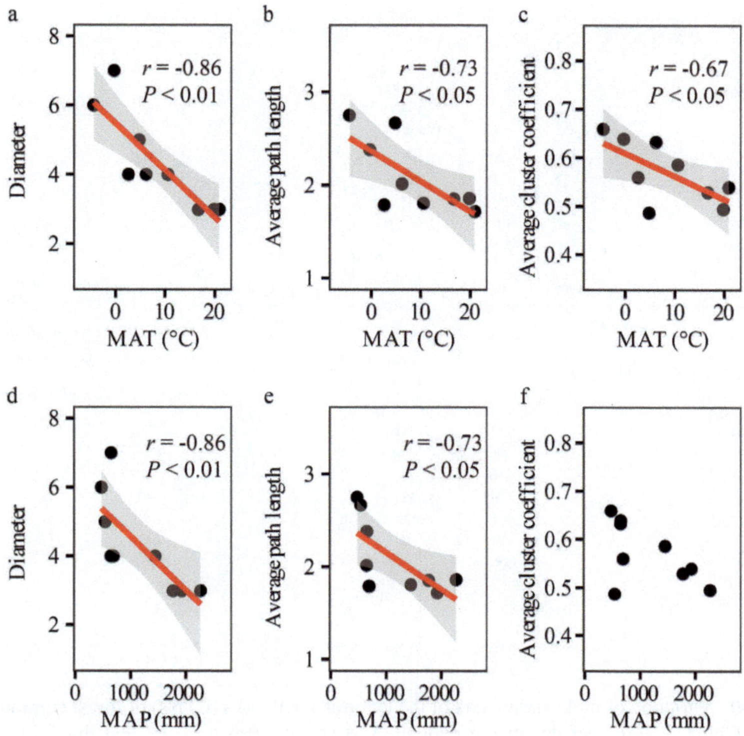

Fig. 5.9 Relationships between general parameters of leaf trait networks (LTNs) and the environment in forest communities in eastern China. The red lines were fitted using linear regression ($p < 0.05$). Shaded areas indicate 95% confidence intervals. *MAT* mean annual temperature, *MAP* mean annual precipitation

multidimensional adaptation strategies and explores their responses to global changes. The global LES includes six functional traits related to leaf chemistry, structure, and physiology: leaf lifespan, specific leaf area, leaf nitrogen content, leaf phosphorus content, photosynthetic rate, and dark respiration rate. According to the LES theory, fast-growing species have high photosynthetic rates and nitrogen concentrations, while slow-growing species have high specific leaf areas and long leaf lifespans (Wright et al. 2004a, b; Shipley et al. 2006a, b). As these six leaf functional traits are closely related to plant photosynthesis and productivity, quantifying the relationships among these functional traits is of great significance in understanding plant environmental adaptation and productivity optimization.

In nature, plants optimize photosynthesis by adjusting multiple leaf functional traits and their relationships, achieving a balance between maintenance costs and resource gains. Therefore, the complex relationships among various functional traits may form a large network (Fig. 5.11b), and the general and node parameters of functional trait networks adapt to the environment. At a global scale, leaf functional traits form nodes, while the relationships between pairs of functional traits form the

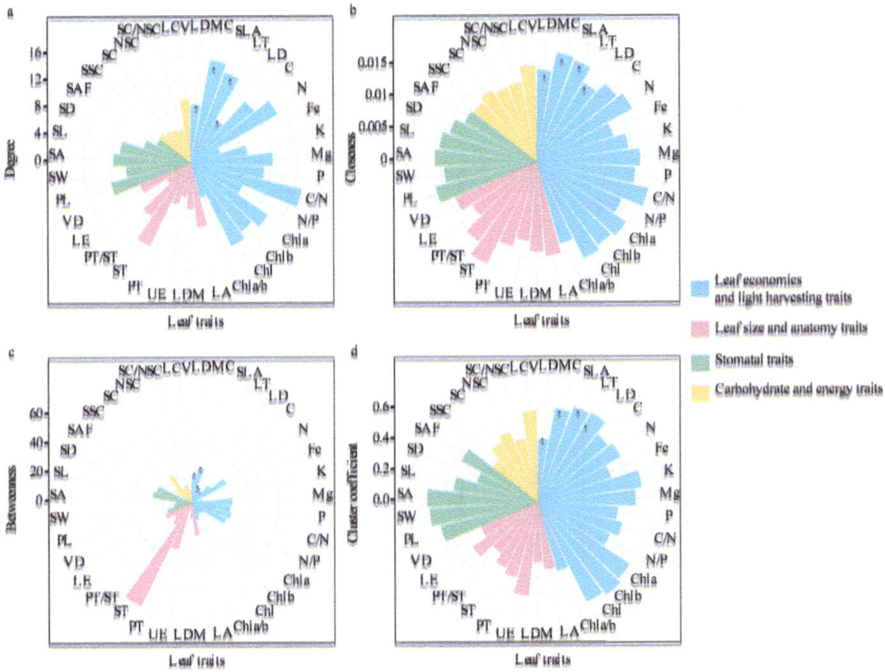

Fig. 5.10 Variation of node parameters of the leaf traits networks (LTNs) of forest communities in eastern China. *LDMC* leaf dry matter content, *SLA* specific leaf area, *LT* leaf thickness, *LD* leaf tissue density, *C* leaf carbon concentration, *N* leaf nitrogen concentration, *Fe* leaf iron concentration, *K* leaf potassium concentration, *Mg* leaf magnesium concentration, *P* leaf phosphorus concentration, *C/N* the ratio of carbon to nitrogen concentration, *N/P* the ratio of nitrogen concentration and phosphorus concentration, *Chl a* leaf chlorophyll-a concentration, *Chl b* leaf chlorophyll-b concentration, *Chl* total chlorophyll concentration, *Chl a/b* the ratio of Chl a to Chl b, *LA* leaf area, *LDM* leaf dry mass, *UE* upper epidermal cells width, *PT* palisade tissue thickness, *ST* sponge tissue thickness, *PT/ST* the ratio of the palisade to spongy tissue thickness, *LE* lower epidermal cell width, *VD* vessel diameter, *PL* stomatal pore length, *SW* stomatal width, *SA* stomatal area, *SL* stomatal length, *SD* stomatal density, *SAF* stomatal area fraction, *SC* starch concentration, *SSC* soluble sugar concentration, *NSC* non-structural carbohydrate concentration, *SC/NSC* the ratio of SC to NSC, *LCV* leaf caloric value

edges of the network (Fig. 5.11b). The width and length of the edges are important core parameters for this functional trait network (Fig. 5.11c). Moreover, owing to the evident differences in the relationships among leaf functional traits in different regions (direction or strength), there is a need to establish a new leaf functional trait network, based on the weighted relationships between functional traits, to accurately express and quantify the interdependencies among various leaf functional traits.

Based on these considerations, Li et al. (2021a, b, c) used global leaf functional trait data for over 2000 plant species, which were previously used to establish the LES, to study the global variation patterns of leaf functional trait networks. This study expanded and verified the recently developed plant functional trait theoretical

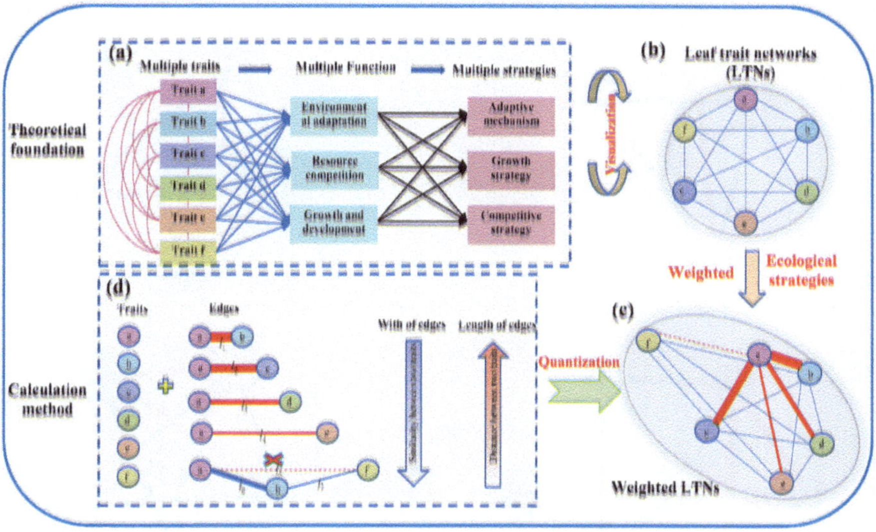

Fig. 5.11 Theoretical basis and method for calculating leaf trait networks (LTNs). Integrated LTNs help capture complex relationships between different traits and explore adaptive strategies in plants. The complex relationships among plant traits are visualized as plant trait networks (PTNs; panels **a** and **b**). The LTN in the figure can be represented as a set of nodes (circles) connected by edges (lines) (panel **d**). Different colored circles represent different leaf traits. As the distance and intensity between traits are different, accurately quantifying the relationship between traits can help understand the adaptation strategies of plants. Therefore, the width of the edge (red line) represents the strength of the relationship between traits, and the length of the edge represents the distance between traits, which is used to weigh the non-weighted trait network (panel **b**) and obtain the weighted trait network (panel **c**)

system and methodology. The main objectives were as follows: (1) construct a global-scale leaf economic trait network and reveal the complex relationships among functional traits, (2) investigate the differences in the interdependencies among leaf functional traits among different plant functional groups and life forms, and (3) identify key functional traits among the six leaf economic traits.

The researchers first used the described methods and parameters to construct a functional LTN using the six leaf functional traits of over 2000 global plant species (Wright et al. 2004a, b). The LTNs visualized the complex relationships among all functional traits and quantified their differences through network parameters (Li et al. 2021a, b, c). The experimental results showed that the LTNs of different life forms and growth forms exhibited significant differences at a global scale, indicating the distinct growth and survival strategies of different types of plants (Fig. 5.12).

Further analysis revealed that the functional trait relationships in broadleaf trees were more closely connected than those in needle leaf trees, suggesting that broadleaf trees may be more efficient at performing functions such as absorption and material transfer. The functional trait relationships in shrubs were more closely

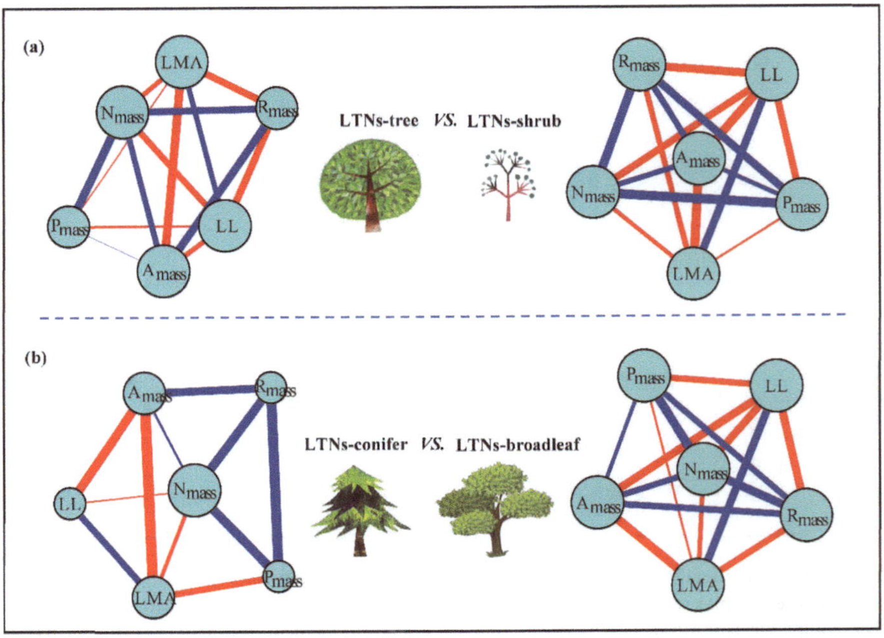

Fig. 5.12 Leaf trait networks (LTNs) of different growth and life forms at a global scale. *LL* leaf lifespan, *LMA* specific leaf weight, N_{mass} leaf nitrogen content, P_{mass} leaf phosphorus content, A_{mass} photosynthetic rate, R_{mass} dark respiration rate

connected than those in trees, likely owing to the need for effective cooperation among multiple functional traits to achieve multiple functions with limited resources (Fig. 5.13). In addition, leaf nitrogen concentration and leaf lifespan had the highest centrality in the LTNs. Therefore, variations in these two traits may have an impact on overall plant phenotype and, therefore, should be prioritized in future studies.

5.5.3 Other Applications of PTNs

The functional differences between liana vines and trees have been demonstrated (Liu et al. 2022a, b, c), although research on how their multiple functional traits are integrated is scarce. With the development of functional trait correlation networks, researchers have used network analysis to distinguish their differences based on the topological relationships among multiple functional traits. Recently, researchers have tried using publicly available datasets including 17 functional traits of 16 liana species and 16 tree species, including 12 leaf functional traits, three stem functional traits, and two branch functional traits (Medina-Vega et al. 2021). The results showed that petiole length was a highly connected trait in the tree trait network, while phosphorus was a highly connected trait in the liana trait network

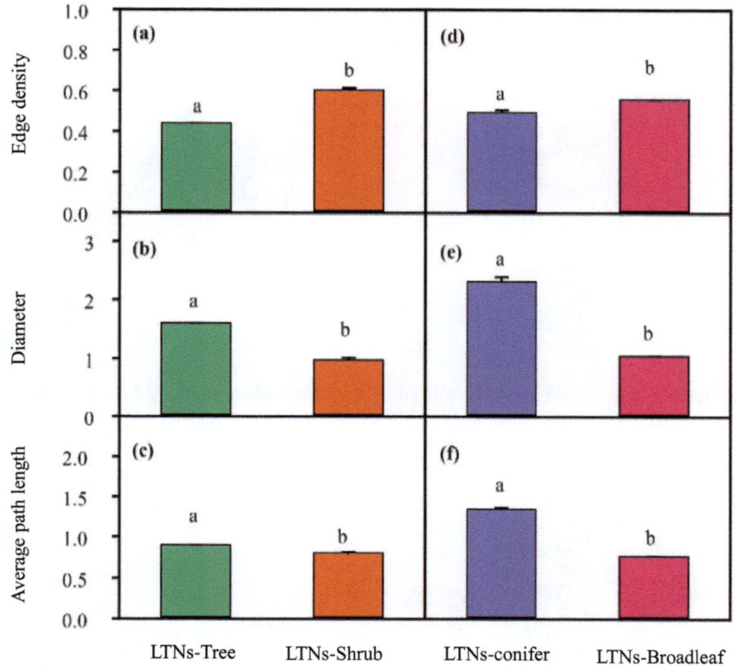

Fig. 5.13 Differences in the leaf trait networks (LTNs) of different growth types and life forms

(Fig. 5.14). Compared to trees, lianas in both wet and dry forest types had high modularity in their trait networks, indicating that environmental changes had a less impact on lianas. If plant functional traits are reclassified into economic traits, hydraulic traits, and mechanical traits, the economic traits of lianas exhibited high connectivity, while the hydraulic traits of trees exhibited high connectivity. Therefore, increasing the connectivity of specific key functional traits in plants should be cost-effective when resource conditions are poor (Liu et al. 2022a, b, c). The number and composition of functional trait modules change with life forms and environmental conditions, and plants can typically choose different combinations of traits to adapt to local environments. Therefore, these existing studies demonstrate that the variations in plant functional trait networks can enhance plant adaptability and provide a new perspective for explaining the phenomenon of increased liana abundance under global change.

The interactions among multiple elements also support various biochemical reactions and environmental adaptations in plants, with different mineral elements being responsible for different dimensions of function. While the development of ecological stoichiometry has provided a good approach for studying element interactions, it remains challenging to reflect overall changes in these interactions and characterize the multidimensional element structure. Theoretically, complex networks provide a framework for describing multidimensional systems. Therefore,

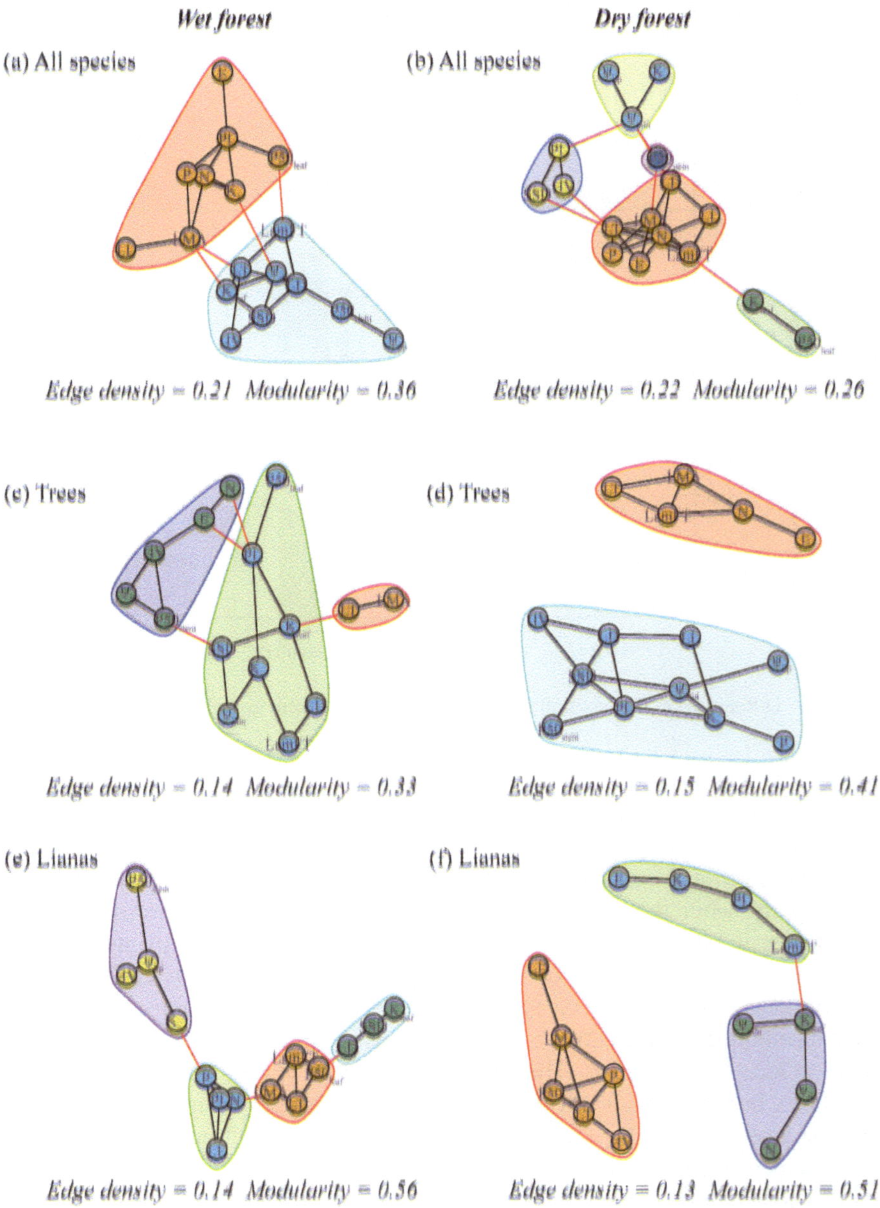

Fig. 5.14 Trait correlation networks of 17 plant traits for all species in wet (panel **a**) and dry forests (panel **b**), for trees in wet (panel **c**) and dry forests (panel **d**), and for lianas in wet (panel **e**) and dry forests (panel **f**). For full abbreviation definitions, refer to Liu et al. (2022a, b, c)

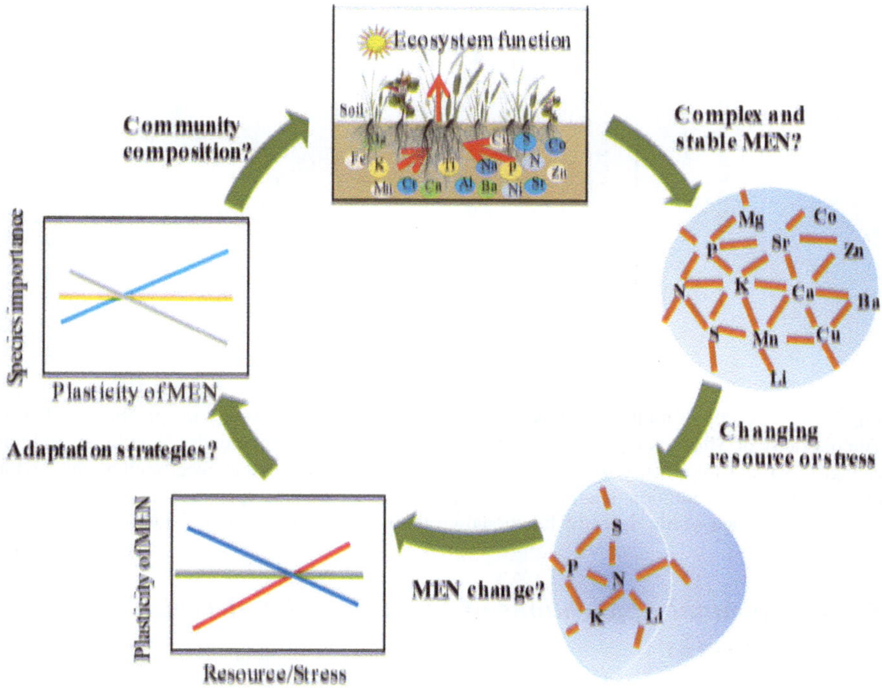

Fig. 5.15 Theoretical framework for the multielement network (MEN). Complex networks comprise interdependent multiple elements. MENs are expected to be relatively stable and species specific in any particular environment. When resource availability or environmental stress changes, MENs may alter accordingly, and the plasticity of MENs may be an effective strategy for adapting to change. MENs provide a plausible explanation for community composition

the MEN hypothesis was proposed (Fig. 5.15). This suggests that the high plasticity of MENs allows plants to respond rapidly to environmental changes, thereby increasing their competitiveness and giving them a high advantage in plant communities, resulting in high relative biomass (Zhang et al. 2021a, b, c, d, e).

Researchers measured 18 elements from five species and long-term biomass measurements from simulated nitrogen (N) deposition plots in Inner Mongolia, China. They used gap analysis and matrix-to-matrix methods to explore the characteristics of the MEN and validate the MEN hypothesis (Fig. 5.15). Their experiment showed that plant MENs exhibit uniqueness, and different species can be identified using MENs under different nitrogen addition gradients, where different species respond differently to simulated nitrogen addition (Zhang et al. 2021a, b, c, d, e). If a plant's MEN could rapidly respond to N addition (i.e., higher plasticity), its relative biomass increased or remained unchanged in the plant community. Conversely, if the plasticity of a plant's MEN was low, relative biomass decreased (Fig. 5.16). This study successfully applied the MEN approach in controlled global change research, which may show promise as a potential approach for predicting changes in community structure and function under global climate change in the future.

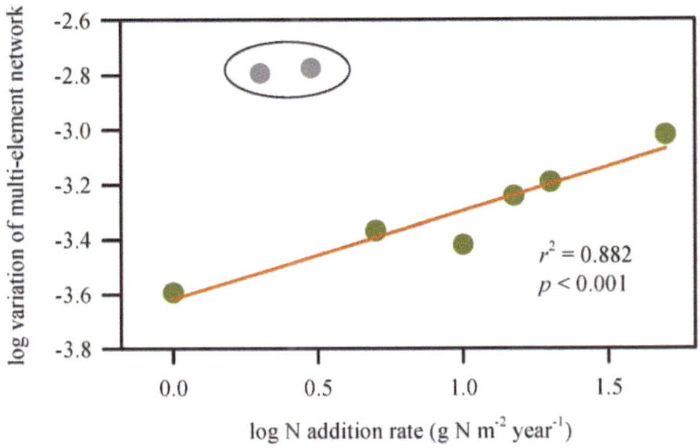

Fig. 5.16 Plasticity of multielement networks (MENs) determines changes in species importance. The plasticity of MENs is represented by the slope of the MEN change under N addition, and the change in species importance is represented by the slope of the relative aboveground biomass with increasing rate of N addition. High plasticity exhibited by a particular species in an MEN enhances species importance within the same functional group in communities under N addition

5.5.4 Potential Applications of Plant Functional Trait Networks

PTNs provide a comprehensive approach for studying plant functional traits, expanding from single traits to multiple traits and introducing new methods to elucidate the multidimensional responses and adaptation mechanisms of plants to environmental and resource changes. Based on various considerations, we believe that PTNs have significant potential in various fields (Li and He 2024), including but not limited to the following (Fig. 5.17):

1. Biogeographical pattern: PTNs could be used for spatiotemporal variability and determinants of plant functional trait networks within natural communities on a macroscale. By employing this approach, we can gain insights into the multidimensional adaptive strategies of plants in response to diverse environmental changes (Li et al. 2022a, b, c, d).
2. Comparative network dynamics: PTNs could help in identifying disparities in functional trait networks across diverse vegetation types, functional groups, and life forms. This line of inquiry also aims to predict patterns of change in these networks at the regional or global scale. The knowledge gained can be applied to quantify and forecast complex natural communities or ecosystems (Li et al. 2022a, b, c; Liu et al. 2022a, b, c).
3. Integration of aboveground and belowground traits: This involves delving into the interplay between leaf and root functional trait networks, examining their structural similarities and differences, and assessing whether their environmental responsiveness varies. This approach offers a novel and effective method to investigate the complex connections between aboveground and belowground

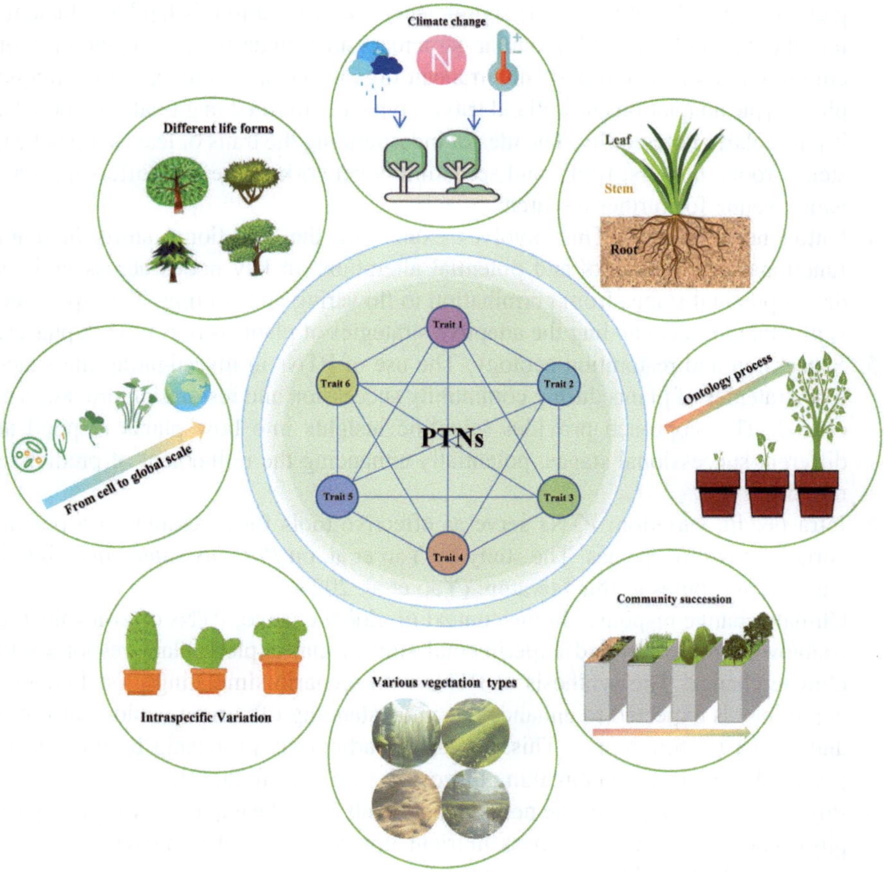

Fig. 5.17 Potential applications of plant functional trait networks (PTNs). The analysis of PTNs provides a high-definition quantification of multidimensional trait correlations, and it is expected to provide valuable insights across diverse ecological research domains. The figure shows a schematic representation of a PTN in the center of the diagram, and the eight panels around it represent the potential applications of PTNs: (**a**) The versatile framework of PTN allows for a comprehensive exploration of plant traits across various scales, from cellular and tissue levels to the intricate dynamics of individual organisms, communities, and even the expansive patterns of biogeography on a global scale. (**b**) PTNs offer a robust comparative tool for examining the distinct trait networks among different life forms, such as trees, shrubs, conifers, and broadleaf trees. (**c**) PTNs are instrumental in uncovering the responses of plants and communities to pressing global changes, such as extreme precipitation events, nitrogen deposition, and climate warming. (**d**) PTNs could reveal the integration of both aboveground and belowground traits and provide a holistic view of plant characteristics. (**e**) PTNs are also crucial for investigating the variations in ontogenetic processes. (**f**) PTNs are used for studying how plants adapt to their environment throughout different successional stages. (**g**) The framework of PTNs is useful in comparing the trait networks of various vegetation types, such as forests, grasslands, deserts, and wetlands. (**h**) The framework can also be used to examine intraspecific variations within plant species

parts of plants. Ye et al. (2024) significantly contributed to this field by elucidating the hierarchical and modular structures that underlie the integration of complex traits of aerial and subterranean organs. Their identification of unique phenotypic adaptations in wetland leaves, typically masked in global trait models, is particularly noteworthy. The idea of incorporating the traits of leaves, branches, stems, roots, flowers, fruits, and seeds into a network framework offers a promising avenue for further research.

4. Ontogenetic process: This involves examining the directional shifts in plant functional trait networks and potential alterations in key nodes across various developmental stages from germination to flowering and fruiting. This approach is pivotal to understanding the adaptive strategies of plants across growth phases.
5. Succession and restoration ecology: The use of PTNs in investigating the adaptive strategies of plants during community succession and restoration processes is crucial. This approach provides scientific insights into how plants respond at different successional stages, potentially enhancing the refinement of predictive models.
6. Intraspecific variation: PTNs serve as effective tools for assessing intraspecific variations within species. The study by Yao et al. on *Populus euphratica* significantly contributes to this literature (Yao et al. 2024).
7. Climate change response: In the context of global change, PTNs offer a valuable framework for controlled experimental studies that explore plant responses to climate change. The synthesis of PTNs with scenario-simulating global change conditions is expected to enhance our understanding of current ecological states and forecast future trends. This domain is particularly promising because of its potential to yield groundbreaking discoveries (Wei et al. 2023).
8. Physiological processes: The network approach could be expanded to encompass physiological processes such as nutrient uptake rates and photosynthetic efficiency in plants and provides an opportunity to achieve a comprehensive understanding of plant biology (Li et al. 2021a, b, c).

Overall, PTNs represent a new field in plant trait research. We predict that in the future, the methodological innovations within PTNs could surpass the path analyses paradigm of the 1990s and the structural equation modeling (SEM) paradigm of the 2000s. In terms of applications, PTN research is not limited to the seven areas listed above. Indeed, the core concepts of plant functional trait networks can be applied to various fields of traditional ecosystem network structures, such as food chains, food webs, relationships between different plant organs, and relationships between different biological groups. Regardless of future progress, novel ideas, techniques, and parameters introduced by multidimensional network analysis will enhance our understanding of the complexity of the natural world.

Notably, researching plant functional trait networks requires systematic and complementary plant functional trait data, which can be challenging to obtain. For example, obtaining PTNs from the renowned TRY database (https://www.try-db.org) is currently difficult. Therefore, developing methods to obtain datasets that meet the requirements of PTNs and rapidly expanding data sources through field surveys and controlled experiment significant challenges. To tackle this, in the future, widely distributed and long-term field ecological stations, biodiversity plots, nature reserves, and national parks could be used to rapidly construct systematic databases.

5.6 Summary

Plants possess multiple functional traits that demonstrate multidimensional adaptation pathways and multifunctionality; they are closely connected, enabling plants to successfully survive and adapt to various stressors, disturbances, and even extreme disasters in the process of evolution. In the presence of complex systems, traditional correlation analysis, redundancy analysis, path analysis, and SEM may not effectively reveal the multidimensional adaptation mechanisms of plants. Furthermore, various hypotheses have been proposed in research on plant environmental adaptation and functional optimization, such as the minimum limiting factor law, niche differentiation theory, niche complementarity theory, and economic spectrum theory; however, integrating these theories remains a major challenge for ecologists. The development of PTNs, which utilize network ideology, can help us to in elucidating the complex relationships among multiple plant functional traits. This chapter has discussed the theoretical basis, construction methods, core parameters, and ecological significance of PTNs and outlined their potential applications in exploring plant adaptation strategies at different scales, community succession, and responses to global change. Currently, establishing systematic and complementary plant functional trait databases is a significant hurdle in promoting the theoretical development of PTNs. As efforts to address this issue increase, future applications of PTNs are expected to expand significantly.

References

Ackerly D (2004) Functional strategies of chaparral shruabout to seasonal water deficit and disturbance. Ecol Monogr 74:25–44

Anderson P (1999) Complexity theory and organization science. Organ Sci 10:216–232

Armbruster WS, Pelabon C, Bolstad GH et al (2014) Integrated phenotypes: understanding trait covariation in plants and animals. Philos Trans R Soc B Biol Sci 369:20130245

Barabasi A, Albert R (1999) Emergence of scaling in random networks. Science 286:509–512

Bruelheide H, Dengler J, Purschke O et al (2018) Global trait-environment relationships of plant communities. Nat Ecol Evol 2:1906–1917

Chave J, Coomes DA, Jansen S et al (2009) Towards a worldwide wood economics spectrum. Ecol Lett 12:351–366

Cornelissen JHC, Lavorel S, Garnier E et al (2003) A handbook of protocols for standardised and easy measurement of plant functional traits worldwide. Aust J Bot 51:335–380

Deng Y, Jiang Y, Yang Y et al (2012) Molecular ecological network analyses. BMC Bioinformatics 13:113–113

Díaz S, Kattge J, Cornelissen JHC et al (2016) The global spectrum of plant form and function. Nature 529:167–171

El Gamal A, Kim YH (2011) Network information theory. Cambridge University Press, Cambridge

Elser JJ, Fagan WF, Denno RF et al (2000) Nutritional constraints in terrestrial and freshwater food webs. Nature 408:578–580

Enquist BJ, Brown JH, West GB (1998) Allometric scaling of plant energetics and population density. Nature 395:163–165

Feng K, Peng X, Zhang Z et al (2022) iNAP: an integrated network analysis pipeline for microbiome studies. iMeta 1:e13

Fleischmann RD, Adams MD, White O et al (1995) Whole-genome random sequencing and assembly of Haemophilus influenzae Rd. Science 269:496–512

Flores-Moreno H, Fazayeli F, Banerjee A et al (2019) Robustness of trait connections across environmental gradients and growth forms. Glob Ecol Biogeogr 28:1806–1826

Freschet GT, Cornelissen JHC, van Logtestijn RSP et al (2010) Evidence of the 'plant economics spectrum' in a subarctic flora. J Ecol 98:362–373

Guimera R, Amaral LAN (2005) Functional cartography of complex metabolic networks. Nature 2005(433):895–900

He NP, Liu CC, Tian M et al (2018a) Variation in leaf anatomical traits from tropical to cold-temperate forests and linkage to ecosystem functions. Funct Ecol 32:10–19

He NP, Liu CC, Zhang JH et al (2018b) Opportunities and challenges for plant trait research: from organs to communities. Acta Ecologic Sinica 38:6787–6796

He NP, Zhang JH, Liu CC et al (2018c) Patterns and influencing factors of traits in forest ecosystems: synthesis and perspectives on the synthetic investigation from the north-east transect of eastern China (NETEC). Acta Ecol Sin 38(18):6359–6382

He NP, Liu CC, Piao SL et al (2019a) Ecosystem traits linking functional traits to macroecology. Trends Ecol Evol 34:200–210

He L, Chen JM, Liu J et al (2019b) Diverse photosynthetic capacity of global ecosystems mapped by satellite chlorophyll fluorescence measurements. Remote Sens Environ 232:111344

He NP, Li Y, Liu CC et al (2020a) Plant trait networks: improved resolution of the dimensionality of adaptation. Trends Ecol Evol 35:908–918

He NP, Liu CC, Xu L et al (2020b) Ecosystem traits open new insights into macroecology: opportunity and challenge. Acta Ecol Sin 40(8):2507–2522

He P, Gleason SM, Wright IJ et al (2020c) Growing-season temperature and precipitation are independent drivers of global variation in xylem hydraulic conductivity. Glob Chang Biol 26: 1833–1841

Hetherington AM, Woodward FI (2003) The role of stomata in sensing and driving environmental change. Nature 424:901–908

Jeong H, Tombor B, Albert R et al (2000) The large-scale organization of metabolic networks. Nature 407:651–654

Kleyer M, Trinogga J, Cebrianpiqueras MA et al (2018) Trait correlation network analysis identifies biomass allocation traits and stem specific length as hub traits in herbaceous perennial plants. J Ecol 107:828–842

La Riva EGD, Olmo M, Poorter H et al (2016) Leaf mass per area (LMA) and its relationship with leaf structure and anatomy in 34 mediterranean woody species along a water availability gradient. PLoS One 11:e0148788

Lavorel S, Garnier E (2002) Predicting changes in community composition and ecosystem functioning from plant traits: revisiting the holy grail. Funct Ecol 16:545–556

Li Y, He NP (2024) Innovations and prospectives of multidimensional trait integration. New Phytol 244:337–340

Li L, McCormack ML, Ma C et al (2015) Leaf economics and hydraulic traits are decoupled in five species-rich tropical-subtropical forests. Ecol Lett 18:899–906

Li X, He N, Xu L et al (2021a) Spatial variation in leaf potassium concentrations and its role in plant adaptation strategies. Ecol Indic 130:108063

Li Y, Li Q, Xu L et al (2021b) Plant community traits can explain variation in productivity of selective logging forests after different restoration times. Ecol Indic 131:108181

Li Y, Liu CC, Xu L et al (2021c) Leaf trait networks based on global data: representing variation and adaptation in plants. Front Plant Sci 12:710530

Li X, Li MX, Xu L et al (2022a) Allometry and distribution of nitrogen in natural plant communities of the Tibetan plateau. Front Plant Sci 13:845813

Li Y, Hou J, Xu L et al (2022b) Variation in functional trait diversity from tropical to cold-temperate forests and linkage to productivity. Ecol Indic 138:108864

Li Y, Liu CC, Sack L et al (2022c) Leaf trait network architecture shifts with species-richness and climate across forests at continental scale. Ecol Lett 25:1442–1457

Li J, Chen X, Niklas KJ et al (2022d) A whole-plant economics spectrum including bark functional traits for 59 subtropical woody plant species. J Ecol 110:248–261

Liljeros F, Edling C, Amaral LAN et al (2001) The web of human sexual contacts. Nature 411:907–908

Liu CC, Li Y, He NP et al (2022a) Differential adaptation of lianas and trees in wet and dry forests revealed by trait correlation networks. Ecol Indic 135:108564

Liu CC, Sack L, Li Y et al (2022b) Contrasting adaptation and optimization of stomatal traits across communities at continental-scale. J Exp Bot 73:6405–6416

Liu H, Ye Q, Simpson KJ et al (2022c) Can evolutionary history predict plant plastic responses to climate change? New Phytol 235:1260–1271

Medina-Vega JA, Bongers F, Poorter L et al (2021) Lianas have more acquisitive traits than trees in a dry but not in a wet forest. J Ecol 109:2367–2384

Moles AT, Ackerly DD, Tweddle JC et al (2006) Global patterns in seed size. Glob Ecol Biogeogr 16:109–116

Moles AT, Warton DI, Warman L et al (2009) Global patterns in plant height. J Ecol 97:923–932

Monaco A, Monda A, Amoroso N et al (2018) A complex network approach reveals a pivotal substructure of genes linked to schizophrenia. PLoS One 13:e0190110

Mongiovì M, Sharan R (2013) Global alignment of protein-protein interaction networks. Methods Mol Biol 939:21–34

Murren CJ (2002) Phenotypic integration in plants. Plant Spec Biol 17:89–99

Newman MEJ, Girvan M (2004) Finding and evaluating community structure in networks. Phys Rev E 69:026113

Oliveras I, Bentley L, Fyllas NM et al (2020) The influence of taxonomy and environment on leaf trait variation along tropical abiotic gradients. Front For Global Change 3:18

Osnas JLD, Katabuchi M, Kitajima K et al (2018) Divergent drivers of leaf trait variation within species, among species, and among functional groups. Proc Natl Acad Sci USA 115:201803989

Poorter H, Anten NPR, Marcelis LFM (2013) Physiological mechanisms in plant growth models: do we need a supra-cellular systems biology approach? Plant Cell Environ 36:1673–1690

Poorter H, Lambers H, Evans JR (2014) Trait correlation networks: a whole-plant perspective on the recently criticized leaf economic spectrum. New Phytol 201:378–382

Proulx SR, Promislow DEL, Phillips PC (2005) Network thinking in ecology and evolution. Trends Ecol Evol 20:345–353

Reich PB (2014) The world-wide 'fast-slow' plant economics spectrum: a traits manifesto. J Ecol 102:275–301

Reich PB, Walters MB, Ellsworth DS (1997) From tropics to tundra: global convergence in plant functioning. Proc Natl Acad Sci USA 94:13730–13734

Roumet C, Birouste M, Picon-Cochard C et al (2016) Root structure-function relationships in 74 species: evidence of a root economics spectrum related to carbon economy. New Phytol 210: 815–826

Sack L, Buckley TN (2020) Trait multi-functionality in plant stress response. Integr Comp Biol 60: 98–112

Sack L, Scoffoni CJ (2013) Leaf venation: structure, function, development, evolution, ecology and applications in the past, present and future. New Phytol 198:983–1000

Sack L, Cowan PD, Jaikumar N et al (2003) The hydrology of leaves: co-ordination of structure and function in temperate woody species. Plant Cell Environ 26:1343–1356
Sack L, Scoffoni C, John GP et al (2013) How do leaf veins influence the worldwide leaf economic spectrum? Review and synthesis. Ecology 4:4053–4080
Scheiner SM, Kosman E, Presley SJ et al (2017) Decomposing functional diversity. Methods Ecol Evol 8:809–820
Shipley B, Lechowicz MJ, Wright I et al (2006a) Fundamental trade-offs generating the worldwide leaf economics spectrum. Ecology 87:535–541
Shipley B, Vile D, Garnier E (2006b) From plant traits to plant communities: a statistical mechanistic approach to biodiversity. Science 314:812–814
Sterner RW, Elser JJ (2002) Ecological stoichiometry: the biology of elements from molecules to the biosphere. Princeton University Press, Princeton
Strogatz SH (2001) Exploring complex networks. Nature 410:268–276
Violle C, Navas ML, Vile D et al (2007) Let the concept of trait be functional! Oikos 116:882–892
Wang C, Wang J (2011) Spatial pattern of the global shipping network and its hub-and-spoke system. Res Transp Econ 32:54–63
Wang J, Mo H, Wang F et al (2011a) Exploring the network structure and nodal centrality of China's air transport network: a complex network approach. J Transp Geogr 19:712–721
Wang L, Wang N, You X et al (2011b) Analysis of chain saw selective felling operations on damage rate of residual trees during winter time in a mixed conifer-broad-leaved forest in China. For Prod J 61:283–289
Wang RZ, Huang WW, Chen L et al (2011c) Anatomical and physiological plasticity in Leymus chinensis (Poaceae) along large-scale longitudinal gradient in Northeast China. PLoS One 6: e26209
Wang H, Harrison SP, Prentice IC et al (2018a) The China plant trait database: toward a comprehensive regional compilation of functional traits for land plants. Ecology 99:500–500
Wang S, Wang X, Han X et al (2018b) Higher precipitation strengthens the microbial interactions in semi-arid grassland soils. Glob Ecol Biogeogr 27:570–580
Wang RL, Wang Q, Liu CC et al (2018c) Changes in trait and phylogenetic diversity of leaves and absorptive roots from tropical to boreal forests. Plant Soil 432:389–401
Wang RL, Wang QF, Zhao N et al (2018d) Different phylogenetic and environmental controls of first-order root morphological and nutrient traits: evidence of multidimensional root traits. Funct Ecol 32:29–39
Watts DJ, Strogatz SH (1998) Collective dynamics of 'small-world' networks. Nature 393:440–442
Wei B, Zhang D, Wang G et al (2023) Experimental warming altered plant functional traits and their coordination in a permafrost ecosystem. New Phytol 240(5):1802–1816
Wright JP, Suttongrier AE (2012) Does the leaf economic spectrum hold within local species pools across varying environmental conditions. Funct Ecol 26:1390–1398
Wright IJ, Reich PB, Westoby M et al (2004a) The worldwide leaf economics spectrum. Nature 428:821–827
Wright IJ, Groom PK, Lamont BB et al (2004b) Leaf trait relationships in Australian plant species. Funct Plant Biol 31:551–558
Wright IJ, Dong N, Maire V et al (2017) Global climatic drivers of leaf size. Science 357:917–921
Yan Z, Hou X, Han W et al (2018) Effects of nitrogen and phosphorus supply on stoichiometry of six elements in leaves of *Arabidopsis thaliana*. Ann Bot 123:441–450
Yao S, Wang J, Huang W et al (2024) Adaptation strategies of Populus euphratica to arid environments based on leaf trait network analysis in the mainstream of the Tarim river. Forests 15:437
Ye Z, Mu Y, Van Duzen S et al (2024) Root and shoot phenology, architecture, and organ properties: an integrated trait network among 44 herbaceous wetland species. New Phytol 244:436–450
Yin Q, Wang L, Lei M et al (2018) The relationships between leaf economics and hydraulic traits of woody plants depend on water availability. Sci Total Environ 621:245–252

Zhang JH, Li MX, Xu L et al (2021a) C:N:P stoichiometry in terrestrial ecosystems in China. Sci Total Environ 795:148849

Zhang JH, Ren TT, Yang JJ et al (2021b) Leaf multi-element network reveals the change of species dominance under nitrogen deposition. Front Plant Sci 12:580340

Zhang Y, He NP, Li MX et al (2021c) Community chlorophyll quantity determines the spatial variation of grassland productivity. Sci Total Environ 801:149567

Zhang ZH, Hou JH, He NP (2021d) Predictability of functional diversity depends on the number of traits. J Resour Ecol 12:332–345

Zhang Y, Migliavacca M, Penuelas J et al (2021e) Advances in hyperspectral remote sensing of vegetation traits and functions. Remote Sens Environ 252:112121

Chapter 6
Plant Community Traits and Their Spatial Variation and Influencing Factors

Many previous studies have demonstrated that understanding the functional traits of organisms (plants, animals, and microorganisms) can help clarify mechanisms of biodiversity changes in nature and accurately predict the impact of plant species on ecosystem functions and services from the individual to community levels. However, owing to the limitations of the traditional definitions of functional traits, most studies have focused on organ or species levels. Studies on functional traits at the community, ecosystem, regional, and global scales have used the simple arithmetical averaging method to scale up from species to community levels. The structure and composition of plant species are highly complex in natural communities, particularly forests. Significant differences exist in the relative contributions of different plant species among different regions. Therefore, using traditional methods involving simple arithmetic means in large-scale studies on functional traits is challenging. Accordingly, researchers have developed a new concept of plant community traits, systematically introduced the theoretical basis, and presented a normative derivation method of specific plant community traits to expand these functional traits into complex natural communities. In this study, plant community traits are defined as the functional or quantitative characteristics of plants at the community level and are expressed as the intensity (or density) normalized per unit land area. Furthermore, plant community traits may be further developed as two-dimensional characteristics considering the definition of "density" or "intensity." For traits related to productivity or adaptation, the density and intensity of functional traits per unit land area should be considered as intrinsic efficiency ($Trait_{effi}$) and quantity traits ($Trait_{quan}$), respectively, which are more ecologically significant and practical at community, regional, and global scales. $Trait_{effi}$ can be calculated using simple arithmetic, geometric, and community-weighted means, considering data availability in natural communities, although they may not be accurate to some extent. $Trait_{quan}$ requires comprehensive or matched information on species composition and the relative contribution of each species to the natural communities. Based on this, we introduced the spatial variability of the two-dimensional characteristics of plant community traits and their influencing factors on a regional scale. These included the spatial

N. He et al., *Plant Functional Ecology*,
https://doi.org/10.1007/978-981-96-8264-5_6

variations of nitrogen and phosphorus content and storage and their influencing factors in plant communities on the Qinghai–Tibet Plateau, and sulfur content and storage and its influencing factors in typical plant communities in China. These new parameters of plant community traits may enhance the integration of different studies in the future, owing to a relatively unified land area, by incorporating functional traits, remote sensing, modeling, or flux.

Traditionally, plant functional traits are genetic parameters that are relatively stable and measurable and can affect the growth, reproduction, and survival of plant species at the individual level (Violle et al. 2007; He et al. 2019a, b). Plant functional traits, such as leaf size and leaf nitrogen and phosphorus contents, are the result of interactions between species and the environment during evolution and development. Therefore, these traits reflect the adaptation strategies of plants to the external environment and the adverse effects of changing environment by adjusting internal physiological and morphological traits to some extent. In 1987, the United Kingdom Ecological Society launched its flagship journal of *Functional Ecology*, and scientists proposed functional ecology as a new subdiscipline of ecology in its inaugural publication (Calow 1987). Herein, plant functional traits refer to a research approach that is not confined to a specific discipline with apparent boundaries but uses plant functional traits to explore, validate, or develop key ecological problems or hypotheses. For example, researchers can shift their focus from exploring species–environment interactions across various environmental conditions to examining plant functional trait–environment interactions at the species, plant functional group, or community scales. With the rapid development of global climate change research in the past two decades, the concepts of plant functional traits and types have been combined to analyze, simulate, and evaluate the impact of climate change on ecosystem functions to a large extent. With the development of new concepts and assessment methods for plant functional traits and the expansion of their applications, related scientific topics have been involved in all aspects of ecological research. The most critical areas are the relationships between plant functional traits and the environment and ecosystem functions, the influence of the environment on plant functional traits, and how plant functional traits reflect the ecosystem processes and functions. Scientists should establish a close relationship among the environment, plant functional traits, and ecosystem functions at different scales.

In functional ecology (or functional trait ecology), most plant functional traits vary to some degree along environmental gradients, and the functional trait–environment interactions may be derived from intraspecific or interspecies trait variation, serving as the basis of related studies. On a large scale, researchers usually pay more attention to interspecific variation than that to intraspecific variation. The variations in functional traits, such as mean and variance, among different plant communities have been widely used to predict the changes of ecosystem function under environmental change scenarios, while the changes within communities can reflect the potential resilience of a specific community to disturbances. Furthermore, spatial variations in the spectrum distribution of plant functional traits along environmental gradients may reveal the regulation of species aggregation in communities, which depend on the strength of environmental influences (Ackerly and Cornwell 2007;

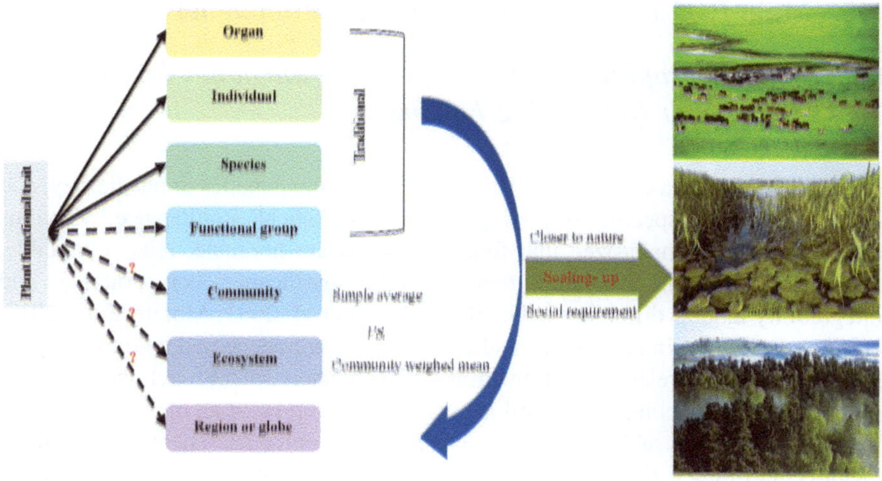

Fig. 6.1 Current development and requirements of plant functional traits to scale up from organ to community levels. Traditionally, the definition of plant functional traits is limited to organ or species levels

Keddy 1992), and help us link the theory of community assemblage to the biodiversity–ecosystem function model (Naeem and Wright 2003). Researchers can comprehensively describe the variations in the mean and variance in functional traits under different environmental conditions by evaluating the interactions between functional traits and the environment, significantly promoting the development of ecology based on functional traits (Moles et al. 2009; Zanne et al. 2010).

Researchers give varying importance to different functional trait–environment interactions. Although there are many studies on leaf traits (or the leaf economic spectrum) and their relationship with the environment, there are few studies on the variations in root functional traits along the environmental gradient (Funk et al. 2017). Most previous studies on functional traits are limited to the organ, individual, or population levels (Fig. 6.1) owing to the restriction of traditional characterizations of functional traits at the species level, and there are relatively few studies on functional traits at the community scale (He et al. 2018a, b). In natural communities, information on species composition is complex, and its relative distribution is spatially variable. For example, natural forest communities have significant differences in the community structure and species composition among different regions or forest types. Therefore, using traditional methods in large-scale studies on functional traits is difficult, considering their significance and accuracy. Furthermore, there is a need to develop the new concept of plant community traits, to promote the study of functional traits in complex natural communities and the application of new technologies in macro ecosystems, particularly high-resolution remote sensing technologies (He et al. 2019a, b, 2020a, b, c).

6.1 Main Research Progresses on Plant Functional Traits

6.1.1 Biogeographical Patterns of Plant Functional Traits and Their Influencing Factors

The classic definition of functional traits originated from genetics or physiology, and each trait has a specific physiological and ecological significance. Previous researchers have emphasized that functional traits are heritable, relatively stable, and measurable, which enhance their scientific significance, comparability, and integrability among different studies (Cornelissen et al. 2003; Moretti et al. 2017; He et al. 2019a, b). In large-scale studies, ecologists are concerned with the biogeographic patterns of related functional traits. Theoretically, plant functional traits are partly regulated by climatic and soil factors and may represent an important strategy for plant adaptation and response to changing environments. Therefore, the main factors controlling different functional traits and spatial patterns differ under different environments. As mentioned above, most large-scale studies on functional traits did not consider intraspecific trait variation but focused on interspecific variation at the spatial scale owing to measurement difficulties. Wright et al. (2004a, b) collected six functional physiological and ecological trait data of 2548 species from 219 families in 175 sites worldwide. They found that these six functional traits exhibit universal relationships and then developed the leaf economic spectrum theory. Plant functional traits were arranged orderly along a continuous spectrum, one end of which represented a rapid investment-return strategy with low specific leaf weight, short leaf life span high nitrogen content, photosynthetic rate, and respiratory rate. The other end represented a slow investment-return strategy with long life, large leaf weight low nitrogen content, photosynthetic rate, and respiratory rate. Díaz et al. (2007) further developed the leaf economic spectrum at the species level using six functional traits, including height, leaf area, and leaf nitrogen concentration of 46,000 mature plants, and found that the tree height-seed quality dimension and leaf economic spectrum dimension were independent. Reich and Oleksyn (2004) analyzed the data of leaf nitrogen and phosphorus contents of 1280 plants worldwide, observed a correlation between leaf nitrogen and phosphorus contents with temperature and latitude gradients at a global scale, and subsequently developed the T-Plant Physiology hypothesis.

Moles et al. (2007) integrated the seed quality data of 11,481 species worldwide and found a 320-fold difference in seed quality from the equator to 60° latitude. Furthermore, the change in seed quality was not linearly correlated with latitude, and this spatial pattern may be owing to changes in plant life forms and vegetation types. They further found that the global maximum tree height had apparent spatial variation along latitudinal gradient, and there was a 2.4-fold maximum sudden change in tree height at the tropical edge, confirming the transition of the life history strategy between tropical and temperate plant species (Moles et al. 2009). Chave et al. (2009) established the wood economic spectrum by analyzing the correlations among trunk traits such as wood density, mechanical strength, and branching characteristics at a global scale. Kunstler et al. (2016) analyzed the maximum growth rate, density, height, and specific leaf area (SLA) of 3 million individual trees in 140,000 sample plots worldwide and found that the relationship between functional traits and competitive exhibited good consistency at regional and global scales, providing novel insights into predicting the interactions among tree species. Joswig et al. (2022) found that two main axes can capture the primary plant morphology and function pattern: the first axis reflected the size spectrum of whole plants and plant organs, and the second axis corresponded to the leaf economic spectrum, which should be the result of balancing leaf persistence and plant growth potential. In summary, plant functional traits are not independent but are closely related at large scales, which is an important basis for the multidimensional adaptation and functional optimization of plant species in nature (He et al. 2020a, b, c; Maynard et al. 2022).

6.1.2 Relationships Between Plant Functional Traits and the Environment

Functional traits are the results of long-term plant responses and adaptations to changing environments, leading to the relationships between plant functional traits and the environment. Theoretically, the adaptation to the environment leads to differences in plant physiological and ecological traits, which occur along the basic resource axis among different plant communities in nature. Plant species with fast resource acquisition strategies are clustered at one end of the resource axis, whereas those with highly conservative resource strategies are clustered at the other end (Wright et al. 2004a, b; Reich 2014). The resource axis may represent various environmental factors, and the scale effect of these environmental factors is evident. Generally, climatic factors are crucial for the spatial variations in plant functional traits at large scales; however, at small or local scales, topographic and soil factors are crucial in regulating the variations of these functional traits (McGill 2010). Through network control experiments, Liu et al. (2015a, b) revealed that the higher the ambient temperature, the more arid and stronger the solar radiation, the higher the leaf mass and nitrogen contents per unit area, the shorter the leaf lifespan, and the weaker the photosynthetic capacity. Furthermore, Blonder et al. (2018) reported that the paleoclimate may influence the components of plant functional traits through slow combinatorial dynamics, indicating that plant functional traits are affected by the environment and phylogenetic history of the plants, even in multidimensional approaches (Miles and Dunham 1992; Song et al. 2016; He et al. 2020a, b, c; LeRoy et al. 2020).

In North America, Šímová et al. (2018) revealed that the diversity of functional traits is not limited to increased environmental pressure; however, the optimal value of some functional traits under specific environments may be influenced by the effect of environmental filtering, leading to interactions with species composition within natural communities on a large scale. Owing to the complexity of species composition and community structure, simple averages for these observed species are difficult to scientifically reflect community-scale functional traits and their spatio-temporal variation, considering carbon, nitrogen, phosphorus, and their stoichiometric ratios (Zhang et al. 2018a, b). Wang et al. (2016) systematically explored the latitudinal pattern of leaf size and SLA of forest communities using a biomass weighting method in combination with community structure and biomass allometric equations for the first time and found that climatic factors and soil nitrogen content influence leaf size and SLA at the biome scale. In 2018, *Functional Ecology*, the flagship journal of the United Kingdom Ecological Society, dedicated a special feature to presenting the systematic findings of nine papers authored by Professor Nianpeng He and Guirui Yu. They demonstrated spatial variations in several functional traits at the species, functional group, and community levels from tropical to temperate forests along a 3700 km transect, considering the community-weighted mean (CWM) and detailed information on community composition (He et al. 2018b, c; Liu et al. 2018a, b; Zhang et al. 2018a, b). Furthermore, they revealed significant correlations between functional traits and ecosystem productivity in natural communities, which has been widely considered a major challenge for ecologists (Reichstein et al. 2014). These related studies have attracted interest from researchers investigating community-scale functional traits and their quantitative relationship with ecosystem functions and have opened the prelude to expanding the study of plant community traits at different scales in ecology (Liu et al. 2023; Liu et al. 2024).

6.1.3 Requirements and Theoretical Basis for the Expansion of Functional Traits from Organs to Communities

Many previous studies have aimed to explore the spatial patterns of functional traits at the community scale and their influencing factors. However, due to the traditional definition of functional traits focusing on the level of organ or individual level, previous macroscale-related studies on functional traits are considered incorrect and insignificant. According to a review of *Oikos* by Violle et al. (2007), plant traits have specific adaptations or functional optimizations to changing environments at the organ, individual, population, functional group, and community levels. Although the concept of functional traits has been widely accepted and the study has been cited in over 1300 publications in the past decade, there is still no clear definition of functional traits at the community level. Therefore, we speculated that as a living organic system, plant communities may optimize their function and adapt to specific environments by continuously adjusting their spatial structure or species composition based on the adaptation and evolution of functional traits of plant species (Fig. 6.1). Thus, functional traits may be essential for ecosystem function optimization and environmental adaptation at the community scale. Therefore, understanding functional traits at the community scale may lead to a breakthrough in understanding the bottleneck of some topical issues in ecology, such as "why the structure of plant community cannot explain the ecosystem function variations at different scales?" "how the structure of plant community explains ecosystem adaptation to the environment, as well as resistance and recovery from disturbances?" Overall, the functional traits of plant communities are the primary properties of plant communities in nature and regulate multiple functions of terrestrial ecosystems (He et al. 2019a, b, 2023; Yan et al. 2023a, b).

The ecological significance of plant functional traits at the organ level is well-known and accepted (Violle et al. 2007; Wang et al. 2016; He et al. 2018a, b, c, 2019a, b). However, most studies on functional traits are limited to the organ, individual, or population levels owing to the constraints of traditional definitions of functional traits at the organ level (Fig. 6.1), and there are few studies on functional traits at the community scale. Additionally, studies on functional traits at community, ecosystem, regional, and global scales are primarily conducted using the single arithmetic mean method to scale up from the organ to community levels (Zhang et al. 2018a, b; Wieczynski et al. 2019), typically falling into the trap of "simple arithmetic mean at species level being equal to community." Considering carbon, nitrogen, and phosphorus contents in plant organs, the differences between simple arithmetic mean and community weighted mean methods are apparent in transects in China and globally, even significantly affecting their biogeographic patterns and their controlling factors (Zhang et al. 2018a, b, 2021a, b, c, d, e; Zhao et al. 2022b). Future studies should avoid using the "simple arithmetic mean" to ensure that studies on functional traits at the community level, including the background of model parameters, mimic nature, and reality (Lu et al. 2017; He et al. 2019a, b). There is an urgent need to develop new concepts and scientific methods to study plant community traits.

Additionally, the development of plant functional traits at the community level can promote the application of rapidly developing technologies in macroecology, such as remote sensing and flux observations and model simulation, thereby bridging research on traditional functional traits and macroecology and promoting studies on plant functional traits by developing new methods and expanding their vision. For example, with the rapid development of remote sensing technologies, some parameters observed using remote sensing, such as SLA, leaf nitrogen content, and spectral characteristics, closely correlate with the functional traits of plant communities or among themselves. Therefore, these new technologies and parameters provide numerous data sources and novel ideas for future ecological research. Additionally, functional traits at the plant community scale should contribute to research on traditional functional traits in macroecology to expand from organ to community levels, broadening the application of traditional functional traits. The parameters of functional traits at the community level can provide validation, optimize parameters, facilitate comparisons for remote sensing, flux observations, and model simulations, and enhance the accuracy and depth of macroecological research by establishing a three-dimensional sky–space–ground observation system (Fig. 6.2). Most of the ecological problems faced by humans must be solved at ecosystem, watershed,

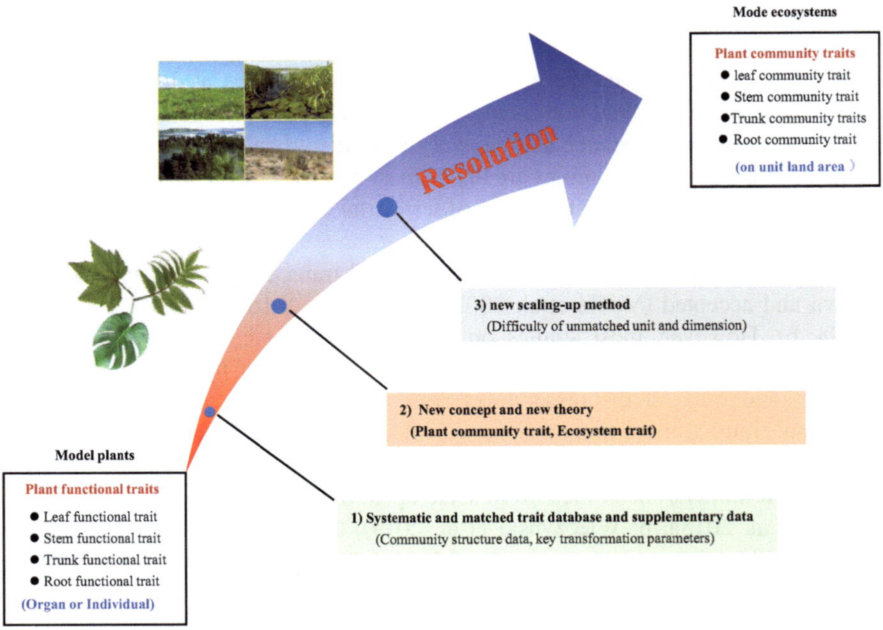

Fig. 6.2 Theory and challenges in expanding the plant functional traits from individuals to communities

regional, and global scales. Therefore, there is need to resolve the scientific problems related to inconsistency in dimension and scale among traditional functional traits and macroecological research and to develop novel concepts and methods for understanding functional traits at the community level. There is a need to develop different disciplines and promote socioeconomic development at regional or global levels.

6.2 Definition and Connotation of Plant Community Traits

6.2.1 Definition and Specifications of Plant Community Traits

As previously mentioned, plant community traits are measurable functional traits standardized by unit land area at the community scale, such as density or emphasis per unit area, which can reflect adaptations to the environment, reproduction, and productivity optimization within these communities (He et al. 2018a, b, c, 2019a, b, 2023). This definition can resolve the problem of dimensional or spatial scale mismatch that may arise during transformation or when conducting studies on functional traits across different disciplines (Fig. 6.3). For any specific community,

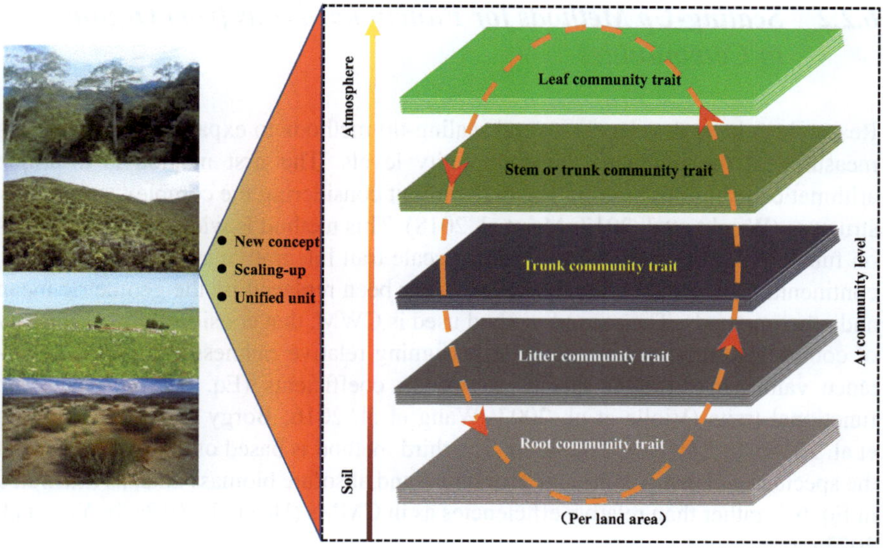

Fig. 6.3 Concepts of plant community traits and its primary components

the parameters of plant community traits comprise a series of different plant functional traits with specific effects or interactions that complete multiple plant community-level functions. The morphological, physiological, and functional elements of leaf, stem/branch, tree, and root communities are the primary units for studying plant functional traits at the community level (Fig. 6.3).

Here, any component of plant community traits should conform to the following principles (He et al. 2019a, b, 2020a, b, c): first, the functional traits of any plant community, such as leaf area, dry matter, stomatal density, can be transformed into functional traits based on unit land area, such as leaf area expanded to leaf area index, leaf dry weight expanded to leaf biomass, leaf stomatal density expanded to stomatal number per unit land area, and leaf carbon, nitrogen, and phosphorus contents can be expanded to carbon, nitrogen, and phosphorus densities, respectively. Second, any plant community trait can be measured or derived, and in principle, they are derived using information on species composition in the plant community, such as the CWM method combined with leaf SLA, traditional CWM, or other weighted average methods based on species-level data. Third, any plant community trait should reflect the adaptation, reproduction, or productivity of the plant community to the changing environment, which implies that it should have clear ecological significance (He et al. 2019a, b, 2023). In this conceptual framework, plant community traits consist of various functional traits at the community level that interact and influence each other within plant communities and jointly respond and adapt to external environments or disturbances.

6.2.2 Scaling-Up Methods for Functional Traits from Organs to Communities

Researchers have developed several scaling-up methods to expand functional traits measured at the organ to the community levels. The first method is a simple arithmetic mean, illustrated in Eq. 6.1, without considering the complex community structure (Wright et al. 2017; Ma et al. 2018). This method is widely used in studies on functional traits, particularly in large-scale trait integrative analyses of national, continental, and global traits; however, it has been replaced by the geometric mean and other methods. The second method used is CWM, that considers the complexity of community structure in nature, by assigning relative richness or relative importance values to different species as weight coefficients (Eq. 6.2) derived from functional traits (Violle et al. 2007; Wang et al. 2016; Borgy et al. 2017; Zhang et al. 2018a, b; Yan et al. 2023a, b). The third method is based on functional traits at the species level using community structure and absolute biomass data, as illustrated in Eq. 6.3, rather than relative efficiencies as in CMMs (He et al. 2019a, b; Yan et al. 2023a, b).

$$\mathrm{Trait_{SAM}} = \frac{\sum_i^n N_i}{n} \tag{6.1}$$

where SAM is the species arithmetic mean, and n is the number of species observed in the community.

$$\mathrm{Trait_{CWM}} = \sum\nolimits_i^n P_i \cdot \mathrm{Trait}_i \tag{6.2}$$

where n is the number of species in the community. P_{i} is the species relative leaf biomass i in the community, and Trait_i is the functional trait of the ith species.

$$\mathrm{Trait_{PLA-CWM}} = \sum\nolimits_i^n B_i \cdot \mathrm{Trait}_i \tag{6.3}$$

where n is the number of species in the community. B_{i} is the biomass density of species i per land area (PLA), and Trait_i is the functional trait of species i.

Equation 6.3 can extend the functional traits to the community on the land area. Because the community has its specific corresponding area, it denotes the unit land area; however, the unit is not forced to convert to per unit land area as illustrated in Eqs. 6.1 and 6.2. For example, g kg^{-1}, mm^2 mg^{-1}, and cm^{-2} are nitrogen and phosphorus contents, SLA, and leaf stomatal density units, respectively. The three methods mentioned above can be regarded as the means of the species level, which can be used to explore higher-level processes and community interrelations to explore the interaction, competition, and coexistence among individuals and species on a large scale in the future.

6.2.3 *Two-Dimensionality of Plant Community Traits*

According to the primary definition of plant community traits, most functional traits have two dimensions at the community level: density and intensity per unit area (He et al. 2019a, b, 2023). We further developed the two-dimensionality of plant community traits to operate more conveniently and reveal the underlying productivity mechanisms in natural communities. Practically, the density parameters of specific functional traits per land area are designated as efficiency traits ($Trait_{effi}$), whereas the intensity parameters of functional traits per land area are considered quantity traits ($Trait_{quan}$) (Zhang et al. 2021a, b, c, d, e; He et al. 2023; Yan et al. 2022). For example, plant communities with high leaf nitrogen content or large SLA should have high $Trait_{effi}$. Thus, these plant communities can have high productivity per unit of nitrogen content or leaf area during the same period (Fig. 6.4). $Trait_{effi}$ is calculated using Eqs. 6.1 and 6.2, and $Trait_{quan}$ is calculated using Eq. 6.3.

The primary plant community traits are community-level functional traits standardized on land area. The new concept of community traits originates from traditional community-level traits and has internal interrelations. The main differences are summarized as follows: first, the functional traits of plant communities have two-dimensional characteristics ($Trait_{effi}$ vs. $Trait_{quan}$), thereby expanding the new accumulation parameters per unit area (Table 6.1). Therefore, the new concept of plant community traits and their two-dimensionality may be used to reveal increasingly complex ecological problems, such as predicting the spatiotemporal variation in ecosystem productivity, including gross primary (GPP) and net primary productivities (NPP) (Zhang et al. 2021a, b, c, d, e; Yan et al. 2022; He et al. 2023). Second, plant community traits require systematic data. For example, the absolute quantity of a specific functional trait is calculated using the data of accurately matched species composition and the absolute biomass of each plant species, while the functional traits should be replaced by relative parameters such as relative richness and importance values. Furthermore, if the functional traits at the traditional

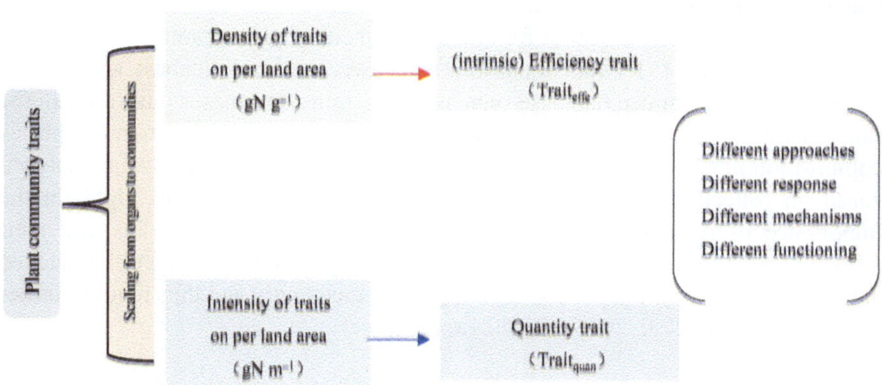

Fig. 6.4 Plant community trait concept and its two-dimensionality of efficiency vs quantity traits

Table 6.1 Two-dimensionality of plant community traits on leaves

	Efficiency trait ($Trait_{i\text{-Effi}}$)[a]	Quantity trait ($Trait_{i\text{-Quan}}$)
1	Leaf chlorophyll concentration (Chl, %)[b]	$Chl_{\text{-Com}}$ (g m^{-2})
2	Leaf nitrogen concentration (N, %)	$N_{\text{-Com}}$ (g m^{-2})
3	Leaf phosphorus concentration (P, %)	$P_{\text{-Com}}$ (g m^{-2})
4	Specific leaf area (SLA, cm^2/g)	Leaf quantity or leaf area index(g m^{-2} or cm^2/m^2)
...	...	....

[a]Efficiency and quantity traits are transformed to the data of plant species and their relative or absolute contribution to natural communities (He et al. 2019a, b, 2020a, b, c)
[b]$Trait_{effi}$ is calculated using Eqs. 6.1 and 6.2, and $Trait_{quan}$ is calculated using Eq. 6.3

community level meet the requirement of accurate and matched data, they could be converted into two-dimensional plant community traits using important parameters, such as leaf area or biomass per species. Plant community traits are not limited to functional traits per unit land area but can be derived from two or more plant community traits, such as nitrogen: phosphorus ratio (ratio of nitrogen to phosphorus contents per unit land area) and SLA (ratio of leaf area to leaf biomass per unit land area). The derived plant community traits were numerically identical to the functional traits at the community level. Additionally, the efficiency traits of the plant community and the newly expanded quantity traits formed a new system of "efficiency traits vs. quantitative traits," which provides novel insights into the expansion of subsequent research on functional traits in natural communities.

6.3 Spatial Variations in Plant Community Traits and Its Influencing Factors at the Biome Scale

According to the definition of plant community traits and their derivation methods, matching data between functional traits and the relative contribution of each species to the communities are essential for the development of plant community traits. As described in Chap. 1, for functional trait databases, the ideal database should be a systematic and integrated database such as China_traits. Databases must contain the functional traits of most plant species and the corresponding information on species composition in each community. However, most large-scale studies were analyzed based on collected data, such as the TRY database. These studies highlight the importance of recognizing spatial variation in plant functional traits and their influencing factors. However, owing to methodological limitations and unmatched data, the results carry a relatively high degree of uncertainty and thus require further validation (Zhang et al. 2018a, b, 2021a, b, c, d, e).

6.3.1 Spatial Variation and Regulation Mechanism of Functional Traits in Chinese Forests

Nine typical forest communities were selected from the 3700 km north–south transect of eastern China, covering the main forests in China, from tropical rain forests to cold-temperate coniferous forests (Fig. 6.5 and Table 6.2). Based on the field survey data of key functional traits of all plant species and the matched data of community structures (China_traits), various plant functional traits were deduced from organ to community levels, and the corresponding plant community traits were calculated using Eqs. 6.1, 6.2, and 6.3 (He et al. 2018b, 2019a, b). The detailed parameters of plant functional traits are designated as leaf morphological traits (i.e., leaf size, mass, and SLA), leaf chlorophyll content, nonstructural carbohydrate content, stomatal morphological traits (i.e., stomata size, length, and density), vein traits, and anatomical traits (i.e., adaxial, abaxial epidermis, leaf, palisade tissue, and spongy tissue thicknesses) and the content of 15 functional elements (including carbon, nitrogen, phosphorus, sulfur, potassium, sodium, calcium, and magnesium) in the leaf, branch, stem, and root (Wang et al. 2016; He et al. 2018a, b, c; Li et al. 2016, 2018a, b, 2022a, b, c, d; Liu et al. 2018a, b; Zhang et al. 2018a, b).

Previous studies have revealed latitudinal variation and identified the main factors influencing functional traits. These studies were the first to analyze the scales from organ, individual, functional group, and community levels at the biome scale.

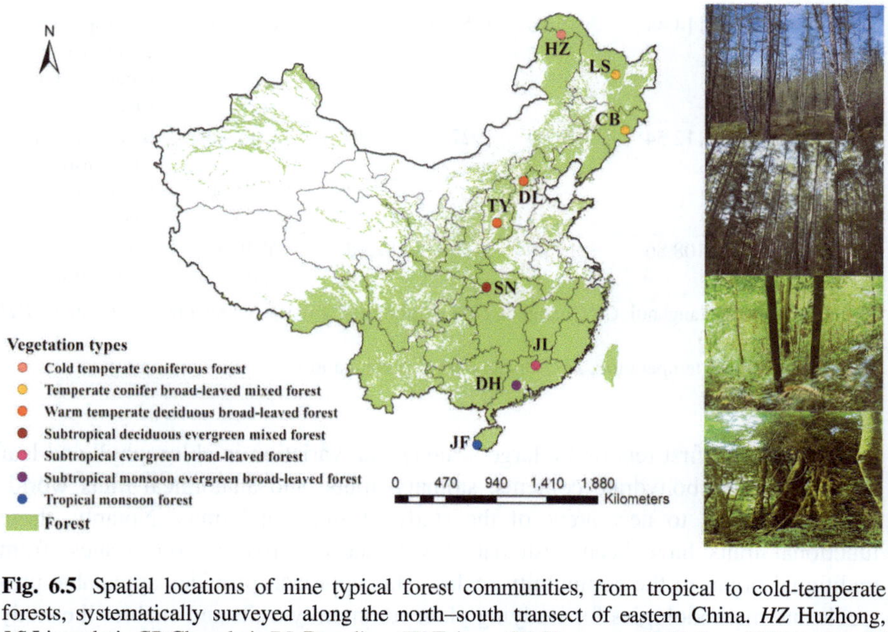

Fig. 6.5 Spatial locations of nine typical forest communities, from tropical to cold-temperate forests, systematically surveyed along the north–south transect of eastern China. *HZ* Huzhong, *LS* Liangshui, *CB* Changbai, *DL* Dongling, *TY* Taiyue, *SN* Shennong, *JL* Jiulian, *DH* Dinghu, *JF* Jianfeng

Table 6.2 Selected information on the nine representative forest communities along the north–south transect of eastern China

Site	Latitude (°)	Longitude (°)	MAT[b] (°C)	MAP (mm)	de Martonne aridity index	Soil type	Forest type
HZ[a]	51.78	123.02	−4.40	481.60	86.00	Gray forest soil	Cold-temperate coniferous forest
LS	47.19	128.90	−0.3	676.00	69.69	Dark brown soil	Temperate conifer broad leaf mixed forest
CB	42.40	128.09	2.60	691.00	54.84	Dark brown soil	Temperate conifer broad leaf mixed forest
DL	39.96	115.42	4.80	539.07	36.42	Brown soil	Warm temperate deciduous broad-leaved forest
TY	36.70	112.08	6.20	662.00	40.86	Cinnamon soil	Warm temperate deciduous broad-leaved forest
SN	31.32	110.50	10.60	1330.00	64.56	Yellow brown soil	North subtropical deciduous evergreen mixed forest
JL	24.58	114.44	16.70	1954.00	73.18	Red soil	Subtropical evergreen broad-leaved forest
DH	23.17	112.54	20.90	1927.00	62.36	Lato soil	South subtropical monsoon broad-leaved forest
JF	18.74	108.86	19.80	2449.00	82.18	Yellow soil	Tropical monsoon forest

[a]*HZ* Huzhong, *LS* Liangshui, *CB* Changbai, *DL* Dongling, *TY* Taiyue, *SN* Shennong, *JL* Jiulian, *DH* Dinghu, *JF* Jianfeng
[b]*MAT* mean annual temperature, *MAP* mean annual precipitation

Specifically, we first reported a large-scale spatial variation in chlorophyll and leaf nonstructural carbohydrate contents, stomatal traits, and anatomical traits worldwide, expanding to new areas of the study of functional traits. Notably, these functional traits have been systematically deduced across various scales, from traditional organ to the community scales. This approach provides applicable protocols and methodologies for studying plant community traits and enables the study of scaling up plant community traits. It is necessary to pointed out that most of these plant community traits involved in these previous studies belong to the category of efficiency traits, and few being relative to quantity traits.

Traditionally, studies of functional trait function are limited to the species level or are conducted under controlled experiments owing to the lack of systematic data on functional traits and matching species composition in natural communities; thus, their conclusions need to be verified using more large-scale data from natural communities (Borgy et al. 2017; He et al. 2018a, b, c). Based on the detailed survey data of China_traits and the integrating method of "community structure, allometric equation, specific area", Liu et al. (2018a, b) conducted a comprehensive analysis of survey data of China_traits and used the integrating method of "community structure, allometric equation, and specific area," and found that stomatal density at the plant community level explained 51% of spatial variations in water-use efficiency in natural communities. Similarly, there were quantitative correlations among SLA, leaf anatomical structure, chlorophyll content, and GPP (He et al. 2018a, b, c; Li et al. 2018a, b). These studies have established correlations between functional traits and productivity in natural communities and provided a reference for subsequent research on the quantitative interrelations between plant functional traits and ecosystem functions in natural communities (Reichstein et al. 2014). *Functional Ecology*, the flagship journal of the United Kingdom Ecological Society, reported the relative results in 2018, as s special feature of nine papers (He et al. 2018b, c; Liu et al. 2018a, b; Zhang et al. 2018a, b). Although some plant community traits, particularly those of the leaves, had significant positive correlations with GPP in this study, the explanatory degree of each trait was lower than that expected, suggesting that the coordination of multiple functional traits is an important mechanism for plant productivity optimization and that any single functional trait should not be exaggerated to determine productivity in natural communities (Li et al. 2022a, b, c, d).

6.3.2 Spatial Variations in Nitrogen and Its Influencing Factors in Qinghai–Tibet Plateau Natural Communities

Nitrogen (N) is an essential element for constructing amino and nucleic acids and chlorophyll and is important for plant growth and survival. Nitrogen deficiency limits productivity in most terrestrial ecosystems. Theoretically, productivity in natural communities is affected by the amount of leaf nitrogen per unit land area [leaf nitrogen content (mg g^{-1}) and the amount of leaf nitrogen per unit land area [leaf nitrogen intensity, (g m^{-2})]. According to the two-dimensionality of plant community traits, the highest community productivity can be obtained when the product of the leaf nitrogen content and leaf nitrogen intensity is optimum (Yan et al. 2022; He et al. 2023). Therefore, nitrogen variation and distribution among different organs of the plant community are the basis for understanding the underlying mechanisms of community structure stability and function optimization (Fig. 6.6).

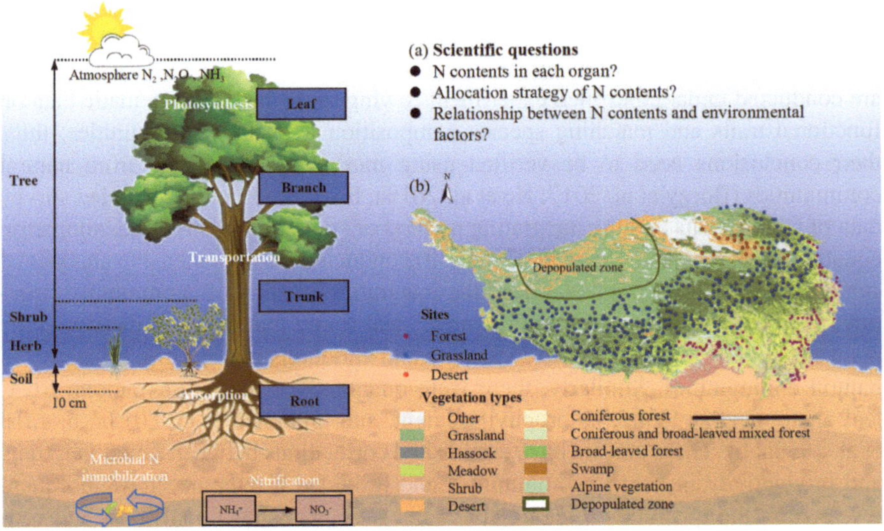

Fig. 6.6 Theoretical framework for the allometry of nitrogen allocation in natural plant communities in the Tibetan Plateau (TP). (**a**) Overview of scientific questions of the study; (**b**) spatial distribution of sampling plots on the TP. Depopulated parts of the TP were not sampled owing to concerns about traffic, climate, and safety of researchers

Researchers have surveyed the data of leaves, branches, stems, and roots in 2040 natural plant communities using the 0.5° grid sampling method and evaluated the spatial variations in nitrogen content, density, and storage in the Qinghai–Tibet Plateau and their driving factors. Obtaining information about the different vegetation types on the Qinghai–Tibet Plateau using remote sensing can be challenging because of the significant variations in altitude ranging from 800 to 5400 m. a. s. l. Therefore, we used the random forest algorithm to predict the spatial variations in nitrogen density on the Qinghai–Tibet Plateau and plotted a spatial distribution map of the nitrogen density of vegetation on the Qinghai–Tibet Plateau with a 1 km resolution (Li et al. 2022a, b, c, d).

The nitrogen content and density of leaves and roots were higher than those of branches and trunks, and leaves and roots had the strongest activities, consistent with the hypothesis that nitrogen accumulates in more active organs at the community level. Among the different vegetation types (Fig. 6.7), the nitrogen content of plant communities dominated by herbaceous plants was higher than that of communities dominated by woody plants. According to the productivity–longevity hypothesis, trees have a long lifespan and can use relatively small amounts of nutrients for long-term photosynthesis, whereas herbaceous plants need to accumulate sufficient nutrients in a short period to enhance productivity. However, the distribution of nitrogen density in different vegetation types contrasts with that of the spatial nitrogen content. The results of the allometric models revealed that nitrogen content and density are distributed in different plant organs. Furthermore, vegetation type and

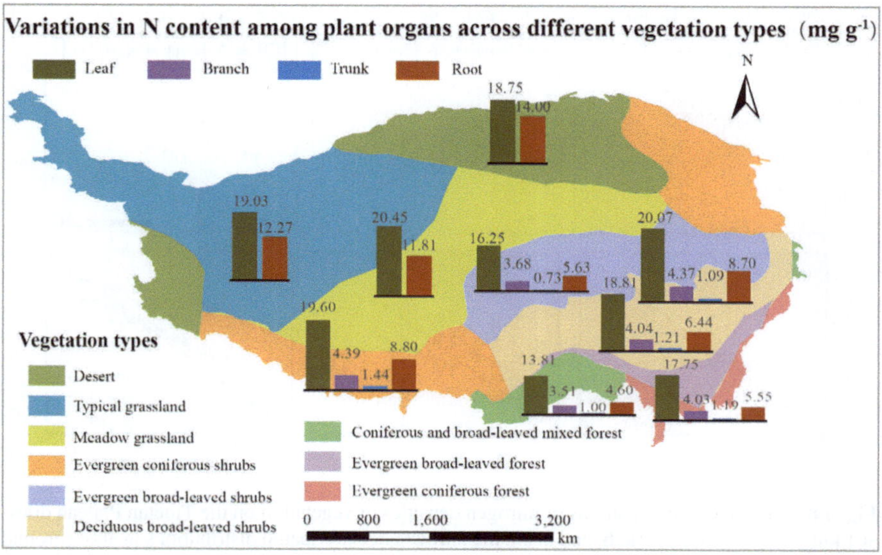

Fig. 6.7 Variations in nitrogen content among plant organs across different vegetation types. Vegetation zoning is based on dominant vegetation type considering vegetation types in China

environmental factors marginally affect the nitrogen content, and the distribution correlation is relatively conservative. The path analysis results revealed that the nitrogen content of the vegetation community has high internal stability and was less affected by environmental factors. The results of the analysis of nitrogen content alone verified the allometric hypothesis that the distribution of nutrients among organs is independent of environmental changes. However, the nitrogen density was greatly affected by environmental factors, primarily driven by radiation factors. Generally, the nitrogen density was high in the southeastern forest and low in the northwestern grassland. Based on field-surveyed data, grid sampling, and machine algorithms, the random forest model explained 69% of the spatial variations in nitrogen density in the Qinghai–Tibet Plateau, and a spatial distribution map of vegetation nitrogen density was drawn at a 1 km resolution (Fig. 6.8).

Overall, the TP has a large vegetative nitrogen pool. The average nitrogen content, density, and total storage of the vegetation were 8.48 mg g^{-1}, 31.97 g m^{-2}, and 50.62 Tg, respectively. These findings and spatial data on the nitrogen content and density of plant communities provide basic parameters for future remote sensing and regional nitrogen cycle models.

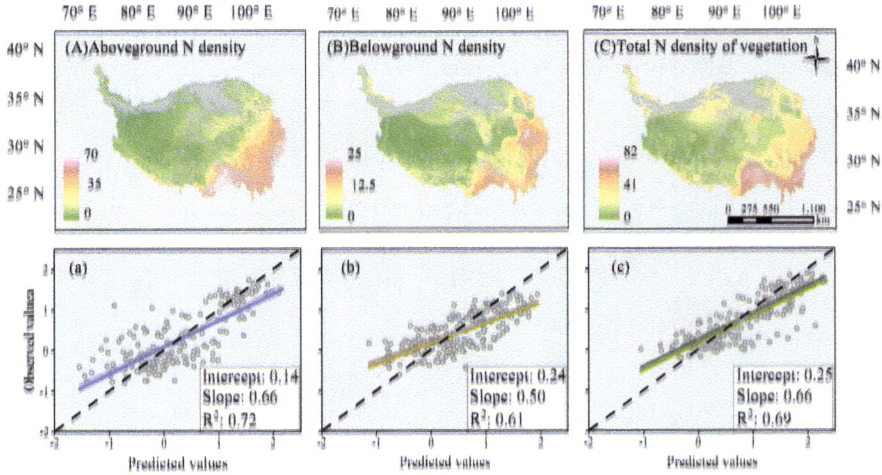

Fig. 6.8 Predicted spatial patterns of nitrogen densities of vegetation on the Tibetan Plateau drawn at 1 km resolution. Panels (**a**, **b**, and **c**) are the nitrogen density spatial distributions in aboveground, belowground, and total vegetation, respectively; panels (**d**, **e**, and **f**) are the correlations between observed and predicted values, using the random forest model in aboveground, belowground, and total vegetation, respectively

6.3.3 Spatial Variations in Phosphorus and Its Influencing Factors in the Plant Communities of Qinghai–Tibet Plateau

Phosphorus (P) is an important constituent element of the structures of nucleic acids, proteins, enzymes, ATP, and the phospholipid layer of the cell membrane (Sterner and Elser 2002) and is essential for plant growth and development. Furthermore, phosphorus participates in important energy metabolism processes, such as starch synthesis, ion absorption, and photosynthesis (Woodrow and Rowan 1979; Mollier and Pellerin 1999). It is also involved in carbohydrate transport in plants and can alter biomass accumulation by regulating the distribution of photosynthates among different organs (Adams et al. 1987; Roblin et al. 1998; Ma 2002). Phosphorus is involved in various plant metabolic processes and can limit productivity in most terrestrial ecosystems. Plants generally adapt to complex and changing environments by altering their morphological structures and physiological functions (Zhao 2022). In nature, the phosphorus content and storage in different plant organs will change with a changing environment, and its spatial distribution can reflect the adaptation and response mechanisms of plants to different environmental conditions to some extent (Mollier and Pellerin 1999).

To reveal the spatial variations in phosphorus and influencing factors in plant communities at the regional scale, we considered the climate gradient and vegetation zonal distribution and adopted a 0.5° standardized grid-sampling method to investigate 2040 plant community plots in the Qinghai–Tibet Plateau (Fig. 6.6). Most of

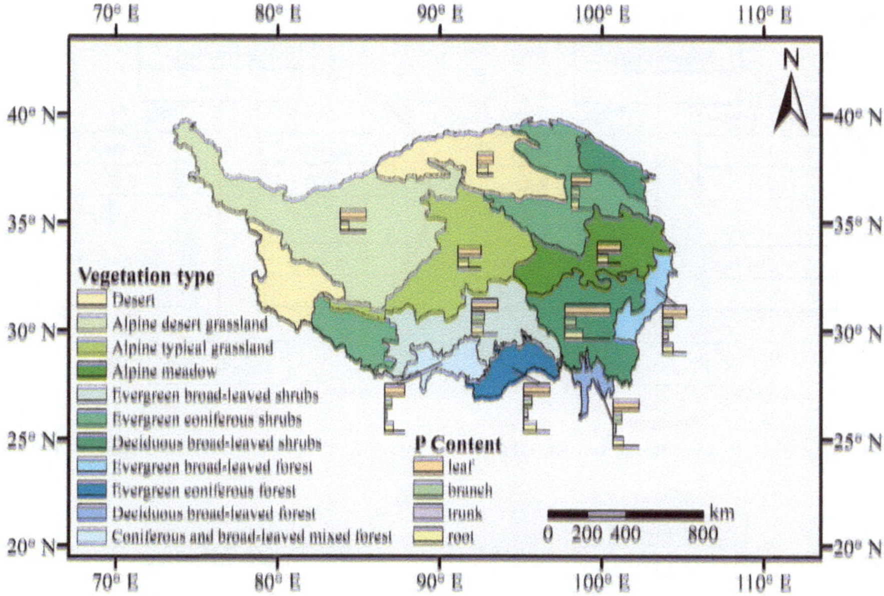

Fig. 6.9 Spatial variations in phosphorus content of plant organs among the different vegetation types

these plots were in areas with minimal human disturbance. Field investigations were conducted during the peak plant growth period from 2019 to 2021, from early July to the end of August. The phosphorus contents in leaves, branches, stems, roots, litter, and 0–10 cm soil layer were measured using standard methods (Li et al. 2022a, b, c, d). The phosphorus content of the vegetation and soil samples was measured using an inductively coupled plasma optical emission spectrometer (Optima 5300 DV, Perkin Elmer, Waltham, MA, USA).

The phosphorus contents in different organs of plants in Qinghai–Tibet Plateau were in the order of leaf (1.71 g kg^{-1}) > root (0.77 g kg^{-1}) > branch (0.63 g kg^{-1}) > stem (0.27 g kg^{-1}), consistent with the functional activities of different organs. Furthermore, the phosphorus density was in the order of branch (1.95 g m^{-2}) > trunk (1.81 g m^{-2}) > root (0.91 g m^{-2}) > leaf (0.63 g m^{-2}). The distribution ratio of above- and belowground phosphorus storage followed the order of forest (2.56) > shrub (1.94) > desert (0.51) > grassland (0.59), and plant communities were dominated by woody plants with a high aboveground phosphorus density. Additionally, phosphorus content, density, and distribution differed among ecosystems and vegetation types (Fig. 6.9).

Path analysis was used to explore the factors influencing phosphorus density and its spatial distribution in plant communities (Fig. 6.10). The results revealed that climatic factors (radiation, precipitation, and temperature), soil properties, and vegetation type (normalized difference vegetation index) jointly explained 60, 46, and 56% of the spatial variations in aboveground, belowground, and overall

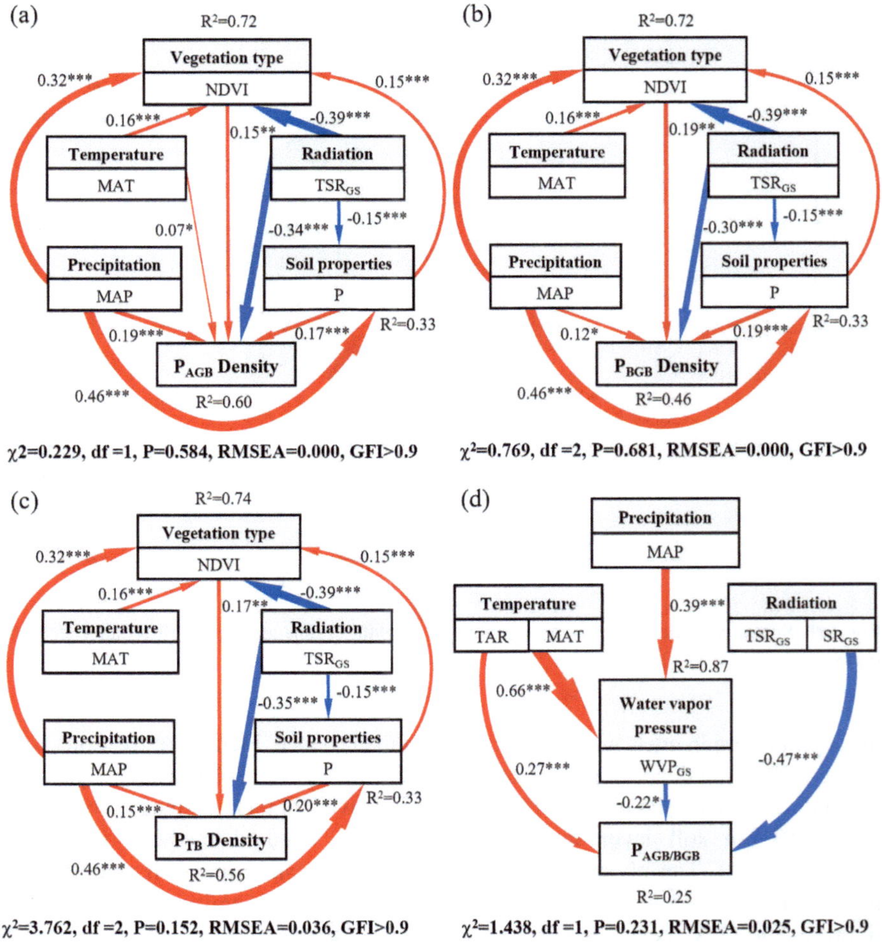

Fig. 6.10 Structural equation models (SEMs) used to explore the direct and indirect effects of environmental factors on P_{AGB} density (**a**), P_{BGB} density (**b**), P_{TB} density (**c**), and $P_{AGB/BGB}$ (**d**) on the TP. *TP* Tibetan Plateau, *SP* total soil phosphorus content, *MAT* mean annual temperature, *TAR* annual temperature range, *MAP* mean annual precipitation, TSR_{GS} total radiation during the growing season, SR_{GS} solar radiation during the growing season, *NDVI* normalized difference vegetation index, *WVP* water vapor pressure, *RMSE* root mean square error. Blue lines and negative values indicate negative effects, whereas red lines indicate positive effects in the SEMs. *, $p < 0.05$; **, $p < 0.01$; ***, $p < 0.001$

phosphorus density of plant communities, respectively, among which radiation (total solar radiation during the growing season, TSRGS) and precipitation (mean annual precipitation, MAP) were the dominant factors affecting phosphorus density of plant communities (Fig. 6.10). Climatic factors (radiation, temperature, vapor pressure, and precipitation) explained 25% of the variations in the above- and belowground phosphorus distribution ratios, among which radiation (TSRGS) and temperature (annual temperature range and mean annual temperature) were dominant. Notably, intense radiation significantly affects vegetation phosphorus density and spatial distribution on the Qinghai–Tibet Plateau. Intense radiation influences plant growth by inhibiting photosynthesis (Strid and Porra 1992; Yao and Liu 2009) and morphological characteristics, such as branch diameter and length (Han et al. 2009), and negatively affects biomass accumulation, thus decreasing phosphorus content in plant communities (Tong et al. 2021). Moreover, intense radiation can limit the activity of soil microbial enzymes, resulting in reduced soil organic matter decomposition and phosphorus content (Sun et al. 2019; Wang et al. 2017a, b, c). Therefore, soil nutrient deficiency reduces plant nutrient uptake and phosphorus storage (Chapin 1980, 2003; Reich and Oleksyn 2004). Furthermore, radiation changed the biomass distribution ratio of phosphorus in plant communities. Because excessive radiation severely damaged the aboveground parts of plants more than the belowground parts, plants on the Qinghai–Tibet Plateau enhance phosphorus storage in their root systems to adapt to high radiation stress (Wang and Liu 2011; Chu et al. 2018). Therefore, ensuring root growth and development to better utilize nutrient resources may be an important adaptive strategy for growth and survival of plants in the Qinghai–Tibet Plateau (Schmid and Weiner 1993).

Generally, the phosphorus content and density of plant communities exhibit significant spatial variation in different vegetation types on the Qinghai–Tibet Plateau, and the phosphorus density and spatial distribution in plant communities are closely correlated with environmental factors. The distinct strong radiation on the Qinghai–Tibetan Plateau is the dominant factor affecting the phosphorus density and the spatial distribution of vegetation through dual effects on the vegetation and the soil environment. Radiation is negatively correlated with precipitation and temperature, which jointly influence plant growth and phosphorus density. Therefore, the decreasing trend of phosphorus density and the relative proportion of aboveground vs. belowground phosphorus distribution in plant communities under strong radiation environments may reflect one of the adaptation strategies of vegetation to a changing environment.

6.3.4 Spatial Variations in Sulfur in Natural Communities and Its Influencing Factors in China

Sulfur (S) is an important macronutrient widely distributed in nature and has essential biological functions in plant catalysis, regulation, and morphogenesis. Therefore, sulfur plays a pivotal role in photosynthesis, stress resistance, and secondary metabolism (Capaldi et al. 2015). Sulfur is primarily taken up by plant roots in the form of sulfate and transported to different plant organs and participates in plant physiological and biochemical activities. The intrinsic requirements of plants primarily drive sulfur uptake, and different organs have different requirements for sulfur, considering their different functions, resulting in significant differences in sulfur accumulation among different organs. Meanwhile, the differences in the distribution of sulfur among different plant organs or regions can reflect the synergy of plant trade-off strategies. From an ecological perspective, the flow and distribution of specific nutrients among plant organs reflect the partitioning of photosynthates among different organs, which is an important mechanism for plants to adapt to various environments and ultimately influences plant growth and development (Poorter et al. 2012; Zhang et al. 2020a, b). Generally, metabolically active organs are allocated limited nutrients to maximize their functions, such as photosynthesis in leaves and nutrient uptake in roots (Zhao et al. 2019, 2022b). To date, most previous large-scale studies have focused on macro elements, such as carbon, nitrogen, and phosphorus; however, information on the biogeographic patterns of sulfur in major plant organs is rare.

Therefore, we conducted a systematic and standardized survey of natural communities in typical terrestrial ecosystems in China to determine the sulfur content and storage of plant communities and their spatial patterns in terrestrial ecosystems. A total of 49,836 plant samples and 1396 soil samples were collected and analyzed from 17,618 natural communities in terrestrial ecosystems in China, including forests (8296), shrubs (1388), grasslands (7371), croplands (235) and others (328). Simultaneously, we constructed a database of sulfur content and density in different plant organs and the topsoil (0–30 cm soil layer) by integrating multiple data sources.

The results revealed that the average sulfur contents in the leaves, branches, stems, and roots of terrestrial ecosystems in China were 2.03 ± 0.01, 0.67 ± 0.01, 0.14 ± 0.00, and 1.21 ± 0.01 g kg^{-1}, respectively. The average sulfur content in the 0–30 cm soil layer was 2.81 ± 0.24 g kg^{-1}. The sulfur content in metabolically active organs, such as leaves, was significantly higher than that in the other organs. Furthermore, desert ecosystems had higher sulfur content than other ecosystems, indicating that plants in arid areas have a stronger accumulative effect on sulfur, improving the adaptation of plants to drought stress (Zhao et al. 2022a). Geographical, climatic, and soil factors and vegetation types influence the spatial variation in sulfur content in plant communities. However, their effect on soil sulfur content is relatively weak, demonstrating that the spatial distribution and partitioning of sulfur in terrestrial ecosystems are complex and strongly affected by environmental changes.

The mean sulfur density in Chinese terrestrial ecosystems was approximately 272.25 × 14.94 × 10^{-2} t hm^{-2}, with those of vegetation and soil being 4.32 × 0.04 × 10^{-2} and 267.93 × 10^{-2} t hm^{-2}, respectively. The sulfur density and storage in different plant community organs were considerably different. Sulfur density (2.18 ± 0.02 × 10^{-2} t hm^{-2})) and storge (12.45 ± 0.31 Tg) of the roots were significantly higher than those of the leaves (0.99 ± 0.01 × 10^{-2} t hm^{-2}, 5.28 ± 0.10 Tg), branches (1.24 ± 0.02 × 10^{-2} t hm^{-2}, 3.24 ± 0.08 Tg), and stems (1.10 ± 0.01 × 10^{-2} t hm^{-2}, 2.59 ± 0.05 Tg). Vegetation and soil were the main components of terrestrial ecosystems, storing 2228.77 ± 121.72 Tg sulfur in China.

Using the random forest model, researchers have mapped the spatial distribution of sulfur density in plant communities (including aboveground, belowground, and

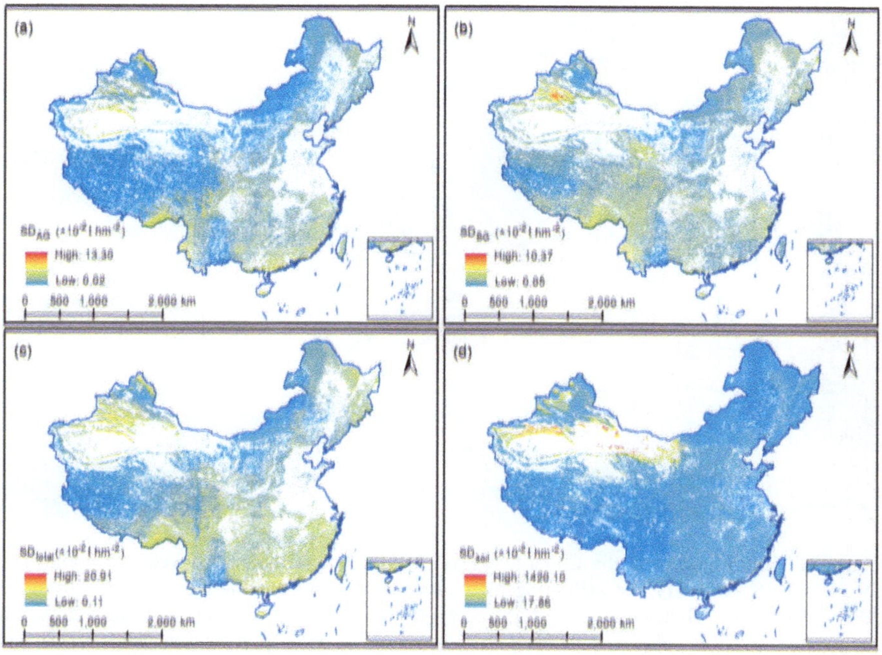

Fig. 6.11 Predicted spatial patterns of sulfur densities at a resolution of 1 km in the vegetation and soils of terrestrial ecosystems in China. (**a**) Aboveground sulfur density (SD_{AG}), (**b**) belowground sulfur density (SD_{BG}), (**c**) total sulfur density of vegetation (SD_{total}), and (**d**) soil sulfur density (SD_{soil}). The white areas in the figures represent cropland (excluding panel **d**), glaciers, lakes, rivers, and human land use (e.g., residential areas, industrial, mining, and transport land)

total vegetation) and surface soil (0–30 cm) of terrestrial ecosystems in China, at 1 km resolution (Fig. 6.11). Overall, the sulfur density of terrestrial ecosystems in China is generally low in the middle latitudes, with high values primarily concentrated in the Northwest and eastern China regions. Northwest China is primarily a desert area, whereas the eastern China region is primarily a forested area.

In summary, this study is the first to systematically reveal the spatial variation in sulfur content and storage in natural plant communities and soils in China and to explore its response to environmental factors at a national scale. The findings of this study broaden our understanding of the biological functions of sulfur and its role in plant–environment interactions, particularly at large scales (Fig. 6.12). Furthermore, these findings will help us understand the adaptive mechanisms of plants and their nutrient utilization when considering global climate change. A matching database of sulfur content and density in different ecological regions, vegetation types, and organs (leaves, branches, stems, and roots) can help modify and improve the regional sulfur cycle model by providing new and reliable parameters at a large scale.

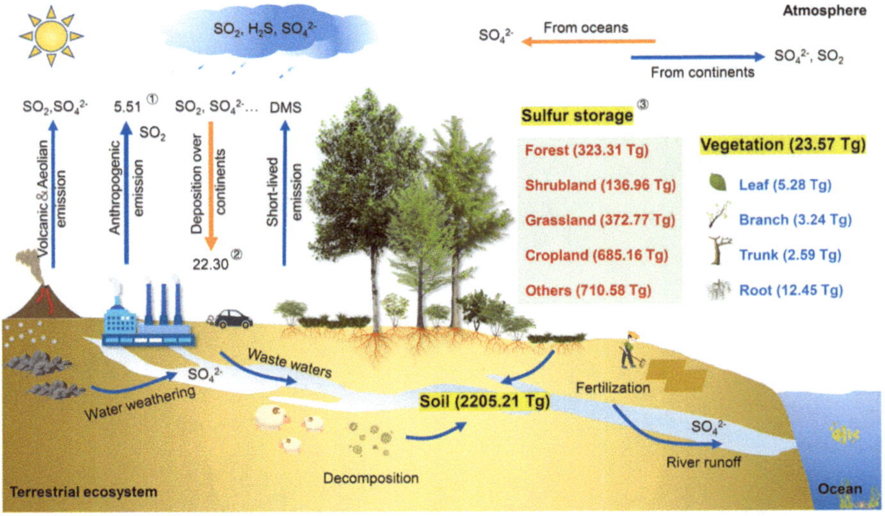

Fig. 6.12 Sulfur storage in the China terrestrial ecosystem and the sulfur cycle illustrating key fluxes at Tg S $yr.^{-1}$. (1) The anthropogenic emissions of sulfur dioxide were obtained from the 2015–2020 Annual Statistic Report on the ecological environment in China. (2) The total sulfur deposition (S_{dep}) flux value was obtained from the multiyear deposition datasets (Zhao 2022), and the average S_{dep} from 2015 to 2020 was used. (3) Sulfur storage values of vegetation and soil in the terrestrial ecosystem were obtained from this study. *DMS* dimethyl sulfide

6.4 Summary

The new concept of plant community traits helps us establish a new research framework on a large scale. It changes our perception considering the traditional methods, where studies of functional traits are limited to the organ or species level and motivate researchers to generate new data on plant community traits on a relatively unified spatial scale and dimension. Furthermore, the results of this study enable the exploration of the interactive interrelations among plants, animals, soil microbes, soil, and climate and reveal the construction and maintenance mechanisms of natural communities from the perspective of functional traits. Additionally, plant community traits consist of various combinations of functional traits standardized by unit land area at the community scale. This resolves the ongoing challenge of plant functional trait data not aligning with the spatial scale of the macroscale observation technology. Therefore, plant community traits can make full use of new data sources and data acquisition technologies, particularly remote sensing, and promote the development of macroecological research. With the rapid development of macroscale high-resolution and new technologies, it is possible to produce more parameters that can be used to explain the structure, process, and functions of ecosystems, such as the LAI, SLA, fluorescence parameters, and community structure parameters. The new data can serve as a solution to various regional environmental problems in different regions, even at the global level. Notably, plant community traits can provide accurate key parameters for ecological models, significantly enhancing their simulation accuracy at large scales (Connolly et al. 2013; Lu et al. 2017). The use of plant community traits in ecological studies is a recent development and thus poses many challenges. The future development of this field requires advancements in its theory and technology, close collaboration with researchers in macroecology and geoscience, and continuous expansion of its application in multiple disciplines of biology, ecology, geoscience, and remote sensing.

References

Ackerly DD, Cornwell WK (2007) A trait-based approach to community assembly: partitioning of species trait values into within-and among-community components. Ecol Lett 10:135–145

Adams MB, Campbell RG, Allen HL et al (1987) Root and foliar nutrient concentrations in *loblolly pine*: effects of season, site, and fertilization. For Sci 33:984–996

Blonder B, Enquist BJ, Graae BJ et al (2018) Late quaternary climate legacies in contemporary plant functional composition. Glob Chang Biol 24:4827–4840

Borgy B, Violle C, Choler P et al (2017) Sensitivity of community-level trait-environment relationships to data representativeness: a test for functional biogeography. Glob Ecol Biogeogr 26: 729–739

Calow P (1987) Towards a definition of functional ecology. Funct Ecol 1:57–61

Capaldi FR, Gratão PL, Reis AR et al (2015) Sulfur metabolism and stress defense responses in plants. Trop Plant Biol 8:60–73

Chapin FS (1980) The mineral nutrition of wild plants. Annu Rev Ecol Syst 11:233–260

Chapin FSIII (2003) Effects of plant traits on ecosystem and regional processes: a conceptual framework for predicting the consequences of global change. Ann Bot 91:455–463

Chave J, Coomes DA, Jansen S et al (2009) Towards a worldwide wood economics spectrum. Ecol Lett 12:351–366

Chu R, Chen NL, Han GJ et al (2018) Effects of enhanced UV-B radiation on growth and physiological characteristics of Phragmites australis. Acta Sci Circumst 38(5):2074–2081

Connolly J, Bell T, Bolger T et al (2013) An improved model to predict the effects of changing biodiversity levels on ecosystem function. J Ecol 101:344–355

Cornelissen JHC, Lavorel S, Garnier E et al (2003) A handbook of protocols for standardised and easy measurement of plant functional traits worldwide. Aust J Bot 51:335–380

Díaz S, Lavorel S, de Bello F et al (2007) Incorporating plant functional diversity effects in ecosystem service assessments. Proc Natl Acad Sci 104:20684–20689

Funk JL, Larson JE, Ames GM et al (2017) Revisiting the holy grail: using plant functional traits to understand ecological processes. Biol Rev 92:1156–1173

Han C, Liu Q, Yang Y (2009) Short-term effects of experimental warming and enhanced ultraviolet-B radiation on photosynthesis and antioxidant defense of *Picea asperata* seedlings. Plant Growth Regul 58:153–162

He NP, Liu CC, Tian M et al (2018a) Variation in leaf anatomical traits from tropical to cold-temperate forests and linkage to ecosystem functions. Funct Ecol 32:10–19

He NP, Liu CC, Zhang JH et al (2018b) Opportunities and challenges for plant trait research: from organs to communities. Acta Ecologic Sinica 38:6787–6796

He NP, Zhang JH, Liu CC et al (2018c) Patterns and influencing factors of traits in forest ecosystems: synthesis and perspectives on the synthetic investigation from the north-east transect of eastern China (NETEC). Acta Ecol Sin 38(18):6359–6382

He NP, Liu CC, Piao SL et al (2019a) Ecosystem traits linking functional traits to macroecology. Trends Ecol Evol 34:200–210

He L, Chen JM, Liu J et al (2019b) Diverse photosynthetic capacity of global ecosystems mapped by satellite chlorophyll fluorescence measurements. Remote Sens Environ 232:111344

He NP, Li Y, Liu CC et al (2020a) Plant trait networks: improved resolution of the dimensionality of adaptation. Trends Ecol Evol 35:908–918

He NP, Liu CC, Xu L et al (2020b) Ecosystem traits open new insights into macroecology: opportunity and challenge. Acta Ecol Sin 40(8):2507–2522

He P, Gleason SM, Wright IJ et al (2020c) Growing-season temperature and precipitation are independent drivers of global variation in xylem hydraulic conductivity. Glob Chang Biol 26: 1833–1841

He NP, Yan P, Liu CC et al (2023) Predicting ecosystem productivity based on plant community traits. Trends Plant Sci 28:43–53

Joswig JS, Wirth C, Schuman MC et al (2022) Climatic and soil factors explain the two-dimensional spectrum of global plant trait variation. Nat Ecol Evol 6:36–50

Keddy PA (1992) Assembly and response rules: two goals for predictive community ecology. J Veg Sci 3:157–164

Kunstler G, Falster DS, Coomes DA et al (2016) Plant functional traits have globally consistent effects on competition. Nature 529:204–207

LeRoy CJ, Hipp AL, Lueders K et al (2020) Plant phylogenetic history explains in-stream decomposition at a global scale. J Ecol 108:17–35

Li NN, He NP, Yu GR et al (2016) Leaf non-structural carbohydrates regulated by plant functional groups and climate: evidences from a tropical to cold-temperate forest transect. Ecol Indic 62: 22–31

Li Y, He NP, Hou JH et al (2018a) Factors influencing leaf chlorophyll content in natural forests at the biome scale. Front Ecol Evol 6:64

Li Y, Liu CC, Zhang JH et al (2018b) Variation in leaf chlorophyll concentration from tropical to cold-temperate forests: association with gross primary productivity. Ecol Indic 85:383–389

Li X, Li MX, Xu L et al (2022a) Allometry and distribution of nitrogen in natural plant communities of the Tibetan plateau. Front Plant Sci 13:845813

Li Y, Hou J, Xu L et al (2022b) Variation in functional trait diversity from tropical to cold-temperate forests and linkage to productivity. Ecol Indic 138:108864

Li Y, Liu CC, Sack L et al (2022c) Leaf trait network architecture shifts with species-richness and climate across forests at continental scale. Ecol Lett 25:1442–1457

Li J, Chen X, Niklas KJ et al (2022d) A whole-plant economics spectrum including bark functional traits for 59 subtropical woody plant species. J Ecol 110:248–261

Liu B, Li H, Zhu B et al (2015a) Complementarity in nutrient foraging strategies of absorptive fine roots and arbuscular mycorrhizal fungi across 14 coexisting subtropical tree species. New Phytol 208:125–136

Liu J, Zhang X, Song F et al (2015b) Explaining maximum variation in productivity requires phylogenetic diversity and single functional traits. Ecology 96:176–183

Liu CC, He NP, Zhang JH et al (2018a) Variation of stomatal traits from cold-temperate to tropical forests and association with water use efficiency. Funct Ecol 32:20–28

Liu J, Milne RI, Cadotte MW et al (2018b) Protect third pole's fragile ecosystem. Science 362: 1368–1368

Liu CC, Sack L, Li Y et al (2023) Relationships of stomatal morphology to the environment across plant communities. Nat Commun 14:6629

Liu CC, He NP, Li Y et al (2024) Current and future trends of plant functional traits in macro-ecology. Chin J Plant Ecol 48:21–40

Lu XJ, Wang YP, Wright IJ et al (2017) Incorporation of plant traits in a land surface model helps explain the global biogeographical distribution of major forest functional types. Glob Ecol Biogeogr 26:304–317

Ma Z (2002) Regulation of root hairs by phosphorus: mechanism and significance. The Pennsylvania State University, Philadelphia

Ma ZQ, Guo DL, Xu XL et al (2018) Evolutionary history resolves global organization of root functional traits. Nature 555:94–97

Maynard DS, Bialic-Murphy L, Zohner CM et al (2022) Global relationships in tree functional traits. Nat Commun 13:3185

McGill BJ (2010) Matters of scale. Science 328:575

Miles DB, Dunham AE (1992) Comparative analyses of phylogenetic effects in the life-history patterns of iguanid reptiles. Am Nat 139:848–869

Moles AT, Warton DI, Warman L et al (2009) Global patterns in plant height. J Ecol 97:923–932

Moles AT, Ackerly DD, Tweddle JC, Dickie JB, Smith R, Leishman MR, Mayfield MM, Pitman A, Wood JT, Westoby M (2007) Global patterns in seed size. Glob Ecol Biogeogr 16:109–116. https://doi.org/10.1111/j.1466-8238.2006.00259.x

Mollier A, Pellerin S (1999) Maize root system growth and development as influenced by phosphorus deficiency. J Exp Bot 50:487–497

Moretti M, Dias ATC, de Bello F et al (2017) Handbook of protocols for standardized measurement of terrestrial invertebrate functional traits. Funct Ecol 31:558–567

Naeem S, Wright JP (2003) Disentangling biodiversity effects on ecosystem functioning: deriving solutions to a seemingly insurmountable problem. Ecol Lett 6:567–579

Poorter H, Niklas KJ, Reich PB et al (2012) Biomass allocation to leaves, stems and roots: meta-analyses of interspecific variation and environmental control. New Phytol 193:30–50

Reich PB (2014) The world-wide 'fast-slow' plant economics spectrum: a traits manifesto. J Ecol 102:275–301

Reich PB, Oleksyn J (2004) Global patterns of plant leaf N and P in relation to temperature and latitude. Proc Natl Acad Sci USA 101:11001–11006

Reichstein M, Bahn M, Mahecha MD et al (2014) Linking plant and ecosystem functional biogeography. Proc Natl Acad Sci USA 111:13697–13702

Roblin G, Sakr S, Bonmort J et al (1998) Regulation of a plant plasma membrane sucrose transporter by phosphorylation. FEBS Lett 424:165–168

Schmid B, Weiner J (1993) Plastic relationships between reproductive and vegetative mass in Solidago Altissima. Evolution 47:61–74

Šímová I, Violle C, Svenning JC et al (2018) Spatial patterns and climate relationships of major plant traits in the New World differ between woody and herbaceous species. J Biogeogr 45:895–916

Song G, Li Y, Zhang J et al (2016) Significant phylogenetic signal and climate-related trends in leaf caloric value from tropical to cold-temperate forests. Sci Rep 6:1–10

Sterner RW, Elser JJ (2002) Ecological stoichiometry: the biology of elements from molecules to the biosphere. Princeton University Press, Princeton

Strid Å, Porra RJ (1992) Alterations in pigment content in leaves of *Pisum sativum* after exposure to supplementary UV-B. Plant Cell Physiol 33:1015–1023

Sun J, Liu B, You Y et al (2019) Solar radiation regulates the leaf nitrogen and phosphorus stoichiometry across alpine meadows of the Tibetan plateau. Agric For Meteorol 271:92–101

Tong R, Cao Y, Zhu Z et al (2021) Solar radiation effects on leaf nitrogen and phosphorus stoichiometry of Chinese fir across subtropical China. Forest Ecosystems 8:62

Violle C, Navas ML, Vile D et al (2007) Let the concept of trait be functional! Oikos 116:882–892

Wang HX, Liu WZ (2011) Effects of enhanced UV-B radiation on biomass and contents of camptothecin and 10-hydroxy-camptothecin in Camptotheca acuminata. Plant Sci J 29(6): 712–717

Wang RL, Yu GR, He NP et al (2016) Latitudinal variation of leaf morphological traits from species to communities along a forest transect in eastern China. J Geogr Sci 26:15–26

Wang RL, Wang QF, Zhao N et al (2017a) Complex trait relationships between leaves and absorptive roots: coordination in tissue N concentration but divergence in morphology. Ecol Evol 7:2697–2705

Wang R, Gibson CD, Berry TD et al (2017b) Photooxidation of pyrogenic organic matter reduces its reactive, labile C pool and the apparent soil oxidative microbial enzyme response. Geoderma 293:10–18
Wang H, Prentice IC, Davis TW et al (2017c) Photosynthetic responses to altitude: an explanation based on optimality principles. New Phytol 213:976–982
Wieczynski DJ, Boyle B, Buzzard V et al (2019) Climate shapes and shifts functional biodiversity in forests worldwide. Proc Natl Acad Sci 116:587–592
Woodrow IE, Rowan KS (1979) Change of flux of orthophosphate between cellular compartments in ripening tomato fruits in relation to climacteric rise in respiration. Funct Plant Biol 6:39–46
Wright IJ, Reich PB, Westoby M et al (2004a) The worldwide leaf economics spectrum. Nature 428:821–827
Wright IJ, Groom PK, Lamont BB et al (2004b) Leaf trait relationships in Australian plant species. Funct Plant Biol 31:551–558
Wright IJ, Dong N, Maire V et al (2017) Global climatic drivers of leaf size. Science 357:917–921
Yan P, Li MX, Yu GR et al (2022) Plant community traits associated with nitrogen can predict spatial variability in productivity. Ecol Indic 140:109001
Yan P, Fernández-Martínez M, Van Meerbeek K et al (2023a) The essential role of biodiversity in the key axes of ecosystem function. Glob Chang Biol 29:4569–4585
Yan P, Zhang JH, He NP et al (2023b) Functional diversity and soil nutrients regulate the interannual variability in gross primary productivity. J Ecol 111:1094–1106
Yao X, Liu Q (2009) The effects of enhanced ultraviolet-B and nitrogen supply on growth, photosynthesis and nutrient status of Abies faxoniana seedlings. Acta Physiol Plant 31:523–529
Zanne AE, Westoby M, Falster DS et al (2010) Angiosperm wood structure: global patterns in vessel anatomy and their relation to wood density and potential conductivity. Am J Bot 97:207–215
Zhang JH, Zhao N, Liu CC et al (2018a) C:N:P stoichiometry in China's forests: From organs to ecosystems. Funct Ecol 32:50–60
Zhang JH, He NP, Liu CC et al (2018b) Allocation strategies for nitrogen and phosphorus in forest plants. Oikos 127:1506–1514
Zhang JH, He NP, Liu CC et al (2020a) Variation and evolution of C:N ratio among different organs enable plants to adapt to N-limited environments. Glob Chang Biol 26:2534–2543
Zhang Y, Li Y, Wang RM et al (2020b) Spatial variation of leaf chlorophyll in northern hemisphere grasslands. Front Plant Sci 11:1244
Zhang JH, Li MX, Xu L et al (2021a) C:N:P stoichiometry in terrestrial ecosystems in China. Sci Total Environ 795:148849
Zhang JH, Ren TT, Yang JJ et al (2021b) Leaf multi-element network reveals the change of species dominance under nitrogen deposition. Front Plant Sci 12:580340
Zhang Y, He NP, Li MX et al (2021c) Community chlorophyll quantity determines the spatial variation of grassland productivity. Sci Total Environ 801:149567
Zhang ZH, Hou JH, He NP (2021d) Predictability of functional diversity depends on the number of traits. J Resour Ecol 12:332–345
Zhang Y, Migliavacca M, Penuelas J et al (2021e) Advances in hyperspectral remote sensing of vegetation traits and functions. Remote Sens Environ 252:112121
Zhao N (2022) Advance in molecular mechanisms of adaptive evolution of plants in extreme on the Qinghai-Tibet plateau. Environ Ecol 4(1):65–70
Zhao N, Yu GR, Wang QF et al (2019) Conservative allocation strategy of multiple nutrients among major plant organs: from species to community. J Ecol 108:267–278
Zhao WZ, Xiao CW, Li MX et al (2022a) Variation and adaptation in leaf sulfur content across China. J Plant Ecol 15:743–755
Zhao WZ, Xiao CW, Li MX et al (2022b) Spatial variation and allocation of sulfur among major plant organs in China. Sci Total Environ 844:157155

Chapter 7
Approach to Predict Ecosystem Productivity on Plant Community Traits

Plants play an essential role in terrestrial ecosystems by contributing significantly to productivity, including gross (GPP) and net primary productivity (NPP), which form the basis of material cycle and energy flow in ecosystems. Therefore, improving the accuracy of models that predict ecosystem productivity is important in ecology. Previous studies have primarily used ecological process models on a large-leaf model to predict the NPP spatiotemporal variation. This approach simulates NPP using the unity principle considering the photosynthesis and respiration mechanistic processes, and it can predict spatiotemporal variation in the NPP at the macroscale under a scenario driven by environmental variables. The prediction of spatiotemporal variation in NPP is highly uncertain owing to the limitations of excessive scaling up between test and prediction and the difficulty in accurately obtaining these parameters at the regional or global levels. The regulation of primary productivity using plant functional traits has attracted extensive attention from researchers, and correlations between ecosystem productivity and specific leaf functional traits have been observed in previous studies. However, the underlying mechanisms in which diverse plant functional traits drive various ecosystem functions are unclear. Although previous studies have attempted to use plant functional traits as input parameters in dynamic vegetation models, prediction accuracy falls short of expectations. Previous studies have referenced the output of the classical engine power model and combined it with the two-dimensionality of plant community traits to develop a novel trait-based productivity (TBP) framework. In this chapter, the theoretical basis of the TBP framework and two potential methods of realization are comprehensively discussed. Furthermore, four examples of the application of the novel TBP framework are discussed: (1) using the nitrogen community trait to predict the NPP spatial pattern and temporal dynamics in typical Chinese ecosystems, (2) the leaf chlorophyll community trait to explore the productivity spatial variation and its main influencing factors in Northern China grassland areas, (3) several community traits to predict the spatial variations in ecosystem productivity in China, and (4) several community traits to demonstrate the spatial variation of ecosystem multifunctionality in China. Furthermore, the TBP framework is based

N. He et al., *Plant Functional Ecology*,
https://doi.org/10.1007/978-981-96-8264-5_7

on the two-dimensionality of plant community traits standardized per unit land area. This framework can be integrated with the rapid development of remote sensing and hyperspectral and flux observations, and other advanced technologies to improve our predictive capabilities regarding spatiotemporal variations in ecosystem productivity and multiple functionalities, providing a basis for the development of a new generation of process models in the future.

Plants contribute the most to terrestrial ecosystem productivity, including gross (GPP) and net primary productivity (NPP), which form the basis of material cycling and energy flow in ecosystems. Therefore, improving the accuracy of models that predict ecosystem productivity in ecology is important. Predicting ecosystem processes and functions using plant functional traits is the "holy grail" in ecology. Since the 1990s, ecologists have increasingly employed functional traits-based approaches to understand how plant communities respond to environmental changes and how shifts in plant community structure influence ecosystem processes and functions at different scales (He et al. 2023). Previous studies have demonstrated that almost all functional traits vary along broad environmental gradients, and these variations among different communities can be used to predict changes in ecosystem functions under constant environmental changes (Violle et al. 2007; He et al. 2020a, b, c). However, functional traits are generally measured at the individual or species level, whereas ecosystem-level functions, such as ecosystem productivity, are determined considering unit land area. The scaling up of plant functional traits from the organ or species level to the ecosystem level encompasses complex transmutation challenges.

Scaling these traits up from the species to the ecosystem level, as a tradition, poses inherent transmutation challenges; however, it can potentially increase prediction accuracy. Jensen's inequality emphasizes that, due to nonlinearity, the mean value of a function differs from the function of the mean value (McGill 2019). Previous studies have demonstrated that certain functional traits, including leaf nitrogen concentration and specific leaf area, are strongly related to the net assimilation rate among different species (Field and Mooney 1986; Reich et al. 1999; Reich 2014). These linear correlations inspired ecologists to employ a linear scaling approach to model complex plant communities, such as big-leaf models (which average leaf properties across all leaves) or multilayer models (which consider leaves exposed to sunlight and shade, respectively) to predict ecosystem-level primary productivity (Farquhar 1989; Kull and Jarvis 1995; Luo et al. 2018). Big-leaf and multilayer models linearly scale ecological processes occurring at the organ and species levels to those at the canopy level (Friend 2001; Luo et al. 2018). However, the validity of directly upscaling plant traits remains questionable. As the spatial scale expands, the associated uncertainty between predicted and observed values may increase drastically (McGill 2019). For instance, the optimal temperature for some photosynthetic physiological parameters at the leaf level, like the maximum rate of ribulose-1,5-bisphosphate carboxylase/oxygenase (Rubisco) carboxylation (V_{cmax}) and the maximum rate of photosynthetic electron transport (J_{max}), cannot be applied at the ecosystem level (Huang et al. 2019). In ecosystems along elevation gradients, V_{cmax} and J_{max} measured at the leaf level, did not decrease as elevation increased. However, the productivity at the ecosystem level decreased with elevation (Malhi et al. 2017), emphasizing the need for aligning processes with appropriate scales when predicting ecosystem processes using plant functional traits (McGill 2010). Nevertheless, in most previous studies, this kind of matching was largely absent, often reporting weak correlations between ecosystem processes and specific traits or detected relationships lacking clear causal explanations (Harfoot et al. 2014; Li et al. 2020). Overall, it is not safe to apply this reductionist approach in all cases, and a method should be developed to directly link ecosystem properties (community-level traits) and functions from an emergentist perspective (He et al. 2023).

Previous studies often used the weighted average of species traits within a community as the representation of community-level traits based on the mass ratio hypothesis. To some extent, the community-weighted mean (CWM) trait value is a practical indicator reflecting the average behavior of all species. For instance, a high CWM leaf nitrogen concentration typically indicates a community that is predominantly composed of rapid-growing species with high photosynthetic rates (Wright et al. 2010; Garnier et al. 2016), indicating that the ecosystem has a relatively high production efficiency (Reich et al. 1997; Garnier et al. 2004). However, while the CWM approach provides the photosynthesis rate per unit leaf area or mass, it is not possible to quantify the carbon capture ability per unit land area of a plant community (Yang et al. 2018; Rubio et al. 2021). Meanwhile, the allometric scaling theory of metabolism and the vegetation quantity hypothesis employ biomass as a proxy for community size and have strong empirical support for elucidating productivity-related ecosystem functions (Enquist et al. 2007; Lohbeck et al. 2015; Prado-Junior et al. 2016; Heckman et al. 2020). Natural ecosystems are complex; therefore, it is challenging to adopt the simple reductionist approach of direct biomass-based upscaling. Furthermore, integrating plant traits with species biomass to explain ecosystem function is challenging. Therefore, there is a need to develop a novel method for combining plant traits with ecosystem contextual information.

7.1 Definition of Ecosystem Primary Productivity and Its Classification

Ecosystem primary productivity is the rate at which plants and other photosynthetic organisms produce organic compounds in a specific ecosystem and the basis of the material cycling and energy flow. Therefore, accurate prediction of the primary productivity of ecosystems is key in ecological research. Currently, the primary productivity of an ecosystem is primarily defined by the GPP, NPP, and net ecosystem productivity (NEP) (Yu et al. 2013). GPP is the total amount of organic compounds fixed by plants through photosynthesis per unit time and is the primary productivity of the total ecosystem. Theoretically, GPP determines the initial amount of matter and energy entering terrestrial ecosystems, is the key flux and basis of material cycling and energy flow in ecosystems, driving primary ecosystem processes, such as respiration and growth, and is an important indicator of carbon sink intensity to some extent. NPP refers to the proportion of organic carbon fixed by vegetation, except for that consumed during respiration, which is the basis for vegetation growth and reproduction. Theoretically, NPP reflects the net efficiency of plant fixation and conversion of photosynthates in each ecosystem and largely determines the material and energy available to heterotrophic organisms. NEP is calculated by subtracting the photosynthetic product consumed by heterotrophic organisms during respiration from NPP. NEP is the net photosynthetic output of atmospheric CO_2 entering the ecosystem. When NEP is >0, the ecosystem functions as a net carbon sink without considering carbon loss by fire, wind, and water erosion. When NEP is <0, the ecosystem is a net carbon source.

7.2 Gross and Net Primary Productivity Mechanisms, Processes, and Classical Prediction Approaches

1. **Climate ecological models**

Rosenzweig (1968) and Lieth and Whittaker (1975) independently developed two important global productivity prediction models based on the correlations between ecosystem productivity and climate variables. Rosenzweig (1968) plotted the correlations between the measured NPP values of 24 mature natural forest communities and the annual actual evapotranspiration (AET) and found an apparent linear correlation between AET and NPP after logarithmic transformation; therefore, AET should be used to predict ecosystem NPP. Most previous studies on climate ecological models have been conducted in the United States of America, and their regional conclusions limit the global applicability of this approach to some extent. Lieth and Whittaker (1975) developed a global NPP estimation model using data from >50 field observation sites in different bioregions worldwide. Furthermore, they matched the NPP values with annual temperature and precipitation data from weather stations using linear regression to obtain the first NPP global spatial distribution map, which is known as the classic Miami model. This model has become the foundation for numerous complex climate models, including the high-resolution biosphere model (Esser and Lautenschlager 1994). Nevertheless, the notion that NPP increases with increases in temperature and precipitation does not hold in the tropics, where NPP is not largely dependent on these climatic factors (Clark et al. 2001; Luyssaert et al. 2007). Contrastingly, other environmental variables, like the availability of soil nutrients and water, drive the variations in NPP within the tropics (Aragão et al. 2009; Quesada et al. 2009). Due to the simplicity and operability of climate-based models, many studies still use these traditional models to estimate regional ecosystem productivity at a large scale (Brovkin et al. 2002; Wang et al. 2005).

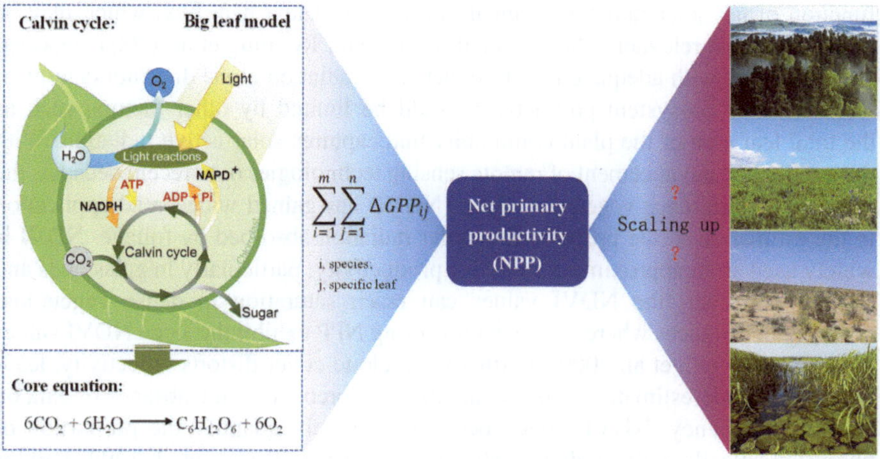

Fig. 7.1 Prediction of ecosystem productivity using the ecological process model based on the large-leaf model

2. **Ecological model of dynamic processes**

Previous studies have widely used the ecological process model based on the big-leaf model to predict the spatiotemporal variations in primary productivity (Fig. 7.1). This approach is based on leaf tissue-level photosynthesis and respiration mechanisms. The unity principle simulates the primary productivity of ecosystems and predicts their variation, which is driven by environmental variables. Practically, this approach is limited by excessive cross-scales between the test and prediction, and obtaining regional parameters through field measurements is challenging. Therefore, the prediction of temporal and spatial variations in ecosystem primary productivity is highly uncertain owing to the lack of regional parameters. Previous studies have developed various complex ecological process models, such as dynamic global vegetation models that simulate ecosystem structures, processes, and functions, such as biogeochemical cycle processes, GPP, and NPP, at different spatiotemporal scales (Cramer et al. 2001). However, these ecosystem process models have high uncertainty in predicting the GPP temporal and spatial variations, owing to the long period between testing and prediction and the difficulty in obtaining accurately measured parameters on a large scale. The effects of nutrients and plant functional traits have been incorporated into models to improve prediction accuracy. However, a substantial increase in model complexity may increase uncertainty in the model output. Despite the challenges, researchers still prefer ecological process models because of their ability to predict primary productivity.

3. **Radiation and remote sensing-based ecological models**

Previous studies have demonstrated that, under optimal irrigation and fertilization, annual crop productivity increases linearly with the absorption of solar energy (Monteith et al. 1977). Therefore, we can assume that, theoretically, NPP is a function of net solar radiation, annual rainfall, and latent heat flux, which may be trained with the relevant data measured in the field. Running et al. (2004) reported that in regions with adequate moisture, net solar radiation alone does not determine NPP. Instead, ecosystem productivity could be limited by other factors, such as the total leaf area of the plant community that captures solar active radiation. With the accelerated advancement of remote sensing technologies over recent decades, the normalized difference vegetation index (NDVI) has gained widespread application in the estimation of the proportion of solar radiation absorbed by foliage. NDVI is widely used as an approximate proxy for productivity, particularly in grasslands and deserts. However, the NDVI values can reach saturation in dense vegetation, resulting in instances where forests with varying NPP exhibit the same NDVI values (Sánchez-Azofeifa et al. 2009). Furthermore, cloud cover distorts reflectivity, leading to NPP underestimation (Zhao et al. 2005). Moreover, in the absence of data on light-use efficiency, NDVI alone does not accurately quantify the proportion of photosynthetically active radiation absorbed by leaves (Ruimy et al. 1999; Jenkins et al. 2007). Therefore, researchers developed a "multilayer" inversion model considering light-use efficiency by combining NDVI with empirical estimates of maximum potential light-use efficiency and other factors limiting photosynthesis and respiration, including, but not restricted to, effective photosynthetic radiation, leaf area index (LAI), temperature, water resource utilization, and atmospheric CO_2 concentration (Zhao et al. 2005; Zhao and Running 2010). Nonetheless, the limited predictive capacity for future primary productivity is a challenge when using these models.

7.3 Theoretical Basis and Approaches for Predicting Productivity Using Plant Community Traits

7.3.1 Theoretical Basis

Plant functional traits are attributes that affect growth, reproduction, and survival (Garnier et al. 2004; Shipley 2010; Garnier et al. 2016) and are often used to study the response of plant species to environmental changes (Reich et al. 1997; Sack et al. 2012; He et al. 2019b). The philosophical conclusion that functional traits influence function provides a crucial theoretical basis for developing a new model for using plant community functional traits to predict ecosystem productivity. Previous studies have investigated how plant functional traits influence ecosystem productivity (Reich 2012; Bahar et al. 2017; Fyllas et al. 2017; Wang et al. 2017a, b, c; Peng et al. 2020), and various studies used plant functional traits in mechanistic analyses and prediction of processes across multiple ecological scales, ranging from organ-level to species-level, extending ultimately to ecosystem-level assessments (McGill et al. 2006; Violle et al. 2007; Sack et al. 2013; Levine 2016; Mittelbach and McGill 2019).

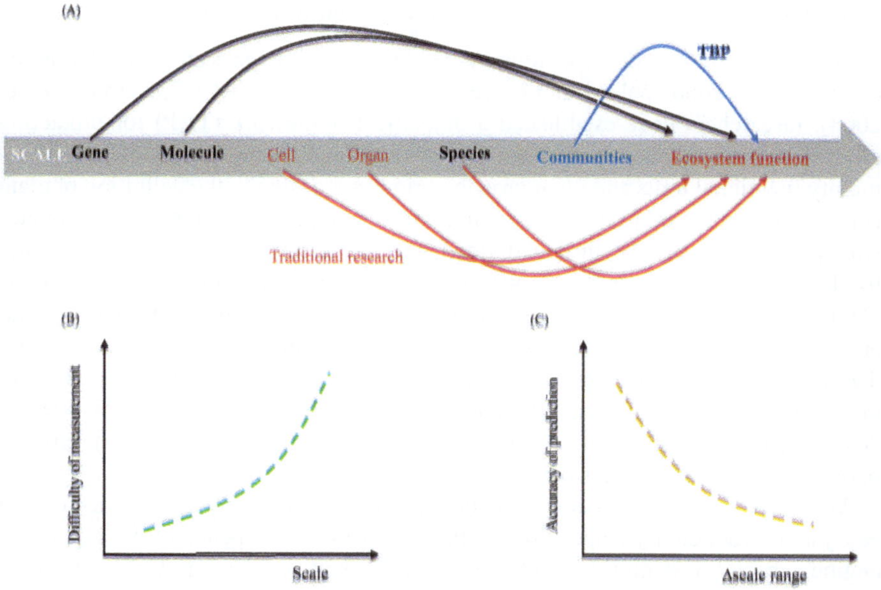

Fig. 7.2 In contrast to previous approaches that directly relate ecosystem-level functions to the traits of genes, molecules, cells, organs, and species (black and red arrows), the trait-based productivity (TBP) framework predicts ecosystem functions from community-level traits (blue arrow) (**a**). With increasing differences in scales considered, the difficulty in measuring traits may sharply increase (**b**), while the accuracy of predicting ecosystem functions may drastically decrease (**c**)

Previous studies have found correlations between ecosystem productivity and some averaged leaf functional traits across species (Bahar et al. 2017; Fyllas et al. 2017); however, effectively linking species traits to whole ecosystem functions is a challenge (He et al. 2019b; Barry et al. 2021). Since current GPP and NPP assessments rely on remote sensing or eddy flux observations per unit area, there is a need to develop a new trait-based model at the ecosystem level (Šímová and Storch 2017; He et al. 2019a; Zhang et al. 2021b). Although attempts have been made to directly upscale individual plant traits to community-scale traits, primarily based on the mass ratio hypothesis, new approaches are required to appropriately incorporate multidimensional community-level traits, which subsequently, along with environmental information, accurately interpret variations in productivity-related ecosystem functions within and among natural ecosystems across spatial scales (Fig. 7.2).

7.3.2 Developing Trait-Based Productivity Based on the Two-Dimensionality of Community Traits

The engine and inverse production models are approaches for developing trait-based productivity (TBP) based on the two-dimensionality of community traits, designed as follows:

1. **Engine model based on two-dimensional community traits**

Inspired by the "engine operating mode" of classical physics and considering the classical production ecology and harvest equations (Monteith 1977; Dermody et al. 2008), researchers have established a theoretical framework (TBP) for predicting ecosystem productivity based on the two-dimensional characteristics of plant community functional traits and plant growth period. According to the definition of plant community functional traits, each community functional trait that closely correlates with productivity can be further divided into two dimensions, which are density [or intrinsic efficiency traits ($\text{Traite}_{\text{efficiency}}$)] and intensity [or quantitative traits ($\text{Trait}_{\text{quantity}}$)], and the detailed description is provided in Chap. 6. Without considering fuel efficiency and friction loss, the engine output energy is theoretically determined primarily by the fuel consumption per unit time, energy released per unit fuel, and working time. By analogy, the GPP of a specific ecosystem is determined by the quantitative and efficiency traits and the length of the growth period of plant communities per unit land area (Fig. 7.3).

We used the engine analogy to illustrate how community-level traits impact ecosystem productivity (Fig. 7.3). Engine power mainly depends on fuel quantity, engine capacity, fuel utilization efficiency, and operating duration. Unlike mechanical engines governed by specific physical laws, natural plant communities are complex. However, this analogy applies to the scaling of productivity from individual and species traits to ecosystem traits. Furthermore, this analogy can be expressed as the classic production ecology equation: *Production* = *resource supply* × *proportion of supply captured* × *production efficiency*. This equation is the harvest equation used in agriculture (Morgan et al. 2005; Dermody et al. 2008): $\boldsymbol{W_h = \varepsilon_c \times \varepsilon_i \times S_t}$. In this study, the energy accumulation in plant biomass ($\boldsymbol{W_h}$) depends on the radiation use efficiency ($\boldsymbol{\varepsilon_c}$), the light interception efficiency ($\boldsymbol{\varepsilon_i}$), and total solar radiation ($\boldsymbol{S_t}$). Similarly, we developed a similar multiplicative framework for TBP, which enabled us to estimate community or ecosystem productivity by incorporating plant community traits, climatic variables, and nutrient inputs (Figs. 7.3 and 7.4, He et al. 2023).

$$P_i = E_{\text{index}} \times \text{Trait}_{\text{capacity}} \times \text{Trait}_{\text{efficiency}} \times L_{\text{gs}} \tag{7.1}$$

where P_i is productivity-related ecosystem functions, such as typical GPP and NPP; E_{index} is the energy index, which represents the energy input per unit time during the growing season, depending on growing season temperature, precipitation, and soil nutrients; $\text{Trait}_{\text{capacity}}$ is the trait capacity, which reflects the capacity of resource uptake of a particular ecosystem; $\text{Trait}_{\text{efficiency}}$ is the trait efficiency, indicating the ecosystem production efficiency; and L_{gs} is the length of the growing season, which can be recognized as the working time of the ecosystem.

In practice, key leaf traits that influence photosynthesis and ecosystem productivity, such as leaf size and specific leaf area, and leaf chlorophyll and nutrient concentration, can be scaled up to $\text{Trait}_{\text{quantity}}$ and $\text{Trait}_{\text{efficiency}}$, representing different dimensions of community-level traits (Table 7.1). Theoretically, each

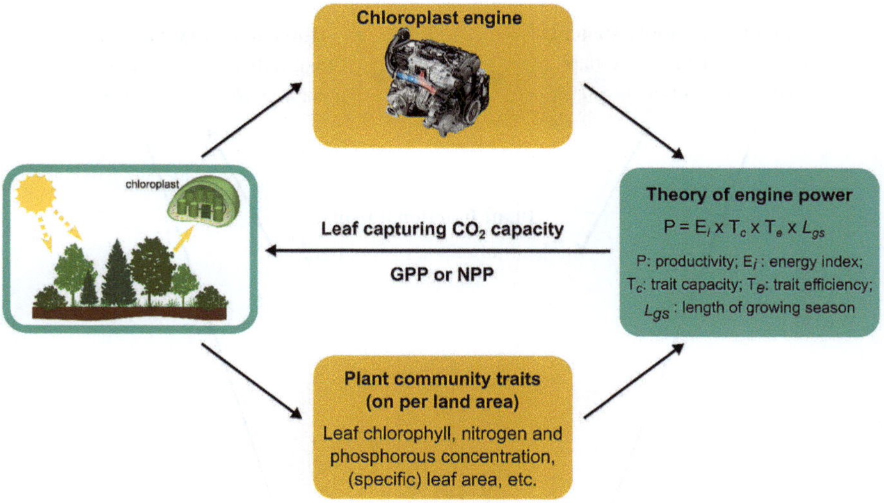

Fig. 7.3 Theoretical framework for predicting natural ecosystem productivity using plant community traits. Different from the complicated process model (Friend 2001), we introduced the concept of emergentism to analogy the whole ecosystem as an integrated engine (He et al. 2019b) and referred to the production ecology (Monteith 1977) or harvest equations (Dermody et al. 2008) to develop a theory of the trait-based productivity. In the provided equation, the energy index $E_i()$ is the energy input during the growing season, dependent on climate and nutrient inputs, trait capacity (T_c) is the resource capture capacity, trait efficiency (T_e) is the resource utilization efficiency of a specific community, and the growing season length (L_{gs}) is the work time of a particular community. *GPP* gross primary productivity, *NPP* net primary productivity

community-level trait from two dimensions can explain a large portion of the spatial variations in GPP or NPP in natural ecosystems, given their strong correlation after long-term adaptation and evolution. However, disentangling the coupling or decoupling between these two-dimensional traits in aggregated leaves poses a considerable challenge for future research.

Furthermore, using the general formula (Eq. 7.1) to explore the spatiotemporal variations in productivity-related ecosystem functions in natural ecosystems requires further development. Taking GPP as an example, the GPP in a natural community can be calculated using the following equation:

$$\text{GPP} = \sum_{i=1}^{n} \text{GPP}_i = \beta_0 + \beta_1 f(z_i) \prod\nolimits_{j=1}^{n} (E_i x_j y_j) \tag{7.2}$$

where i is the specific period, GPP_i is the formation of GPP during the ith period, E_i is the energy input during the ith period, x_j is the capacity of specific Trait_j at the community level ($\text{Trait}_{\text{-capacity}}$; Table 7.1), y_j is the Trait_j efficiency ($\text{Trait}_{\text{-Effiency}}$; Table 7.1), $f(z_i)$ is the growth time in the specific ith period (L_{gs}), β_0 is a constant, and β_1 is the specific coefficients representing the effect of these parameters.

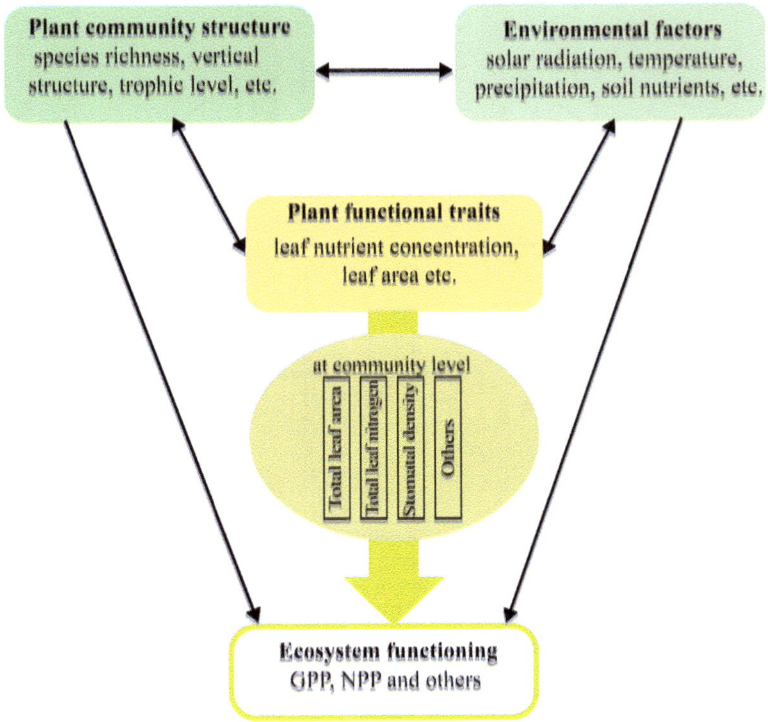

Fig. 7.4 Theory integrating plant traits into the prediction of ecosystem functioning. In this trait-based productivity theory, plant functional traits are core for predicting productivity (expressed at per unit land area). Biotic factors, such as community structure and herbivory, and abiotic factors, such as precipitation, temperature, light, and soil nutrients, can directly affect productivity and interact with and influence traits to indirectly influence gross primary productivity (GPP) and net primary productivity (NPP)

As illustrated in Table 7.1, these traits (Trait_j, 1, 2, 3, … ., and j) at the community level act interactively; therefore, Eq. 7.2 can be transformed into Eq. 7.3, as follows:

$$\text{GPP}_i = \beta_0' + \beta_1' \times f(E_i) \times f(x_j \times y_j) \times f(z_i) \quad (7.3)$$

Most two-dimensional community-level traits related to ecosystem productivity should be included in these equations once we understand their complex relationships and underlying mechanisms.

$$f(x_j) = \gamma_0 + \gamma_1 f(C, S) \quad (7.4)$$

$$f(y_j) = \gamma_0' + \gamma_1' f'(C, S) \quad (7.5)$$

Table 7.1 Selected leaf functional traits at the community level

No.	$Trait_i$ at the organ level	$Trait_i$ at the community level[a]	
		Capacity of $trait_i$ ($Trait_{i\text{-}Capacity}$)	Efficiency of $trait_i$[b] ($Trait_{i\text{-}Efficiency}$)
1	Leaf chlorophyll concentration (Chl, %)	$Chl_{\text{-}Capacity}$ (g m^{-2})	$Chl_{\text{-}Efficiency}$ (%)
2	Leaf nitrogen concentration (N, %)	$N_{\text{-}Capacity}$ (g m^{-2})	$N_{\text{-}Efficiency}$ (%)
3	Leaf phosphorus concentration (P, %)	$P_{\text{-}Capacity}$ (g m^{-2})	$P_{\text{-}Efficiency}$ (%)
4	Leaf size (leaf, cm^2)	Leaf area index (LAI, cm^2 m^{-2})	Leaf area (LA, cm^2)
…	…	….	…

[a]$Trait_{\text{-}Capacity}$ and $Trait_{\_Efficiency}$ represent different dimensions of plant traits at the community level
[b]High $Trait_{\text{-}Capacity}$ indicates high total leaf material input into vegetation photosynthesis at the ecosystem level, reflecting a high capacity of the ecosystem to absorb resources; High $Trait_{\_Efficiency}$ represents a high photosynthetic efficiency, reflecting a high ecosystem production efficiency (Reich et al. 1997; Garnier et al. 2004)

$$f(z_i) = \gamma_0'' + \gamma_1'' f''(C, S) \tag{7.6}$$

where C and S represent the influence of climate (e.g., temperature and precipitation) and soil factors, respectively, on x_j, y_j, and z_i; γ_0 is a constant; γ_1 is the specific coefficients representing the effect of these parameters. Therefore, it is possible to accurately predict the spatial and temporal variations in GPP at different scales if the correlations among abiotic factors on $f(x_j)$, $f(y_j)$, and $f(z_i)$ are determined through numerous field investigations and remote sensing.

2. **Inverse production model of two-dimensional community traits**

The engine model based on two-dimensional community traits underscores the role of community traits in influencing the acquisition and retention of different resources. In a natural ecosystem, production is strongly limited by multiple resources, with the most important in major ecosystems being light, water, and soil nutrients. Thus, we developed a model treating ecosystem productivity as a function of multiple limited resources (Nobel 1999) named the "inverse production equation" (Fig. 7.5), which assumes that resource limitations are approximately separable.

As illustrated in Fig. 7.5, predicted ecosystem productivity is calculated using the following equation:

$$\frac{1}{P} = \frac{1}{P_w \cdot \epsilon_w} + \frac{1}{P_n \cdot \epsilon_n} + \frac{1}{P_i \cdot \epsilon_i} \tag{7.7}$$

where P is ecosystem production (i.e., GPP, NPP, or another production-related variable); P_w, P_n, and P_i represent the maximum production constrained only by water, nitrogen, and irradiance, respectively, that is, potential production when resources are unlimited (e.g., P_w represents the maximum potential production

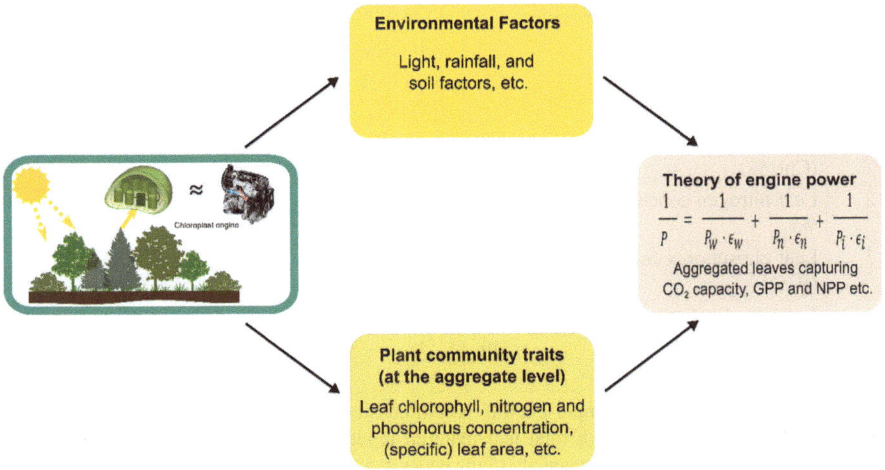

Fig. 7.5 Theoretical framework of the inverse production model. In the equation, P_w, P_n, and P_i indicate the maximum production constrained only by water, nitrogen, and irradiance, respectively. ε_w, ε_n, and ε_i indicate the resource acquisition efficiencies, reflecting the proportion of each available resource that plant community can capture

when nitrogen and light are infinitely available.); ε_w, ε_n, and ε_i indicate the resource acquisition efficiencies, reflecting the proportion of each available resource that a plant community can capture. The inverse production function can be deduced from the classic model of photosynthesis and gaseous exchange, with minor simplifications, providing insights into how traits affect productivity (He et al. 2023).

$$\frac{1}{P} = \frac{1}{c_a - \Gamma} \times \left(\frac{1.6D}{E_o} \left(\frac{E_o}{K(\psi_{\text{soil}} - \psi_{\text{leaf}})} \right) + (rc_a + M) \left(\frac{1}{\chi N_o} \left(\frac{N_o}{N} \right) + \frac{1}{\phi i_o} \left(\frac{1}{f_{\text{par}}} \right) \right) \right) \tag{7.8}$$

where the terms on the right-hand side are expressed in relation to the underlying environmental variables or plant traits.

$$P_w = \frac{E_o(c_a - \Gamma)}{1.6D}, \quad \epsilon_w = \frac{K(\psi_{\text{soil}} - \psi_{\text{leaf}})}{E_o} \tag{7.9}$$

$$P_n = \left(\frac{c_a - \Gamma}{rc_a + M} \right) \chi N_o, \quad \epsilon_n = \frac{N}{N_o} \tag{7.10}$$

$$P_i = \left(\frac{c_a - \Gamma}{rc_a + M} \right) \phi i_o, \quad \epsilon_i = f_{\text{par}} \tag{7.11}$$

where c_a is the ambient CO_2 mole fraction, Γ is the photosynthetic CO_2 compensation point, D is the leaf-to-air water vapor mole fraction difference, E_o is precipitation, K is the plant hydraulic conductance, ψ_{soil} is the soil water potential, ψ_{leaf} is the leaf water potential, r is the average ratio of intercellular to ambient CO_2 concentrations (approximately 0.7 and 0.4 for C_3 and C_4 plants, respectively) (Wong et al. 1979), M is the effective average Michaelis constant for the effect of CO_2 on the photosynthetic rate, N_o is the total ecosystem nitrogen, N is the total plant nitrogen, ϕ is the quantum yield of electrons, i_o is the irradiance above the canopy, and f_{par} is the fraction of i_o that is absorbed by the canopy.

The inverse production model illustrates how to keep the core simplicity of the traditional production-function approach while integrating the independent effects of multiple environmental factors (Fig. 7.5). Notably, the function resolves the limitations of conventional methods by focusing on a single resource like light and separating the three main photosynthetic resource limitations through a process-based framework centered on "acquisition efficiencies" to understand productivity. "Acquisition efficiency" is the proportion of each resource retained by the plant community. For example, ε_w is the ratio of the actual transpiration rate ($K[\psi_{soil} - \psi_{leaf}]$) to the theoretically possible maximum transpiration rate (the rate of water inputs or precipitation, E_o). The inverse production model presents clear and logical hypotheses concerning the integration of various capture traits into the model. For example, hydraulic, light interception, and nutrient capture traits would affect ε_w, ε_i, and ε_n, respectively. The definition of acquisition efficiency reveals the crucial role of plant structural traits in ecosystem productivity. For example, f_{par}, N/N_o, and K are primarily determined by the LAI, belowground carbon investment, and belowground carbon and aboveground water transport, respectively. Numerous plant traits regulating photosynthesis and capturing resources, like leaf chlorophyll and nutrient concentrations, specific leaf area, and root biomass, can be integrated into the model through their effects on equation parameters. The components of Eq. 7.7 may vary considering timescale and should be defined correctly. These timescales could range from minutes to hours or even span longer durations such as from growing seasons to years.

Equation 7.7 represents one among numerous potential methodologies that underscore mechanistic processes. Phenomenological approaches can be applied to the scale from traits to productivity. For instance, Bayesian models of tree growth emphasize how species trait regulate growth responses to environmental changes (Feng et al. 2018; Fortunel et al. 2018). This approach could potentially be extended to predict ecosystem productivity.

7.4 Trait-Based Productivity Application in Predicting Natural Ecosystem Productivity

7.4.1 Prediction of Gross Spatial Variations in Gross Primary Productivity Using Plant Community Trait Associated with Nitrogen

Leaf nitrogen content is a key component of leaf Rubisco, closely related to the photosynthetic rate, and an important part of the leaf economic spectrum (LeBauer and Treseder 2008; Dong et al. 2017a, b). Previous studies have demonstrated a quantitative correlation between leaf nitrogen concentration and photosynthetic rate at the species level (Croft et al. 2017; Smith et al. 2019). Interestingly, dominant species in natural communities often exhibit lower leaf nitrogen content compared to those of rare species (Zhang et al. 2022a, b). Herbaceous species tend to have higher leaf nitrogen content than that of trees (Ghimire et al. 2017). In resource-poor arid regions, the leaf nitrogen content is higher than that of plants in resource-rich regions (Moreno-Martínez et al. 2018), indicating that exploring the relationship between functional traits and ecosystem productivity based solely on leaf nitrogen concentration may be unrepresentative. Therefore, there is a need to resolve the limitation of the traditional assumption of mass ratio when scaling up, calculating the community-weighted-average nitrogen content [efficiency trait (mg g^{-1})] and the total nitrogen density of community leaves per unit land area [quantitative trait (g m^{-2})] and predicting temporal and spatial changes in ecosystem productivity using the combination of efficiency and quantitative traits (He et al. 2019b).

Therefore, researchers used data on leaf nitrogen content and its associated community structure from "systematic and compatible" measurements of 73 typical natural ecosystems in China to test this idea and provide an application case for the TBP framework. The researchers constructed efficiency ($N_{efficiency}$) and quantitative trait datasets ($N_{quantity}$) of the plant community functional trait (leaf nitrogen) (Fig. 7.6).

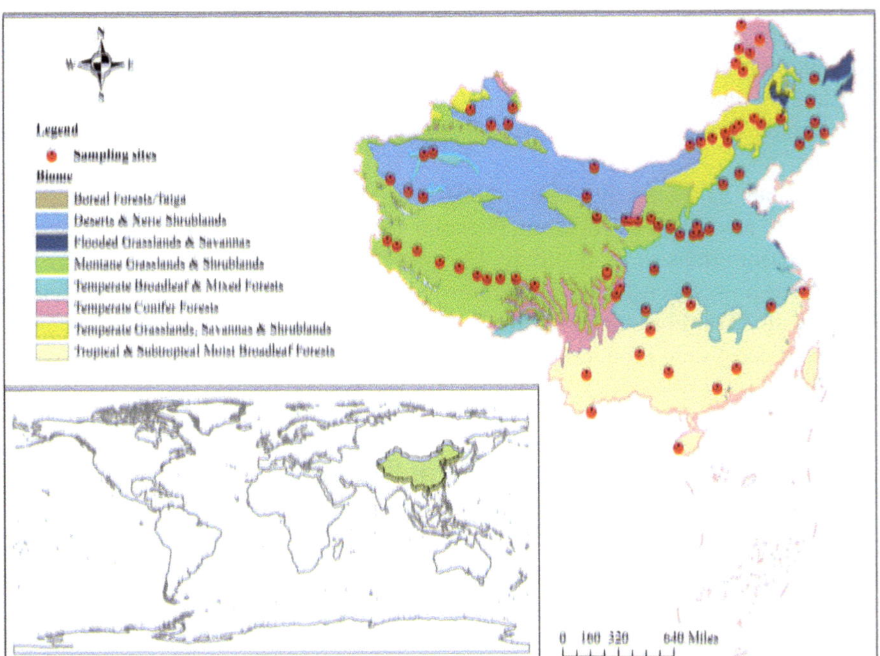

Fig. 7.6 Spatial distribution of study sites across various biomes, including boreal, temperate, tropical forests, and steppe, montane, and desert grasslands

The results revealed that climatic factors primarily regulate the amount and concentration of leaf nitrogen in the community but with a contrasting influence (Yan et al. 2022). Specifically, the first axis of the climate principal component revealed a significant positive effect on the total community leaf nitrogen density but a significant negative effect on the community leaf nitrogen content. In resource-rich areas, the total nitrogen density in community leaves was high, while the nitrogen content in community leaves was low, and contrasting results were observed when the areas with scarce resources. With a self-organizing function, the plant community cannot regulate the leaf nitrogen content as the weighted average of the community but can regulate the total leaf nitrogen density of the community to achieve the best use of resources in the long-term evolution and adaptation process. Additionally, 81% of the spatial variation in ecosystem primary productivity could be explained by combining efficiency ($trait_{efficiency}$) and quantitative traits ($Trait_{quantity}$) and the plant growth period length (Fig. 7.7).

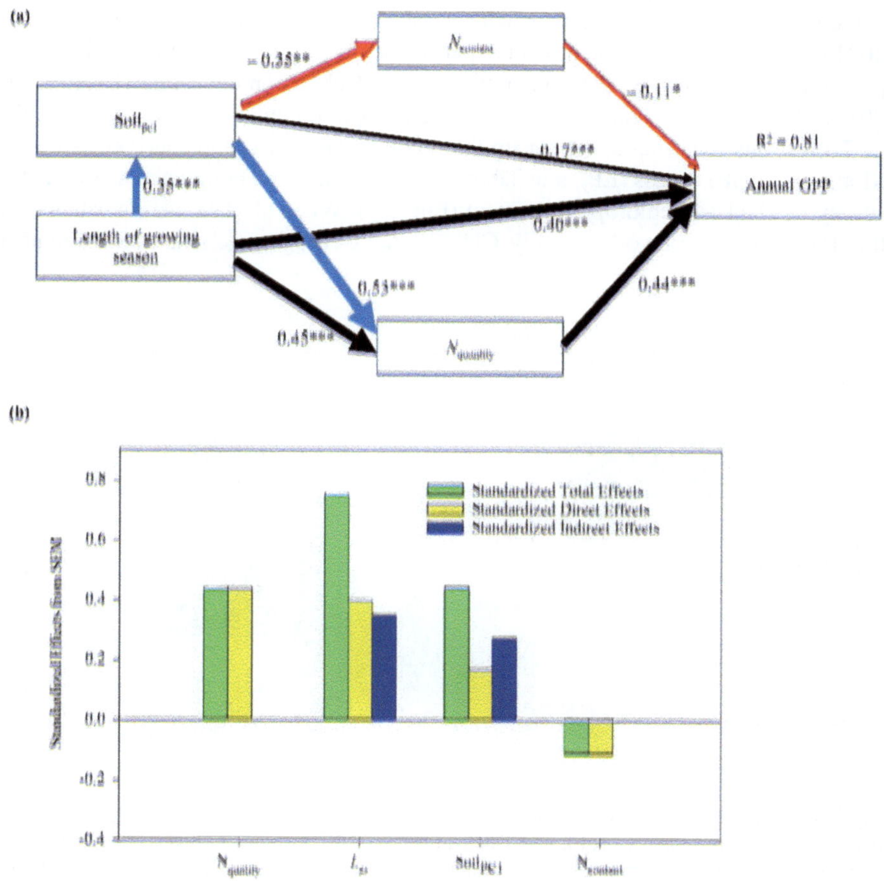

Fig. 7.7 Prediction of spatial variations in ecosystem GPP based on the nitrogen content of plant community functional traits. In **a** and **b**, the nitrogen efficiency trait ($N_{efficiency}$) represents the weighted average of leaf nitrogen concentration in the community (mg g^{-1}), the nitrogen quantitative trait ($N_{quantity}$) represents the total leaf nitrogen density of the community per unit land area (g m^{-2}). *GPP* gross primary productivity

7.4.2 Predicting the Spatial Variations in Grassland Productivity Using Plant Community Traits Associated with Chlorophyll

Chlorophyll, the most abundant pigment in plants, is the molecular basis of plant photosynthesis and a functional trait closely related to productivity (Croft et al. 2017). Previous studies have found that the chlorophyll content is a more reliable predictor of large-scale spatial variation than leaf nitrogen content. Using leaf chlorophyll enables us to verify and expand the application of the TBP framework. According to the TBP framework, when we assume the natural plant community as an engine, theoretically, the engine work output is primarily affected by the fuel consumption per unit time, calorific value of the unit fuel, and engine working time. When leaf chlorophyll was used as the carrier, community productivity was determined using the total leaf chlorophyll quantity [$CHL_{Quantity}$ (mg m^{-2})], chlorophyll content [$CHL_{Efficiency}$ (%)], and growing season length [GSL (months)] (Zhang et al. 2021a). According to classical plant experiments, the higher the chlorophyll content per unit mass of leaves, the higher the saturated photosynthetic rate (Croft et al. 2017); therefore, the community-weighted chlorophyll content is an efficiency trait.

To test this hypothesis, we set up three contrasting grassland transverses in the Mongolian (MP), Loess (LP), and Tibetan (TP) plateaus and conducted a systematic survey of plant leaf chlorophyll content (Fig. 7.8). When all data were attributed to the TBP basic framework, namely $CHL_{Quantity}$, $CHL_{Efficiency}$, and L, we found a

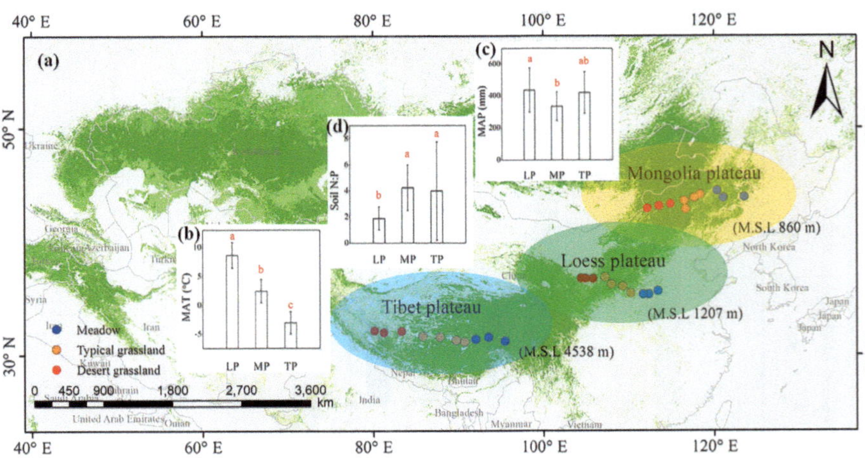

Fig. 7.8 Study sites in Chinese grasslands. Three grassland transects were set up in the Mongolian plateaus (MP), Loess plateaus (LP), and Tibetan plateaus (TP). From east to west, each transect is successively distributed with meadows, typical grasslands, and desert grasslands (**a**). In the MP, LP, and TP regions, the primary limiting constraints to plant growth were mean annual precipitation (MAP, **c**), soil nutrient [soil nitrogen: phosphorus (Soil N: P, **d**)], and mean annual temperature (MAT,**b**), respectively

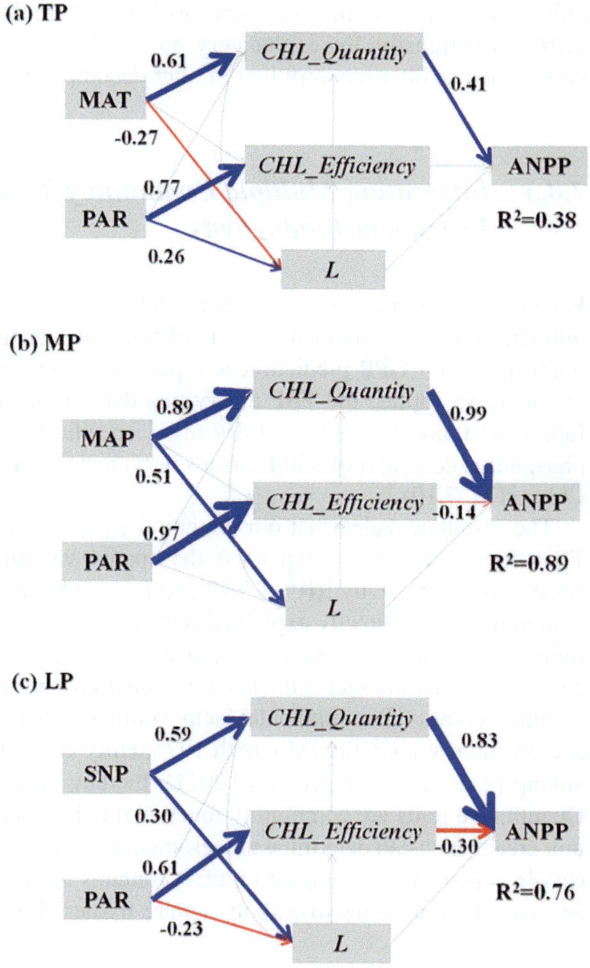

Fig. 7.9 Effects of total leaf chlorophyll content (*CHL_Efficiency*), total leaf chlorophyll quantity (*CHL_Quantity*), and the length of growing season (*L*) on the spatial variations in productivity across the Tibetan (TP) (**a**), Mongolian (MP) (**b**), and Loess plateaus (LP) (**c**). Blue, red, and gray arrows indicate positive, negative, and nonsignificant effects, respectively. *MAT* mean annual temperature, *MAP* mean annual precipitation, *SNP* soil nitrogen to phosphorus ratio, *PAR* photosynthetically active radiation, *ANPP* aboveground net primary productivity

significant positive correlation between the interaction effects of the three traits and the spatial change pattern of grassland productivity (Fig. 7.9). The two-dimensional characteristics of plant community functional traits and L explained 67% of the spatial variation in productivity in the MP, LP, and TP, and 83% of the spatial variation was explained in the TP alpine grassland (Zhang et al. 2021a). The quantitative traits analyzed in this study were related to, but not the same as, those associated with biomass. To be precise, "total traits" should be understood as an extension of the connotation of the "biomass hypothesis." First, there are inconsistencies in the results of the "biomass hypothesis tests." Currently, the biomass hypothesis primarily focuses on tropical forests, and the driving effect of biomass on community function depends on the state of resources and the environment. At different stages of community succession, biomass may have positive or negative effects on community function owing to different limiting factors (Prado-Junior et al. 2016). Additionally, the explanatory power of biomass is high in production processes such as litter but very low in litter decomposition (Lohbeck et al. 2015).

7.4.3 Integrating Multiple Community Traits to Predict Ecosystem Productivity

Plant community productivity does not rely solely on a single functional trait but is influenced by the synergistic effects of multiple functional traits. Integrating plant multi-traits into GPP prediction is a promising but challenging approach. In this research, we applied the TBP theory to a dataset integrating five plant community functional traits (leaf area and dry mass, specific leaf area, and leaf nitrogen and phosphorus contents) of 2500 species in typical forest and grassland ecosystems in China (Fig. 7.10).

The results revealed that our structural equation model (SEM) model based on TBP theory effectively explained the spatial variations in monthly and annual ecosystem productivity [($R^2 = 0.73$ and $R^2 = 0.87$, respectively (Fig. 7.11)]. Plant community traits directly explained 69% of the spatial variation in annual productivity, while environmental factors, including temperature, humidity, and soil nutrients, exerted an indirect influence on annual productivity by modulating the plant community traits. Environmental factors influence plant growth by regulating energy and resource availability (Monteith 1977; Hilty et al. 2021) and carbon distribution among plant organs (Hilty et al. 2021), thereby affecting ecosystem productivity. Quantitative traits are community functional traits standardized by leaf biomass per unit area, which can determine the resource utilization capacity of the ecosystem to a certain extent. As an integrated trait, it contains information about the biomass that an ecosystem allocates to carbon capture tissues (Enquist et al. 2007; Yang et al.

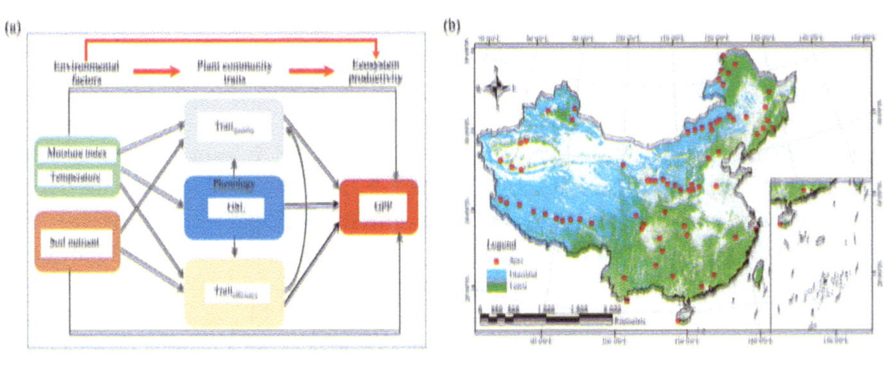

Fig. 7.10 Conceptual models illustrating the effects of environmental factors (climate and soil variables) and plant community traits and phenology on ecosystem productivity (**a**). $Trait_{quantity}$ and $Trait_{efficiency}$ represent the quantity and efficiency traits, respectively; GSL represents the growing season length. The spatial distribution of study sites in Chinese forests and grasslands (**b**)

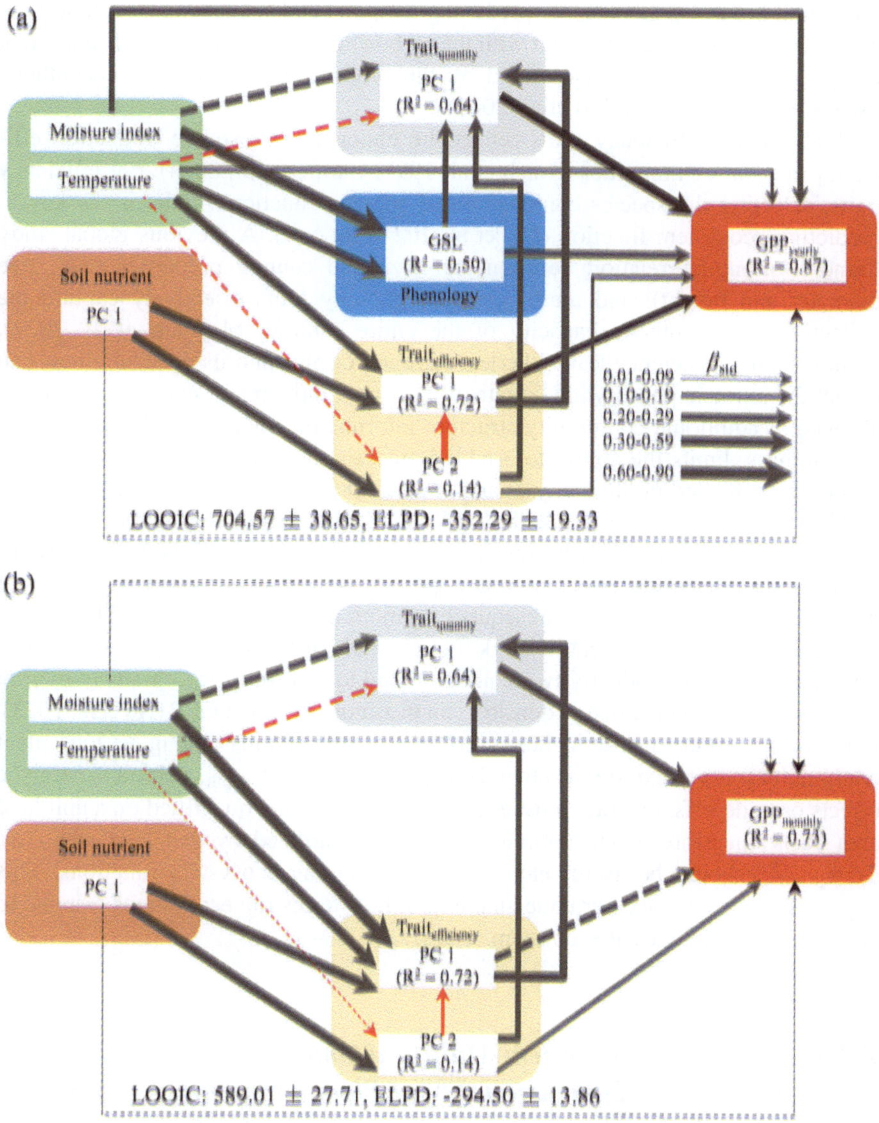

Fig. 7.11 Bayesian piecewise structural equation models elucidating the direct and indirect influences of environmental variables and plant community traits on gross primary productivity (GPP) across all sites: (**a**) yearly GPP (GPP_{yearly}) and (**b**) monthly GPP ($GPP_{monthly}$). $Trait_{quantity}$ and $Trait_{efficiency}$ denote the quantity and efficiency traits, respectively; PC1 and PC2 represent the first two principal components (PC) of the corresponding variables; GSL represents the growing season length. Solid and dotted lines indicate significant and nonsignificant effects, respectively. Red and black arrows denote negative and positive correlations, respectively

2018) and is a key parameter for predicting temporal and spatial changes in productivity. Our findings are consistent with the carbon economy theory, which states that the plant growth rate is determined by the biomass allocated to photosynthetic tissues (Garnier 1991; Rubio et al. 2021).

Additionally, efficiency traits reflect the endogenous resource utilization efficiency of plant communities (Reich et al. 1997; Garnier et al. 2004). For efficiency traits, selecting the species composition of specific functional traits is the key to predicting ecosystem function (Liu et al. 2021a, b, c, d). A previous global study found a strong correlation between leaf size and canopy properties (e.g., tree diameter and height), and these plant morphology traits effectively capture the collective photosynthetic capacity of the entire canopy (Maynard et al. 2022). Plant chemical efficiency traits like leaf nitrogen or phosphorus concentrations can reflect the intrinsic photosynthetic rate of leaves per unit area or mass; however, the absence of additional community structure information, such as total leaf area and leaf biomass, limits the ability to establish underlying relationships between these efficiency traits and productivity (Yang et al. 2018; Rubio et al. 2021; Zhang et al. 2022a, b). When a unit area standardized quantitative trait is considered, its significance in predicting ecosystem productivity is limited because of the lack of importance of the "selection effect of trait type." This finding provides an important theoretical foundation for the subsequent promotion and application of TBP.

There were limitations in this work. For example, the impacts of plant functional traits on ecosystem productivity are not confined to leaf traits. Root traits (e.g., root biomass, nitrogen content, and density) can theoretically affect GPP (He et al. 2023). Furthermore, future research must consider the interactions among various plant community traits to explore whether $Trait_{efficiency}$ and $Trait_{quantity}$ exhibit synergistic effects or trade-offs. Overall, most ecosystem functions are quantified on a unit land area basis; therefore, plant community traits standardized by unit land area (i.e., $Trait_{quantity}$) should be useful indicators for predicting ecosystem functions. This approach could enhance our understanding of how terrestrial ecosystems respond to environmental changes at ecosystem scales.

7.4.4 Predicting Ecosystem Multifunctionality on the Two-Dimensionality of Plant Community Traits

Exploring the multifunctionality of terrestrial ecosystems (EMF) and their regulatory mechanisms is of great interest. Previous studies suggested that the variations in terrestrial ecosystem functions can be mainly represented by three key axes: the ecosystem maximum productivity (GPP_{max}), the ecosystem carbon use efficiency (CUE), and the ecosystem water-use efficiency (WUE). Nevertheless, the roles of plant community traits in driving these fundamental ecosystem functions remain largely unexplored. To quantitatively reveal how biodiversity and plant functional traits affect the EMF (including GPP_{max}, WUE, CUE, and ecosystem light-use efficiency, LUE). A previous study collected data on plant functional traits and biodiversity from 840 plant community plots. Additionally, they collected environmental data on growing season precipitation (GSP), temperature (GST), and soil nutrients from 72 ecosystems in China (Yan et al. 2023).

The results exhibited that biodiversity (species richness and functional and phylogenetic diversity) with plant community traits ($Trait_{quantity}$ and $Trait_{efficiency}$) collectively governed the variation in the EMF. Notably, the effects of biodiversity on the EMF were less important than those of plant community traits. $Trait_{quantity}$ and $Trait_{efficiency}$ were the two most important variables in explaining the variations of multiple ecosystem functions [15.42–46.19% and 14.77–40.08% of the total R^2, respectively (Table 7.2)].

SEM analysis further identified the direct and indirect effects of biodiversity, plant community traits, and environmental factors on the variation in EMF (Fig. 7.12). Environmental factors (GSP and soil nutrients) directly influence EMF and indirectly impact it by modulating various community attributes (biodiversity and functional traits). Plant community traits exerted the highest direct impact on EMF, whereas growing season precipitation had the highest indirect effect on EMF.

7.5 Prospects and Significance of the Trait-Based Productivity Model

In the future, it is imperative to incorporate more plant functional traits to elucidate and predict spatial-temporal variations in ecosystem productivity and improve the accuracy of ecological process models. Environmental factors, including precipitation, temperature, and soil nutrients, can exert direct impacts on productivity and can interact with plant community traits, thereby indirectly influencing productivity. Therefore, researchers can use structural equation models to quantify these indirect effects, further refining the TBP theoretical framework. Moreover, future research should consider the impacts of taxonomy, phylogeny, and functional diversity on ecosystem productivity as they potentially optimize the ecosystem resource absorption efficiency.

The ecosystem structure and species traits determine plant community traits, and the role of ecosystem structure is of primary importance in predicting productivity (Fig. 7.5). To integrate the CWM of leaf traits into prediction equations of ecosystem productivity as demonstrated in Eq. 7.1, the CWM for leaf traits was multiplied by some plant community structural parameters. These could include either the leaf mass index (LMI, total leaf dry matter mass per unit land area) or the LAI (total leaf area per unit land area) (Bréda 2003). Previous analyses of productivity per unit land area across ecosystems generally relied on LAI, but some mass-based leaf traits, when multiplied by LMI, are effective predictors of productivity per unit land area. Similar logic can be applied to other plant traits like root traits. Additionally, intraspecific variations should be incorporated when determining plant community

Table 7.2 Variable importance of biodiversity, plant community traits, and environment in explaining variations in multiple ecosystem functions using hierarchical partitioning

	GPP_{max}[a]		WUE		CUE		LUE	
Predictors	Import. (R^2)	Contrib (%)	Import.(R^2)	Contrib. (%)	Import. (R^2)	Contrib. (%)	Import. (R^2)	Contrib. (%)
GSP[b]	0.12^{**c}	14.83	0.09^{**}	13.34	0.011	9.05	0.18^{***}	21.35
GST	0.02	3.02	0.10^{**}	15.82	−0.010	−7.94	0.04^{*}	5.26
Soil nutrients	0.11^{**}	13.22	0.05	7.19	0.001	1.11	0.05^{*}	5.95
Species richness	0.04^{*}	5.52	0.04	6.27	−0.005	−4.21	0.08^{**}	9.65
Functional diversity	0.10^{**}	11.96	0.08^{*}	11.94	0.022	17.38	0.13^{**}	14.67
Phylogenetic diversity	0.11^{**}	13.53	0.07^{*}	11.06	−0.002	−1.75	0.05^{*}	5.76
$Trait_{efficiency}$ on CWM	0.12^{**}	14.77	0.12^{**}	18.96	0.051^{*}	40.08	0.16^{**}	18.42
$Trait_{quantity}$	0.19^{***}	23.12	0.10^{**}	15.42	0.058^{*}	46.19	0.16^{**}	18.96
Total	0.81	100	0.64	100	0.13	100	0.85	100

[a]GPP_{max}, ecosystem maximum photosynthetic CO_2 uptake (g C m^2 day^{-1}); CUE, ecosystem carbon use efficiency (%); LUE, ecosystem light-use efficiency (g C MJ^{-1}); WUE, ecosystem water-use efficiency (g C L^{-1})

[b]GSP, growing season precipitation; GST, growing season temperature; $Trait_{efficiency}$ on CWM represents the first PC of CWM traits, which explains 53% of the variation in four CWM traits, and "$Trait_{quantity}$ as ecosystem traits" represents the first PC of ecosystem traits, which explains 98% of the variation in five ecosystem traits. *CWM* community weighted mean, *PC* principal component

[c]$^{*} = P < 0.05$; $^{**} = P < 0.01$; $^{***} = P < 0.001$

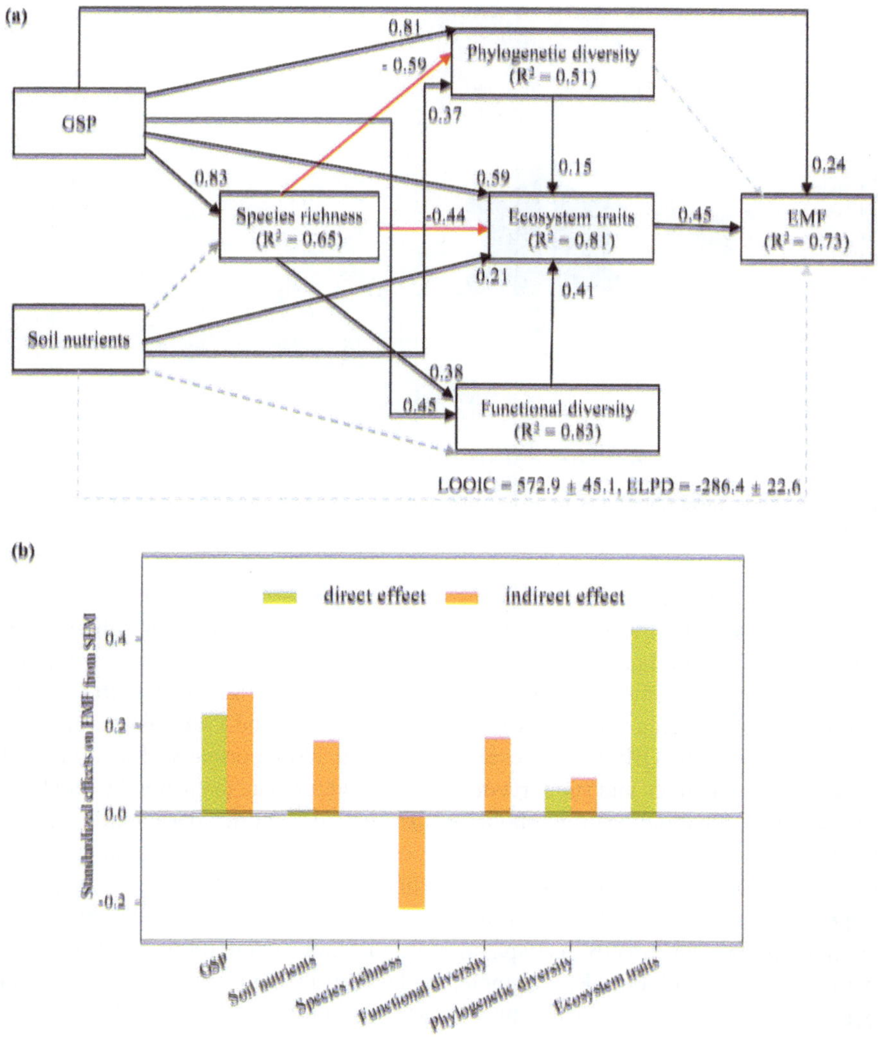

Fig. 7.12 Direct and indirect effects of biodiversity, plant community traits, and environmental factors on ecosystem multifunctionality (EMF), which is associated with the ecosystem water, carbon, and energy cycling. GSP indicates the precipitation of growing seasons. Black arrows signify positive correlations, while red arrows denote negative correlations. Solid gray lines donate significant effects, while dotted gray lines denote nonsignificant effects

traits by accounting for diverse trait values among individuals of a specific species in the calculations. Furthermore, since tissue-level traits may differ depending on age (He et al. 2023), examining the effects of stand age on ecosystem productivity is a promising avenue in future studies.

The TBP framework innovatively divides functional traits at the community level into quantitative and efficiency traits (Fig. 7.13). Quantitative and efficient traits

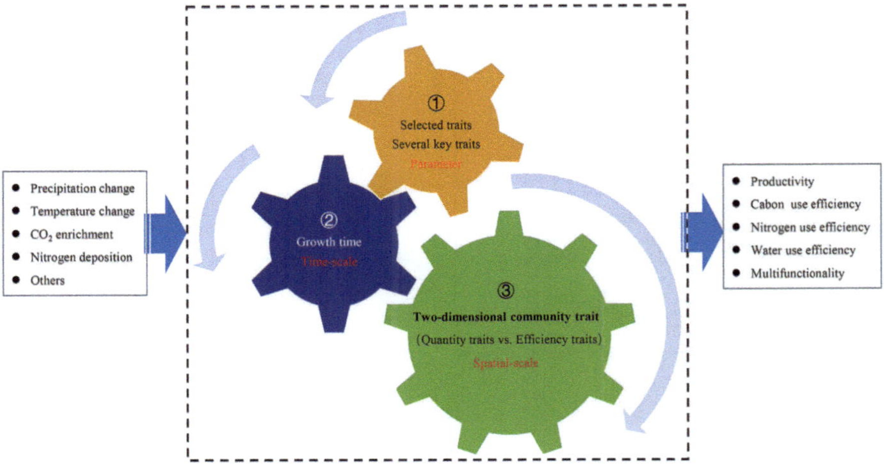

Fig. 7.13 Theoretical framework for using plant functional traits to predict ecosystem functions in future studies

represent the aggregation level of plant community properties, and ecological studies generally ignore the former. Specifically, quantitative traits are the absolute cumulative value per unit land area, and efficiency traits are relative values per unit leaf area or dry matter content. Temporal and spatial variations in primary productivity can be predicted using quantitative and efficiency traits and GSL. To promote the development of the TBP theory, researchers should adopt normative functional trait measurement methods and support community structure data to ensure the reliability of scaling up. Quantitative traits should be obtained in real time, particularly through remote sensing observations, to obtain the corresponding phenological data, that is, GSL (Bruelheide et al. 2018; Moreno-Martínez et al. 2018). Unlike the traditional large-leaf or stratified models, the TBP framework uses physiological parameters for simulation and can directly use spatial scale-matching community functional trait parameters for simulation. For example, the leaf nitrogen content and total nitrogen per unit land area have been substituted into the model (Farquhar et al. 1980; Amthor 1994; Rogers et al. 2017; Yan et al. 2022). Traditional analyses have demonstrated a correlation between productivity and physiological traits at the cellular and leaf levels. However, a previous study highlighted the need to consider the concept of "plant traits per unit land area" when predicting carbon fluxes at the canopy or ecosystem scales (He et al. 2023).

Researchers can obtain increasingly available functional trait data by rapidly assessing the CWM of leaf functional traits (Bruelheide et al. 2018; Moreno-Martínez et al. 2018) and using technologies such as high-resolution remote sensing and light detection and ranging, and research has rapidly expanded to regional and global scales (He et al. 2019b). Therefore, in the current situation of open access to many functional trait datasets and the rapid development of remote sensing technology, the TBP framework that comprehensively focuses on community functional traits has the potential to make full use of various high-tech and multisource data and is expected to become the basis of the formation of new GPP or NPP ecological process models in the future to predict better ecosystem functions and their changes under global climate change and anthropogenic disturbance scenarios (Fig. 7.13).

7.6 Summary

Accurately predicting ecosystem productivity has been a traditional and vital scientific question spanning multiple disciplines, such as ecology and geosciences. After a thorough examination of the strengths and weaknesses of the traditional big-leaf and multilayer models, the researcher referenced the classical engine power output model from physics. This idea was subsequently integrated with the recently developed two-dimensional characteristics of plant community functional traits (efficiency traits and quantitative traits). Further, we developed a TBP framework to predict temporal and spatial changes in ecosystem productivity using plant community traits (He et al. 2023). TBP offers a novel perspective on ecosystem productivity prediction, extending the concept behind the "biomass hypothesis" and applying it to community functional traits. As discussed in this chapter, the TBP framework has an inherent theoretical basis, can be technically extended, and has been verified in many experimental cases. Although the TBP framework is still in its early stages of development, its ability to incorporate many functional traits and remote sensing observation data enables the use of various high-tech and multisource data, further optimizing its application. Therefore, after many subsequent optimizations and improvements, the TBP framework will likely serve as a new core for ecosystem productivity prediction (even ecosystem multifunctional prediction), promoting the development and improvement of a new generation of mechanism process models, thereby bringing innovative impacts and new development opportunities to ecosystem ecology, macroecology, macro-geography, remote sensing science, and other related disciplines.

References

Amthor JS (1994) Scaling CO_2-photosynthesis relationships from the leaf to the canopy. Photosynth Res 39:321–350

Aragão L, Malhi Y, Metcalfe D et al (2009) Above-and below-ground net primary productivity across ten Amazonian forests on contrasting soils. Biogeosciences 6:2759–2778

Bahar NHA, Ishida FY, Weerasinghe LK et al (2017) Leaf-level photosynthetic capacity in lowland Amazonian and high-elevation Andean tropical moist forests of Peru. New Phytol 214:1002–1018

Barry KE, Pinter GA, Strini JW et al (2021) A graphical null model for scaling biodiversity-ecosystem functioning relationships. J Ecol 109:1549–1560

Bréda NJ (2003) Ground-based measurements of leaf area index: a review of methods, instruments and current controversies. J Exp Bot 54:2403–2417

Brovkin V, Bendtsen J, Claussen M et al (2002) Carbon cycle, vegetation, and climate dynamics in the Holocene: experiments with the CLIMBER-2 model. Glob Biogeochem Cycles 16:1139

Bruelheide H, Dengler J, Purschke O et al (2018) Global trait-environment relationships of plant communities. Nat Ecol Evol 2:1906–1917

Clark DA, Brown S, Kicklighter DW et al (2001) Net primary production in tropical forests: an evaluation and synthesis of existing field data. Ecol Appl 11:371–384

Cramer W, Bondeau A, Woodward FI et al (2001) Global response of terrestrial ecosystem structure and function to CO_2 and climate change: results from six dynamic global vegetation models. Glob Chang Biol 7:357–373

Croft H, Chen JM, Luo XZ et al (2017) Leaf chlorophyll content as a proxy for leaf photosynthetic capacity. Glob Chang Biol 23:3513–3524

Dermody O, Long SP, Mcconnaughay K et al (2008) How do elevated CO_2 and O_3 affect the interception and utilization of radiation by a soybean canopy? Glob Chang Biol 14:556–564

Dong SK, Tang L, Zhang SF et al (2017a) Relationship between plant species diversity and functional diversity in alpine grasslands. J Ecol 37:1472–1483

Dong N, Prentice IC, Evans BJ et al (2017b) Leaf nitrogen from first principles: field evidence for adaptive variation with climate. Biogeosciences 14:481–495

Enquist BJ, Kerkhoff AJ, Stark SC et al (2007) A general integrative model for scaling plant growth, carbon flux, and functional trait spectra. Nature 449:218–222

Esser G, Lautenschlager M (1994) Estimating the change of carbon in the terrestrial biosphere from 18 000 BP to present using a carbon cycle model. Environ Pollut 83:45–53

Farquhar GD (1989) Models of integrated photosynthesis of cells and leaves. Philos Trans R Soc Lond B Biol Sci 323:357–367

Farquhar GD, von Caemmerer S, Berry JA (1980) A biochemical model of photosynthetic CO_2 assimilation in leaves of C_3 species. Planta 149:78–90

Feng X, Uriarte M, González G et al (2018) Improving predictions of tropical forest response to climate change through integration of field studies and ecosystem modeling. Glob Chang Biol 24:e213–e232

Field C, Mooney H (1986) Photosynthesis – nitrogen relationship in wild plants. In: On the economy of plant form and function: proceedings of the sixth Maria moors Cabot symposium, evolutionary constraints on primary productivity, adaptive patterns of energy capture in plants, Harvard Forest. Cambridge University Press, Cambridge

Fortunel C, Lask JR, Uriarte M et al (2018) Topography and neighborhood crowding can interact to shape species growth and distribution in a diverse Amazonian forest. Ecology 99:2272–2283

Friend AD (2001) Modelling canopy CO_2 fluxes: are 'big-leaf' simplifications justified? Glob Ecol Biogeogr 10:603–619

Fyllas NM, Bentley LP, Shenkin A et al (2017) Solar radiation and functional traits explain the decline of forest primary productivity along a tropical elevation gradient. Ecol Lett 20:730–740

Garnier E (1991) Resource capture, biomass allocation and growth in herbaceous plants. Trends Ecol Evol 6:126–131

Garnier E, Cortez J, Billes G et al (2004) Plant functional markers capture ecosystem properties during secondary succession. Ecology 85:2630–2637

Garnier E, Navas ML, Grigulis K (2016) Plant functional diversity: organism traits, community structure, and ecosystem properties. Oxford University Press

Ghimire B, Riley WJ, Koven CD et al (2017) A global trait-based approach to estimate leaf nitrogen functional allocation from observations. Ecol Appl 27:1421–1434

Harfoot MB, Newbold T, Tittensor DP et al (2014) Emergent global patterns of ecosystem structure and function from a mechanistic general ecosystem model. PLoS Biol 12:e1001841

He NP, Liu CC, Piao SL et al (2019a) Ecosystem traits linking functional traits to macroecology. Trends Ecol Evol 34:200–210

He L, Chen JM, Liu J et al (2019b) Diverse photosynthetic capacity of global ecosystems mapped by satellite chlorophyll fluorescence measurements. Remote Sens Environ 232:111344

He NP, Li Y, Liu CC et al (2020a) Plant trait networks: improved resolution of the dimensionality of adaptation. Trends Ecol Evol 35:908–918

He NP, Liu CC, Xu L et al (2020b) Ecosystem traits open new insights into macroecology: opportunity and challenge. Acta Ecol Sin 40(8):2507–2522

He P, Gleason SM, Wright IJ et al (2020c) Growing-season temperature and precipitation are independent drivers of global variation in xylem hydraulic conductivity. Glob Chang Biol 26: 1833–1841

He NP, Yan P, Liu CC et al (2023) Predicting ecosystem productivity based on plant community traits. Trends Plant Sci 28:43–53

Heckman RW, Khasanova AR, Johnson NS et al (2020) Plant biomass, not plant economics traits, determines responses of soil CO_2 efflux to precipitation in the C4 grass Panicum virgatum. J Ecol 108:2095–2106

Hilty J, Muller B, Pantin F et al (2021) Plant growth: the what, the how, and the why. New Phytol 232:25–41

Huang M, Piao S, Ciais P et al (2019) Air temperature optima of vegetation productivity across global biomes. Nat Ecol Evol 3:772–779

Jenkins JP, Richardson AD, Braswell BH et al (2007) Refining light-use efficiency calculations for a deciduous forest canopy using simultaneous tower-based carbon flux and radiometric measurements. Agric For Meteorol 143:64–79

Kull O, Jarvis PG (1995) The role of nitrogen in a simple scheme to scale up photosynthesis from leaf to canopy. Plant Cell Environ 18:1174–1182

LeBauer DS, Treseder KK (2008) Nitrogen limitation of net primary productivity in terrestrial ecosystems is globally distributed. Ecology 89:371–379

Levine JM (2016) A trail map for trait-based studies. Nature 529:163–164

Li YQ, Reich PB, Schmid B et al (2020) Leaf size of woody dicots predicts ecosystem primary productivity. Ecol Lett 23:1003–1013

Lieth H, Whittaker RH (1975) Primary productivity of the biosphere. Springer-Verlag, New York

Liu CC, Li Y, Yan P et al (2021a) How to improve the predictions of plant functional traits on ecosystem functioning? Front Plant Sci 12:622260

Liu XR, Chen HX, Sun TY et al (2021b) Variation in woody leaf anatomical traits along the altitudinal gradient in Taibai Mountain, China. Glob Ecol Conserv 26:e01523

Liu C, Li Y, Xu L et al (2021c) Stomatal arrangement pattern: a new direction to explore plant adaptation and evolution. Front Plant Sci 12:655255

Liu R, Yang X, Gao R et al (2021d) Allometry rather than abiotic drivers explains biomass allocation among leaves, stems and roots of Artemisia across a large environmental gradient in China. J Ecol 109:1026–1040

Lohbeck M, Poorter L, Martinez-Ramos M et al (2015) Biomass is the main driver of changes in ecosystem process rates during tropical forest succession. Ecology 96:1242–1252

Luo X, Chen JM, Liu J et al (2018) Comparison of big-leaf, two-big-leaf, and two-leaf upscaling schemes for evapotranspiration estimation using coupled carbon-water modeling. J Geophys Res Biogeo 123:207–225

Luyssaert S, Inglima I, Jung M et al (2007) CO2 balance of boreal, temperate, and tropical forests derived from a global database. Glob Chang Biol 13:2509–2537

Malhi Y, Girardin CAJ, Goldsmith GR et al (2017) The variation of productivity and its allocation along a tropical elevation gradient: a whole carbon budget perspective. New Phytol 214:1019–1032

Maynard DS, Bialic-Murphy L, Zohner CM et al (2022) Global relationships in tree functional traits. Nat Commun 13:3185

McGill BJ (2010) Matters of scale. Science 328:575

McGill BJ (2019) The what, how and why of doing macroecology. Glob Ecol Biogeogr 28:6–17

McGill BJ, Enquist BJ, Weiher E et al (2006) Rebuilding community ecology from functional traits. Trends Ecol Evol 21:178–185
Mittelbach GG, McGill BJ (2019) Community ecology. Oxford University Press
Monteith JL (1977) Climate and the efficiency of crop production in Britain. Philos Trans R Soc Lond B Biol Sci 281:277–294
Monteith JL, Moss CJ, Cooke GW et al (1977) Climate and the efficiency of crop production in Britain. Philos Trans R Soc Lond B Biol Sci 281:277–294
Moreno-Martínez Á, Camps-Valls G, Kattge J et al (2018) A methodology to derive global maps of leaf traits using remote sensing and climate data. Remote Sens Environ 218:69–88
Morgan PB, Bollero GA, Nelson RL et al (2005) Smaller than predicted increase in aboveground net primary production and yield of field-grown soybean under fully open-air [CO_2] elevation. Glob Chang Biol 11:1856–1865
Nobel PS (1999) Physicochemical & environmental plant physiology. Academic Press
Peng Y, Bloomfield KJ, Prentice IC (2020) A theory of plant function helps to explain leaf-trait and productivity responses to elevation. New Phytol 226:1274–1284
Prado-Junior JA, Schiavini I, Vale VS et al (2016) Conservative species drive biomass productivity in tropical dry forests. J Ecol 104:817–827
Quesada M, Sanchez-Azofeifa GA, Alvarez-Añorve M et al (2009) Succession and management of tropical dry forests in the Americas: review and new perspectives. For Ecol Manag 258:1014–1024
Reich PB (2012) Key canopy traits drive forest productivity. Proc R Soc B Biol Sci 279:2128–2134
Reich PB (2014) The world-wide 'fast-slow' plant economics spectrum: a traits manifesto. J Ecol 102:275–301
Reich PB, Walters MB, Ellsworth DS (1997) From tropics to tundra: global convergence in plant functioning. Proc Natl Acad Sci USA 94:13730–13734
Reich PB, Ellsworth DS, Walter MBS et al (1999) Generality of leaf trait relationships: a test across six biomes. Ecology 80:1955–1969
Rogers A, Serbin SP, Ely KS et al (2017) Terrestrial biosphere models underestimate photosynthetic capacity and CO_2 assimilation in the Arctic. New Phytol 216:1090–1103
Rosenzweig ML (1968) Net primary productivity of terrestrial communities: prediction from climatological data. Am Nat 102:67–74
Rubio VE, Zambrano J, Iida Y et al (2021) Improving predictions of tropical tree survival and growth by incorporating measurements of whole leaf allocation. J Ecol 109:1331–1343
Ruimy A, Kergoat L, Bondeau A et al (1999) Comparing global models of terrestrial net primary productivity (NPP): analysis of differences in light absorption and light-use efficiency. Glob Chang Biol 5:56–64
Running SW, Nemani RR, Heinsch FA et al (2004) A continuous satellite-derived measure of global terrestrial primary production. Bioscience 54:547–560
Sack L, Scoffoni C, McKown AD et al (2012) Developmentally based scaling of leaf venation architecture explains global ecological patterns. Nat Commun 3:1–10
Sack L, Scoffoni C, John GP et al (2013) How do leaf veins influence the worldwide leaf economic spectrum? Review and synthesis. Ecology 4:4053–4080
Sánchez-Azofeifa GA, Castro K, Wright SJ et al (2009) Differences in leaf traits, leaf internal structure, and spectral reflectance between two communities of lianas and trees: implications for remote sensing in tropical environments. Remote Sens Environ 113:2076–2088
Shipley B (2010) From plant traits to vegetation structure: chance and selection in the assembly of ecological communities. Cambridge University Press
Šímová I, Storch D (2017) The enigma of terrestrial primary productivity: measurements, models, scales and the diversity-productivity relationship. Ecography 40:239–252
Smith AB, Godsoe W, Rodríguez-Sánchez F et al (2019) Niche estimation above and below the species level. Trends Ecol Evol 34:260–273
Violle C, Navas ML, Vile D et al (2007) Let the concept of trait be functional! Oikos 116:882–892

Wang J, Rich PM, Price KP et al (2005) Relations between NDVI, grassland production, and crop yield in the central great plains. Geocarto Int 20:5–11

Wang RL, Wang QF, Zhao N et al (2017a) Complex trait relationships between leaves and absorptive roots: coordination in tissue N concentration but divergence in morphology. Ecol Evol 7:2697–2705

Wang R, Gibson CD, Berry TD et al (2017b) Photooxidation of pyrogenic organic matter reduces its reactive, labile C pool and the apparent soil oxidative microbial enzyme response. Geoderma 293:10–18

Wang H, Prentice IC, Davis TW et al (2017c) Photosynthetic responses to altitude: an explanation based on optimality principles. New Phytol 213:976–982

Wong S, Cowan I, Farquhar G (1979) Stomatal conductance correlates with photosynthetic capacity. Nature 282:424–426

Wright SJ, Kitajima K, Kraft NJ et al (2010) Functional traits and the growth-mortality trade-off in tropical trees. Ecology 91:3664–3674

Yan P, Li MX, Yu GR et al (2022) Plant community traits associated with nitrogen can predict spatial variability in productivity. Ecol Indic 140:109001

Yan P, Zhang JH, He NP et al (2023) Functional diversity and soil nutrients regulate the interannual variability in gross primary productivity. J Ecol 111:1094–1106

Yang J, Cao M, Swenson NG (2018) Why functional traits do not predict tree demographic rates. Trends Ecol Evol 33:326–336

Yu GR, Zhu XJ, Fu YL et al (2013) Spatial patterns and climate drivers of carbon fluxes in terrestrial ecosystems of China. Glob Chang Biol 19:798–810

Zhang JH, Li MX, Xu L et al (2021a) C:N:P stoichiometry in terrestrial ecosystems in China. Sci Total Environ 795:148849

Zhang JH, Ren TT, Yang JJ et al (2021b) Leaf multi-element network reveals the change of species dominance under nitrogen deposition. Front Plant Sci 12:580340

Zhang JH, Hedin LO, Li MX et al (2022a) Leaf N:P ratio does not predict productivity trends across natural terrestrial ecosystems. Ecology 103:e3789

Zhang M, Gao HR, Mo WY et al (2022b) Variation in leaf vein traits along environmental gradient in loess plateau grasslands. Acta Ecol Sin 42(19):8082–8093

Zhao M, Running SW (2010) Drought-induced reduction in global terrestrial net primary production from 2000 through 2009. Science 329:940–943

Zhao M, Heinsch FA, Nemani RR et al (2005) Improvements of the MODIS terrestrial gross and net primary production global data set. Remote Sens Environ 95:164–176

Chapter 8
Root Community Traits and Their Correlations with Ecosystem Productivity

Roots are crucial plant organs providing physical support and facilitating the uptake of water and essential mineral nutrients, such as nitrogen and phosphorus, thereby supporting plant growth. Previous research has extensively investigated variations in root functional traits and their contributions to individual plant growth, community assembly, and ecosystem functioning. However, conventional conceptual frameworks and measurement techniques for root traits have generally constrained research on the organ, individual, and population levels. Consequently, fundamental ecological questions remain unresolved, including the quantification of the abundance of fine roots in plant communities, the relationship between fine root traits and ecosystem functions, and the feasibility of scaling individual fine root traits to the community level. Although previous studies have attempted to scale root traits from the organ to the ecosystem level, existing methods have scale mismatches between traditional functional traits measured at the organ level and ecosystem-level processes. This methodological limitation has constrained our understanding of physiological mechanisms and ecological processes governing trait–ecosystem function relationships. Therefore, developing more effective methodologies to scale root functional traits from the species to the community levels and clearly establish their relationships with ecosystem functions is critically important. Recently, innovative conceptual frameworks and measurement approaches have been introduced for assessing root traits at the community level, offering promising avenues for establishing linkages between root functional traits and ecosystem functions. In this chapter, we explore the historical context, theoretical foundation, and scientific significance of community-level root functional traits. We comprehensively define the concept of root community traits (RCTs), discuss their implications, explain methods of derivation, and identify key parameters. Furthermore, to empirically validate the utility of RCTs, we conducted field surveys along three contrasting grassland transects across the Inner Mongolia, Loess, and Qinghai–Tibet Plateau.

N. He et al., *Plant Functional Ecology*,
https://doi.org/10.1007/978-981-96-8264-5_8

Through these surveys, we investigated spatial variations in root functional traits and assessed their capacity to predict grassland productivity patterns across large scales. The novel concept of RCTs addresses previous methodological shortcomings by aligning trait measurements with appropriate ecological scales, thereby enhancing the integration of root traits into ecosystem function studies. However, significant challenges remain regarding methodology and data acquisition for field-based investigations of root community traits. Advanced, nondestructive root-sampling technologies are urgently required to expand the datasets on root community traits and to integrate their multidimensional characteristics, thereby improving the prediction accuracy of ecosystem functions.

Understanding and predicting ecological processes and functioning based on plant functional traits are considered as "holy grail" in trait-based ecology (Lavorel and Garnier 2002; He et al. 2019a, b). Numerous studies have incorporated plant functional traits into mechanistic models to predict ecosystem responses to environmental changes (Garnier et al. 2004; Diaz et al. 2007; Suding et al. 2008; van der Plas et al. 2020). However, previous studies indicated that functional traits explain only a limited portion of spatiotemporal variability in ecosystem functioning, possibly due to the neglect of belowground traits in most studies and their indirect relationship with targeted ecosystem processes (Bardgett et al. 2014; van der Plas et al. 2020; Freschet et al. 2021b). From a physiological perspective, plant growth and biomass accumulation primarily depend on the integrated use of aboveground (leaf) and belowground (root) resources (Poorter et al. 2012; Freschet et al. 2015; Weemstra et al. 2021). The roots are the most important plant organs responsible for physical support and nutrient and water uptake, which are essential for plant growth. Consequently, root morphological, structural, physiological, and biochemical traits can significantly influence physiological and ecological processes across various organizational scales, from individual plants to entire ecosystems (Bardgett et al. 2014; Laliberté 2017; McCormack et al. 2017).

Despite increasing recognition of the importance of root traits in ecosystem functioning, uncertainty remains regarding their predictive value, especially at larger ecological scales (McCormack et al. 2017; Freschet et al. 2021b). This uncertainty is largely due to challenges in systematically obtaining comprehensive data on root traits. Fine roots play a crucial role in plant resource uptake; however, sampling these absorptive roots is challenging and laborious (Freschet et al. 2021a). Additionally, methodological complexities complicate the establishment of clear relationships between root traits and ecosystem functions at individual, species, and community levels. Ecosystem function manifests at the community scale through complex interactions of abiotic and biotic factors (Lavorel and Garnier 2002), rather than simply as aggregated outcomes of individual species traits (Reichstein et al. 2014). Therefore, accurately scaling traits from individual or species levels to the community level is essential for elucidating trait-ecosystem function relationships (He et al. 2019a, b; Freschet et al. 2021b). However, using inappropriate methods to scale up plant traits collected at the individual and community levels may lead to

Table 8.1 Brief description and comparison of the four approaches for linking root traits with ecosystem functions

Approach	Sampling and scaling-up methods	Advantages	Disadvantages
Approach I: Root trait of dominant species or community-weighted means	Separately sample and measure root traits of specific species in natural communities Scale up using simple means	Simple and operable for each species Trait data are easy to collect worldwide	Ignores the complexity of natural communities Assumes similar contributions of each species Mismatched units or dimensions with ecosystem functions
Approach II: Pooled-species trait	Pools of plants are sampled over a given soil surface area or soil volume Community-level functional parameters are directly measured	Does not require estimation of root abundance of each species	Ignores the complexity of natural communities Mismatched units or dimensions with ecosystem functions
Approach III: community traits proposed by He et al. (2019a, b)	Separately sample and measure root traits of all species in a natural community Scale up using the data of community structure	Data for root traits being similar to those for natural communities Matches units between root traits and ecosystem functions Integrates data from field traits using different approaches and technologies	Requires systematic measurements of all species in a community Difficult to obtain some basic parameters, such as root biomass and area per unit land area Time-consuming and laborious
Approach IV: Root community traits (per unit land area)	Sampling using large cores or soil blocks in a community Normalize per unit land area	Directly obtains community-level root traits Matches units with ecosystem functions Integrates data from field trait Surveys with novel technologies in macroecology	Destructive sampling Time-consuming and laborious

uncertainties (Violle et al. 2014; He et al. 2019a, b; Freschet et al. 2021b) (Table 8.1). For example, the community-weighted mean (CWM) method has been widely applied to scale functional traits and explore their ecological significance across ecosystems (Garnier et al. 2004; Migliavacca et al. 2021; Yan et al. 2022). While easy to calculate, CWMs are often challenging to interpret (Muscarella and Uriarte 2016; Laughlin et al. 2018), which could be due to mismatched spatial and temporal scales (He et al. 2019a, b; Wang et al. 2021). Recent studies introduced the concept of two-dimensional community traits, distinguishing between "trait efficiency" (CWM) and "trait quantity" (total trait abundance per unit area) (Zhang et al. 2021a, b, c, d, e; Yan et al. 2022). However, quantifying root traits at the community level and effectively linking them to ecosystem functions is challenging.

We propose a novel framework for evaluating root community traits (RCTs) based on per unit land area, which incorporates accurate parameters and practical field methodologies to investigate the multiple roles of roots in community assembly and ecosystem processes, allowing the integration of root traits into studies of ecosystem functioning across multiple scales. In this study, we emphasized the theoretical foundations, clear definitions, and parameter selection of RCTs. Furthermore, through field surveys conducted across grassland transects in the Tibetan (TP), Mongolian (MP), and Loess (LP) plateaus in northern China, we explored spatial variations in RCTs and their correlations with grassland productivity at a large scale.

8.1 Major Progress in Root Trait Variation

Root traits have long been utilized in agricultural systems to develop varieties of crops with superior growth and productivity. More recently, research on root traits in natural communities has increased, particularly regarding nutrient uptake strategies, mechanisms of species coexistence, and the spatial and temporal variations of ecosystem productivity (Eissenstat 1992; Valverde-Barrantes et al. 2017; Ma et al. 2018; Kong et al. 2019; Fort and Freschet 2020). Previous studies have gradually recognized that fine roots (diameter < 2 mm) or absorptive roots (first- to third-order fine roots) exhibit greater resource acquisition at the species level compared to coarse roots (Pregitzer et al. 2002; McCormack et al. 2015). Consequently, classifying roots into distinct functional modules has significantly advanced our understanding of root traits and facilitated the identification of general principles governing root trait variation among plant species, environments, and root orders (McCormack et al. 2015).

Root ecology research has revealed that plant species can regulate the plasticity of their root traits to adapt to environmental conditions by modulating the morphology, architecture, and microbial associations of fine roots through a trade-off between nutrient uptake efficiency and investment costs (Kramer-Walter et al. 2016; Ma et al. 2018). For instance, woody plants exhibit a gradient in the diameter of first-order roots, decreasing from tropical environments (characterized by stable conditions) to desert biomes (characterized by infertile and highly seasonal conditions). This reduction in root diameter is associated with reduced dependence on mycorrhizal associations, thereby increasing carbon-use efficiency and enhancing colonization capabilities in harsh habitats (Ma et al. 2018). Furthermore, the development of comprehensive global databases of root traits, standardized across diverse species, biomes, and environmental gradients, has considerably improved predictive accuracy at multiple ecological scales, contributing significantly to our understanding of belowground root ecology.

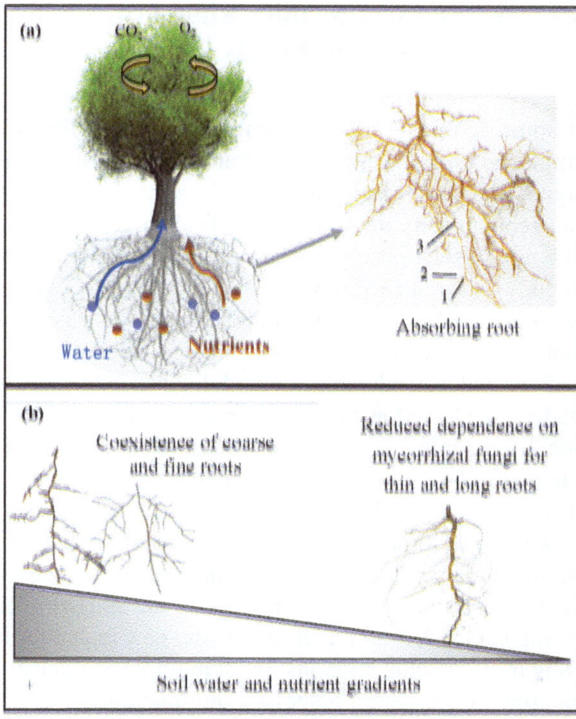

Fig. 8.1 Effect of absorptive roots on plant growth (**a**) and their variation along environmental gradients (**b**)

Roots play essential roles in anchoring plants and facilitating the uptake of water and nutrients (Freschet et al. 2021b), which greatly depend on variations in root traits and their interactions (Fig. 8.1a). Despite their critical ecological functions, most studies on trait–function relationships have focused on aboveground organs, particularly leaves (such as the correlation between leaf nitrogen and chlorophyll contents and photosynthetic capacity), because of the practical difficulties in sampling and measuring root systems and traits, respectively (Reich et al. 1999; Wright et al. 2004a, b; Croft et al. 2017; He et al. 2021). Furthermore, there is ongoing debate and uncertainty regarding root trait–functioning correlations. For example, previous studies have demonstrated that absorptive roots nitrogen content is positively correlated with plant nutrient uptake and respiratory metabolism, which can negatively affect root longevity (Reich et al. 2008; Roumet et al. 2016). However, these relationships were not as strong as those in the leaves and were decoupled in some studies. For instance, by analyzing leaf and root trait data from 10 common species and approximately 13,000 trees in a temperate forest in Michigan, Weemstra et al. (2021) identified two best models that could explain tree growth, including specific leaf area (SLA), root diameter, and soil carbon or nitrogen concentrations. However, the observed relationships and effects on growth were inconsistent with the predictions of the plant resource economics theory. Overall, root trait–function relationships exhibit greater complexity compared to leaf traits, posing significant challenges for incorporating root functional traits into studies of ecosystem processes and functioning.

8.2 Necessity of Scaling Root Traits from Organ to Community Levels

Despite significant advancements in understanding variations in root traits at the intra- and interspecies levels, trait-based research has been primarily limited to organ, individual, and species levels, particularly select model or dominant species and economically important crops (Lynch 2007; Ma et al. 2018; Kong et al. 2019). The rapid development in macroecology, involving technologies and methods such as eddy covariance flux measurements, remote sensing, and ecological modeling, has primarily been oriented toward ecosystem-scale analyses. Effectively integrating root trait research into these macroecological frameworks presents substantial opportunities and considerable methodological challenges (He et al. 2019a, b). Most studies conducted at the organ or species levels limit the application of root functional traits in the parameterization of ecosystem models, resulting in uncertainties when simulating vegetation dynamics and nutrient cycles under global climate change scenarios (Wang et al. 2021). Establishing clear physiological relationships between root traits at individual levels and broader ecosystem functions and integrating these root traits into community-level ecological models is therefore essential (McCormack et al. 2017; Freschet et al. 2021b). Theoretically, we can develop ecological models and interfaces using new techniques in macroecology by quantifying trait–function relationships at multiple ecological scales, enhancing our understanding of ecosystem processes and functions (Fig. 8.2).

Community-weighted mean (CWM) approach is a method widely used in previous studies, where trait values of dominant species are averaged or weighted by species abundances to represent the entire plant community [Approach I (Fig. 8.3), Table 8.1]. Using this method, researchers can identify variations in specific root

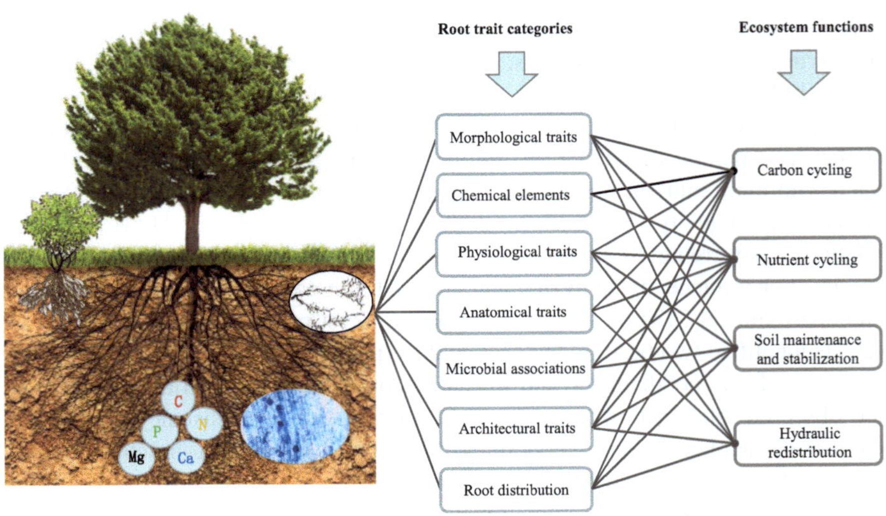

Fig. 8.2 Linkages between root functional traits and ecosystem functions at different scales

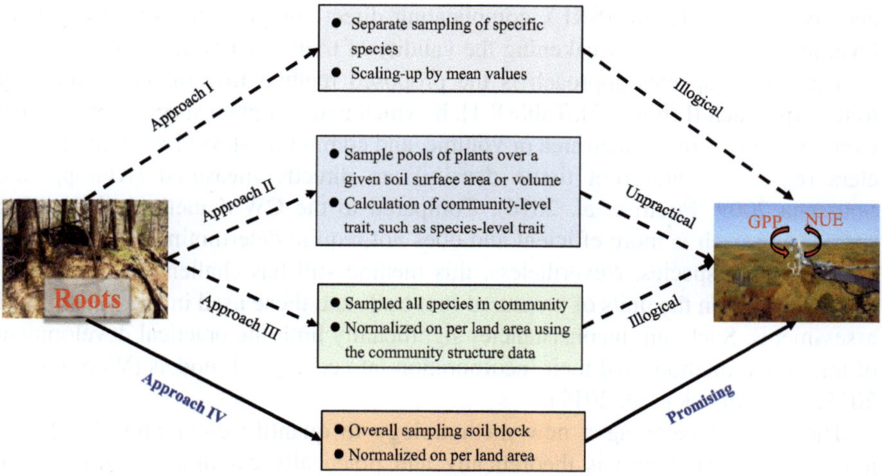

Fig. 8.3 Illustration of the four approaches for linking root traits and forest ecosystem functioning. Approach I: root trait of dominant species or community-weighted means; Approach II: pooled-species trait; Approach III: community traits proposed by He et al. (2019a, b); Approach IV: root community traits (per unit land area). *GPP* gross primary productivity, *NUE* nutrient-use efficiency

traits and their effects on high-level processes and functions to some extent (Lavorel and Garnier 2002; Violle et al. 2007). Typically, this method is used to assess community-level root traits using databases of root traits and community structures for each species. However, three significant practical challenges must be addressed to effectively link community-level root traits with ecosystem functioning at large scales. Firstly, acquiring accurate data on root abundance or biomass for each species in natural communities—a fundamental step in traditional CWM calculations—is complex and labor-intensive. Although aboveground characteristics typically determine plant dominance (Mokany et al. 2008; De Long et al. 2019), the belowground biomass of individual species may not scale proportionately with aboveground features, particularly in ecosystems such as grasslands and shrublands, where the majority of biomass is belowground (Ottaviani et al. 2020a, b). Secondly, these traditional scaling methods often presume linear or approximately equal species contributions to ecosystem functioning (Reichstein et al. 2014), and few previous studies have directly tested and quantified relationships between organ-level root traits and ecosystem functioning, particularly within complex natural ecosystems (van der Plas et al. 2020; Freschet et al. 2021b). Consequently, ecologists continue to face significant methodological challenges studying the integration between root traits and ecosystem functioning analyses (Freschet et al. 2021b). Most importantly, the mismatch between the units of ecosystem functioning and CWM root trait values limits the correlation between them. Functions at the ecosystem level are generally quantified per land area unit using techniques like eddy-flux observations, remote sensing, or ecological modeling. Contrastingly, the units of root traits in the CWMs are consistent with those measured at the organ level, such as root element content and specific root length (SRL), complicating direct comparisons with ecosystem-level measurements and weakening the validity of trait–function relationships.

The pooled-species approach is the proposed method to estimate community traits [Approach II (Fig. 8.3), Table 8.1], in which pools of plant species are sampled over a specified soil surface area or volume, and community-level functional parameters (e.g., SRL and root tissue density) are directly measured (Klumpp and Soussana 2009; Prieto et al. 2016). Compared to the CWM method, the pooled species approach is more efficient and does not require determining the root abundance of each species. Nevertheless, this method still has challenges due to mismatches between the units of measured root traits and those used in ecosystem-level assessments. Such unit inconsistencies significantly limit the practical development of traditional methods and their incorporation into ecological models (Warren et al. 2015; McCormack et al. 2017).

Therefore, developing a new methodology to quantify community-level root traits per unit land area is theoretically and practically essential to improve the application of root traits in ecological models. The novel concept of community traits introduced by He et al. (2019a, b) presents a promising alternative, because it defines plant community traits as the quantitative characteristics of organisms at the community level and is expressed as the intensity or density normalized per unit land area [Approach III (Fig. 8.3), Table 8.1]. For example, previous studies have successfully applied this approach using leaf area or biomass per unit land area to derive leaf physiological parameters of community traits, such as total stomatal number per unit ground surface area, demonstrating strong correlations with ecosystem productivity and water-use efficiency (Liu et al. 2018a; He et al. 2019a, b). Inspired by this conceptual framework, researchers are increasingly re-evaluating sampling strategies in natural ecosystems. Moreover, previous studies have developed the concept of two-dimensional community traits, which regard CWM traits as the efficiency of functional substances within a community (i.e., trait_efficiency). Accordingly, "community traits" emphasize "total amount" per land area and thus can be defined as "trait_quantity" (Zhang et al. 2021a, b, c, d, e; Yan et al. 2022). Leaf trait_quantity measures, such as leaf nitrogen content, chlorophyll, and stomatal density, are often more closely associated with environmental gradients and ecosystem productivity than trait_ efficiency values (Wang et al. 2015; Zhang et al. 2021a, b, c, d, e; Yan et al. 2022). However, owing to methodological difficulties associated with sampling and scaling root traits, the variation in root trait_quantity along environmental gradients and its relation to ecosystem functioning remain poorly understood.

8.3 Definition of Root Community Traits

We previously proposed the use of holistic sampling methods to address methodological limitations inherent in traditional approaches (Wang et al. 2021). Specifically, methods such as root cores or soil block sampling have been employed to directly measure the root morphological, chemical, and physiological traits within a community, with trait values normalized per unit land area [Approach IV (Fig. 8.3)]. The holistic sampling strategy differs significantly from frameworks quantifying "ecosystem traits" or "community traits" as described by He et al. (2019a, b). Typically, deriving functional traits from the organ level to the community level necessitates detailed data on organ mass or area alongside systematic trait measurements and community structure information. Taking leaf functional traits as an example, we require detailed data on the community structure, SLA, leaf chlorophyll, nitrogen, and phosphorus contents of each species. Contrastingly, the RCTs concept proposed by Wang et al. (2021) utilizes holistic sampling methods to directly assess entire root communities without needing intermediate measures of organ mass or area. This direct measurement approach provides trait indicators that are more representative of complex natural ecosystems. Although many previous studies have used root cores to assess root traits, most have only investigated the root biomass per unit land area (Yang et al. 2010; Ma et al. 2021). The RCTs concept encourages researchers to comprehensively assess the spatial and temporal variations in root traits, thereby significantly advancing our understanding of the relationships between root traits and ecosystem functions at large ecological scales.

8.4 Parameter System of Root Community Traits

As previously discussed, RCTs are the quantity or density of root traits normalized per unit land area, representing the absolute accumulation of root trait reserves within a community. Given that ecosystem productivity is typically quantified on a per-unit land area basis, RCTs hold significant potential for improving predictions of ecosystem productivity. Using this framework, we calculated community-level parameters such as root mass density (RMA), root length density (RLA), and root area density (RAA) per unit land area (Table 8.2). Like species-level metrics such as SRL and root area, which play important roles in plant nutrient acquisition, these RCTs can reflect the capacity of root communities to uptake water and nutrients. Therefore, they can better capture variations in ecosystem processes, particularly the belowground cycling of carbon and nutrients (Wang et al. 2021).

In this study, we primarily integrated the quantity of belowground root traits within communities. However, additional root traits—including anatomical and architectural traits and mycorrhizal colonization—also critically influence plant resource uptake strategies and ecosystem functioning. While holistic sampling methods effectively estimate total root biomass, area, and chemical content per unit land area; however, quantifying more intricate traits, such as root branching ratios and mycorrhizal colonization rates per unit land area, remains challenging. Therefore, obtaining accurate community-level data on these complex root architectures and mycorrhizal traits poses a significant methodological challenge. Advancing new technologies to comprehensively quantify diverse community-level root traits, such as chemical, morphological, and anatomical characteristics, is essential. Furthermore, integrating two-dimensional community traits—trait efficiency and trait quantity—is necessary to further enhance predictions of ecosystem function.

Table 8.2 Community-level root traits obtained from different scaling-up approaches

Type	Trait	Definition	Unit	Functional importance	Scaling-up approach
Trait_quantity	RMA	Total root dry mass per unit land area	$g\ m^{-2}$	Nutrient acquisition, carbon allocation to belowground community	Approach III or IV[a]
	RLA	Total root length per unit land area	$m\ m^{-2}$	Nutrient acquisition by community-level roots	Approach III or IV
	RAA	Total root surface area per unit land area	$m^2\ m^{-2}$	Nutrient acquisition by community-level roots	Approach III or IV
	RNA	Root nitrogen content per unit land area	$g\ m^{-2}$	Nutrient acquisition and nitrogen storage by community-level roots	Approach III or IV
Trait_efficiency	RD	Mean root diameter of certain soil depth			Approach I or II[a]
	SRL	Mean root length per unit dry mass of certain soil depth	$m\ g^{-1}$	Root nutrient-uptake efficiency	Approach I or II
	SRA	Mean root surface area per unit dry mass of certain soil depth	$m\ g^{-1}$	Root nutrient-uptake efficiency	Approach I or II
	RTD	Mean root dry mass per unit volume of certain soil depth	$g\ cm^{-3}$	Nutrient acquisition-conservation trade-off, stress tolerance	Approach I or II
	RN	Mean root nitrogen content per unit dry mass of certain soil depth	$g\ kg^{-1}$	Root nutrient-uptake efficiency	Approach I or II

[a]Approach I and II failed to estimate the root traits per unit per unit land area owing to no scale conversion. Approach III could realize scale conversion in theory; however, it could not be operationalized because it was difficult to obtain some basic parameters, such as root biomass and area per unit land area

8.5 Spatial Variations in Root Community Traits and Their Relationships with Grassland Productivity

Roots are important plant organs that provide structural support and facilitate the uptake of water and essential nutrients from the soil, thereby underpinning plant growth and reproduction. Root traits play fundamental roles in maintaining diverse ecosystem processes and functioning, particularly the cycling and transformation of elemental nutrients and mineral/organic compounds across different environmental interfaces. Consequently, exploring the relationships between root traits and different aspects of ecosystem functioning, such as net primary productivity (NPP) and

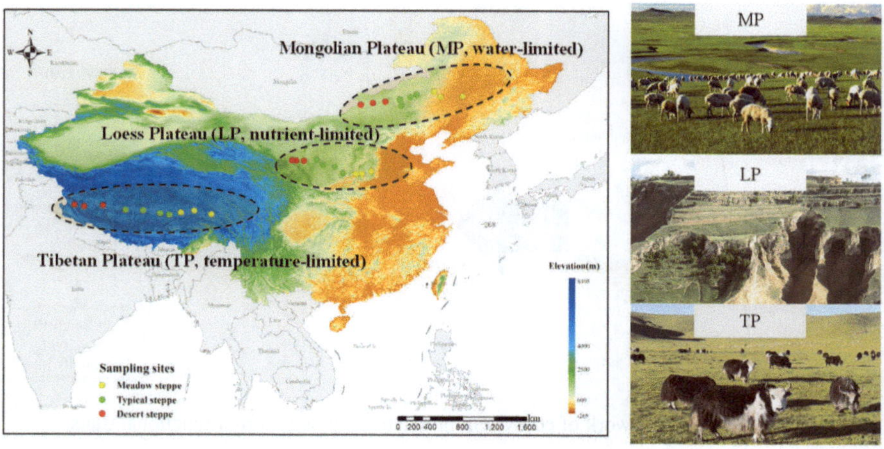

Fig. 8.4 Sampling sites in the three grassland transects along Tibetan (TP), Mongolian (MP), and Loess (LP) plateaus

nutrient-use efficiency (NUE), provides valuable theoretical insights. The concept of two-dimensional community traits has been recently proposed, highlighting that leaf traits such as nitrogen content, chlorophyll, and stomatal density are often more closely correlated with environmental factors and ecosystem productivity compared to traditional CWM trait values (Wang et al. 2015; Zhang et al. 2021a, b, c, d, e; Yan et al. 2022). However, the patterns of root trait quantity variation within plant communities across environmental gradients and their subsequent effects on ecosystem functioning are unclear.

To address these knowledge gaps, we conducted field investigations across 30 sites along three east–west grassland transects in TP (alpine grassland), MP (temperate grassland), and LP (warm-temperate grassland), as illustrated in Fig. 8.4. Different plant growth limiting factors were observed at sampling sites in the three plateau regions (Fig. 8.4), consistent with the findings of previous studies (Bai et al. 2008; Qiu 2008; Liu et al. 2013). Specifically, the TP region, often termed the "Third Pole," represents the highest and most extensive highland region on Earth, characterized by extremely low temperatures (Qiu 2008). Contrastingly, moisture is the most limiting resource for ecosystem productivity in the LP region (Bai et al. 2008). The LP region experiences severe soil erosion, resulting in nutrient-poor soils relative to that of other regions in China (Liu et al. 2013). Thus, these three transects provided optimal settings for examining the ecological adaptations of plant communities to distinct environments. In this study, we investigated spatial variations in root community trait quantities—specifically root mass, length, and area densities—along environmental gradients across the three plateau regions. We further analyzed the relationships between RCTs and ecosystem productivity while investigating how environmental gradients influence these relationships. We hypothesized that root community traits (mass, length, and area) are competitively

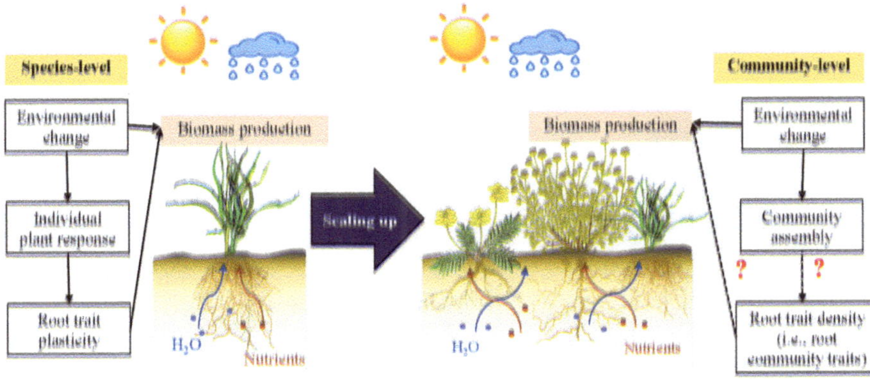

Fig. 8.5 Conceptual framework of ecosystem productivity based on root functional traits

beneficial, predicting that plant communities exhibiting high RCT values would typically inhabit favorable environments and consequently sustain higher biomass productivity (Fig. 8.5).

During the peak growing season (July–September) in 2018, 10 sampling sites were selected on an east–west transect, considering the rainfall gradient on each plateau. These 30 sampling sites span a broad geographical area, representing distinct climates, soil conditions, and vegetation types. Collectively, these transects represented natural zonal grassland vegetation, encompassing three primary vegetation types: meadow, typical, and desert steppes (Fig. 8.4). At each site, we randomly established four plots (1×1 m^2), ensuring they were $\geq$ 10 m apart, resulting in 120 plots. Root samples were collected from each soil core (0–30 cm depth) at every plot and scanned individually at 400 dpi using a scanner (Epson Scanner Perfection V800 Photo, Tokyo, Japan). Subsequently, RMA, RLA, and RAA were quantified for each soil layer.

Across all sampling sites, the mean ± standard error values for RMA, RLA, and RAA were 622.30 ± 57.20 g m^{-2}, 23.45 ± 2.11 km m^{-2}, and 37.11 ± 3.33 m^2 m^{-2}, respectively. Generally, the three RCTs increased from the LP to the MP and TP; however, the increases were not statistically significant [$P > 0.05$ (Fig. 8.6a–c)]. Contrastingly, the three RCTs demonstrated significant variation across different grassland types [all $P < 0.05$ (Fig. 8.6d–f)], with the highest values observed in meadow steppe and the lowest in desert steppes. The RCTs exhibited significant differences between the grassland types within each region. The meadow steppe in the MP region exhibited higher values for RMA, RLA, and RAA compared with those of the corresponding vegetation types in the other regions (Fig. 8.6).

Generally, when data from all sites were combined, RMA increased with mean annual precipitation (MAP), whereas RLA and RAA increased with soil nitrogen content [SN (Fig. 8.7)]. However, the factors influencing root traits differed among regions (Fig. 8.7). Specifically, SN influenced all three RCTs in the TP region; however, in the LP, mean annual temperature (MAT) primarily influenced these traits. Contrastingly, in the MP region, MAP strongly influenced RMA, while SN significantly affected RLA and RAA.

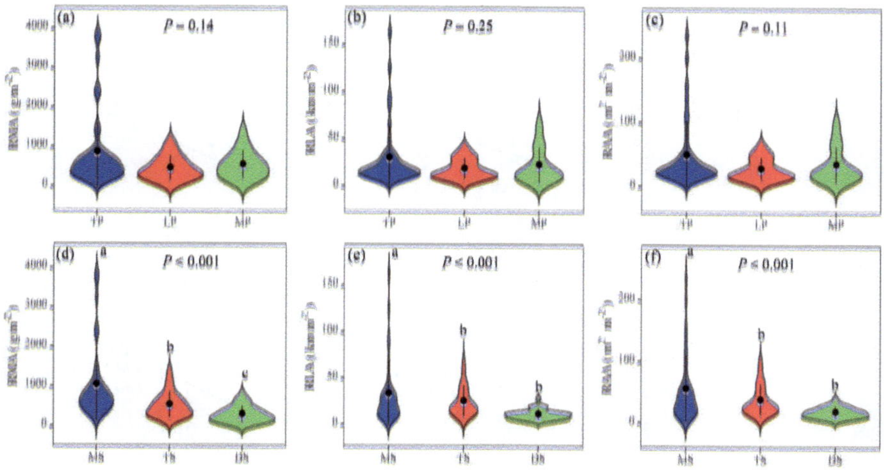

Fig. 8.6 Differences in root community traits (RCTs) among different regions (**a–c**) and grassland types (**d–f**). Different letters denote statistical differences between different grassland types at $P < 0.05$. The RCT parameters include total root dry mass per unit land area (RMA), total root length per unit land area (RLA), and total root area per unit land area (RAA). *TP* Tibetan Plateau, *LP* Loess Plateau, *MP* Mongolian Plateau, *MS* meadow steppe, *TS* typical steppe, *DS* desert steppe

The RCTs exhibited significant correlations with ecosystem productivity (Figs. 8.8 and 8.9). Generally, aboveground NPP (ANPP) increased linearly with increases in RMA, RLA, and RAA [$R^2 = 0.09–0.37$, $P < 0.05$ (Fig. 8.8)] across all sampling sites in the TP and LP regions. However, only the ANPP–RMA correlation was statistically significant in the MP region. Standardized major axis (SMA) tests revealed no significant differences in slope among the different regions [$P > 0.05$ (Fig. 8.8)]. Similarly, SMA tests for common slopes revealed no significant correlations between NPP–RMA and NPP–RAA among the different regions [$P > 0.05$ (Fig. 8.9)]. Contrastingly, considering the NPP–RLA correlation, the slope on the LP was steeper than that on the MP [$P < 0.05$ (Fig. 8.9)].

Piecewise structural equation model (SEM) analyses revealed that when considering combined random effects of "site nested within region" or "site," environmental factors and RCTs explained 79–91% of the total variance in ANPP [i.e., R_c^2 (Fig. 8.10)]. However, the dominant factors controlling regional ANPP variation differed among different datasets. In all sampling site datasets, ANPP variation was primarily driven by environmental factors (particularly MAT and MAP) and root traits. Specifically, MAT ($\beta = 0.53$) and MAP ($\beta = 0.43$) exhibited stronger direct influences on ANPP than other environmental variables. Moreover, MAP indirectly regulated ANPP by influencing RMA ($\beta = 0.06$). SN content mediated RMA and RAA, exerting direct ($\beta = 0.22$) and indirect effects ($\beta = -0.03$) on ANPP. Contrastingly, RMA and RLA had relatively weak effects on the variations in ANPP, with β of 0.21 and −0.18, respectively (Fig. 8.10a).

When analyzed separately by region, environmental influences on ANPP were dominant; however, the direct effects of RCTs on ANPP were not observed in the TP and LP regions (Fig. 8.10b, c). Specifically, in the TP region, SN content was the

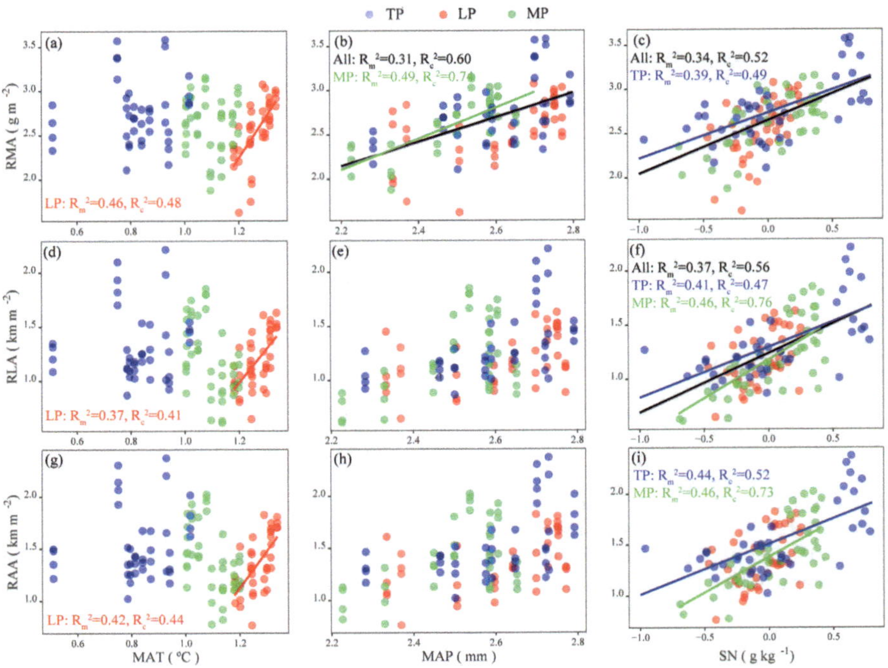

Fig. 8.7 Changes in root community traits along the environmental gradient. *MAT* mean annual temperature, *MAP* mean annual precipitation, *SN* soil total nitrogen concentration, *RMA* total root dry mass per unit land area, *RLA* total root length per unit land area, *RAA* total root area per unit land area, *TP* Tibetan Plateau, *LP* Loess Plateau, *MP* Mongolian Plateau, *MS* meadow steppe, *TS* typical steppe, *DS* desert steppe. Only significant regression lines are presented ($P < 0.05$). R_m^2 represents marginal R^2, and R_c^2 represents conditional R^2. Root traits, MAP, SN, and MAT+10 data were log-transformed

most critical factor influencing ANPP ($\beta = 0.67$), followed by MAT ($\beta = 0.34$); however, MAT for the LP region was $\beta = 0.83$. Contrastingly, in the MP region, ANPP was regulated jointly by MAP, SN, and RCT (indirect effect: $\beta = -0.50$ to 0.88) (Fig. 8.10d). MAP exerted the largest influence on ANPP through direct ($\beta = 0.88$) and indirect effects through RMA ($\beta = 0.25$), whereas SN content influenced ANPP indirectly via RLA ($\beta = -0.33$). Additionally, RMA and RLA had direct positive and negative influences on ANPP ($\beta = 0.35$ and -0.50), respectively.

Plants adapt to changing environments by adjusting fine root morphology, architecture, and microbial associations to optimize resource exploration efficiency per unit of carbon invested (Bardgett et al. 2014; Freschet et al. 2021b). However, the responses of individual plant roots to environmental changes may not always be equal to those at the community level, as community-level root traits—including variations in trait distribution and diversity—also shift due to global change-driven vegetation changes (Bardgett et al. 2014). This study quantitatively assessed community-level root density across three comparable grassland transects, revealing water and soil nitrogen availability as critical factors influencing RCTs across all

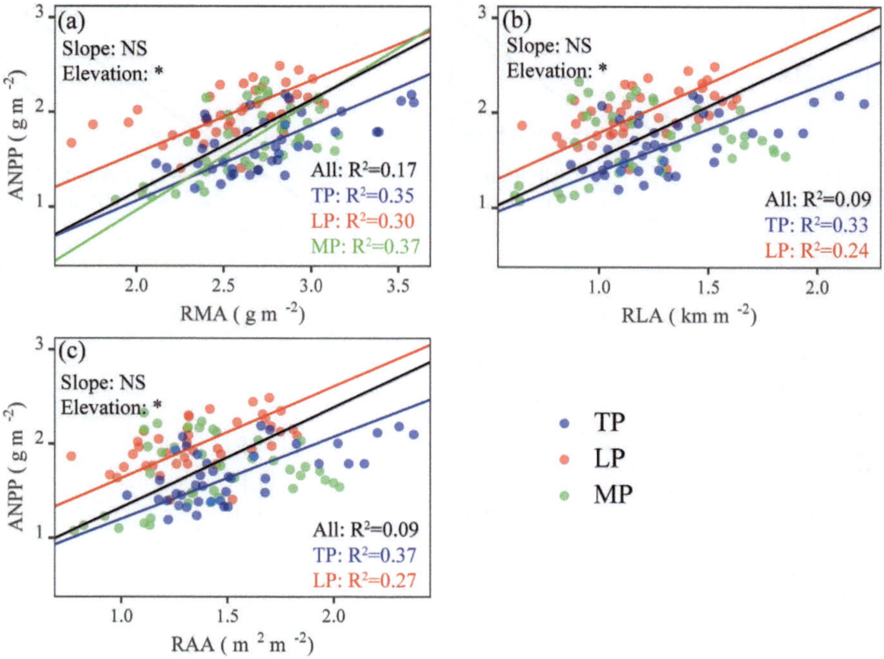

Fig. 8.8 Standardized major axis (SMA) regressions between aboveground net primary productivity (ANPP) and root community traits across different regions. "Slope," difference in SMA slopes; "Elevation," difference in SMA elevations (i.e., y-axis intercept); *, significantly different ($P < 0.05$); NS, not significantly different ($P > 0.05$). All data were log-transformed. *RMA* total root dry mass per unit land area, *RLA* total root length per unit land area, *RAA* total root area per unit land area, *TP* Tibetan Plateau, *LP* Loess Plateau, *MP* Mongolian Plateau, *MS* meadow steppe, *TS* typical steppe, *DS* desert steppe

sampling sites. RMA and RLA linearly increased with increases in MAP and SN (Fig. 8.7). Previous studies indicated that root traits are highly plastic and are strongly influenced by ecological indicators related to nutrient availability, including soil water and nitrogen and phosphorus supply, particularly in arid or semiarid grassland communities (Mokany et al. 2006; Bai et al. 2008; Wang et al. 2020a, b). Water and soil nutrient conditions can directly affect plant root growth and indirectly affect leaf CO_2 assimilation efficiency and carbohydrate allocation patterns (Chapin et al. 2002). Similarly, results from previous studies revealed that nitrogen enrichment and increased precipitation typically increased plant community root biomass and ecosystem biomass productivity (Bai et al. 2013; Wang et al. 2019).

The primary environmental factors influencing RCTs differed among regions; however, they were not the only limiting environmental factors. For example, SN and MAT influenced root traits in TP and LP, respectively. Notably, the Qinghai–Tibet Plateau is a unique ecological region characterized by low temperatures and

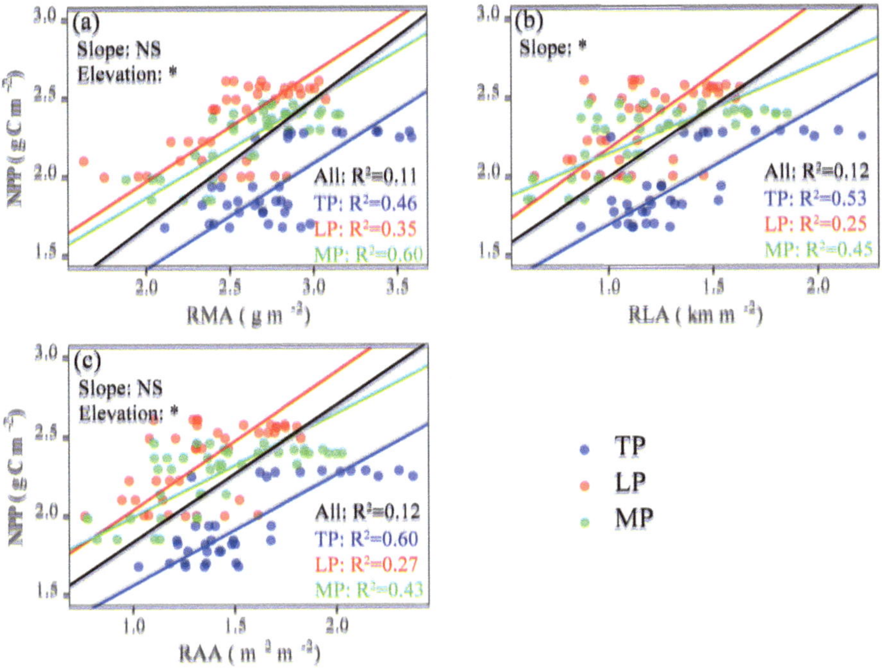

Fig. 8.9 Standardized major axis (SMA) regressions between net primary productivity (NPP) and root community traits (RCTs) across different regions. "Slope," difference in SMA slopes; "Elevation," the difference in SMA elevations (i.e., y-axis intercept); *, significantly different ($P < 0.05$); NS, not significantly different ($P > 0.05$). *RMA* total root dry mass per unit land area, *RLA* total root length per unit land area, *RAA* total root area per unit land area, *TP* Tibetan Plateau, *LP* Loess Plateau, *MP* Mongolian Plateau, *MS* meadow steppe, *TS* typical steppe, *DS* desert steppe. All data were log-transformed. NPP data were obtained from the MODIS NPP product (MOD17A2HGF) with a spatial resolution of 500 m and were averaged from 2000 to 2018

high radiation and is highly sensitive to climate change. (Qiu 2008; Liu et al. 2018b). However, our results demonstrated that temperature had no significant effect on RCTs in the TP region, consistent with findings from previous studies that revealed that precipitation and soil moisture, rather than MAT, are the primary regulators of ecosystem functions in this region. These results may be partly owing to the significant diurnal temperature fluctuations, comparable to seasonal variations, which might enable alpine plant communities to adapt effectively to temperature extremes (Jing et al. 2015). Thus, the expected influence of MAT might be overshadowed. Contrastingly, our results revealed that MAT was the primary factor regulating changes in RCTs in the LP region, although this region is typically characterized by poor soil fertility. This pattern may be explained by the strong positive correlation between SN and MAT ($r = 0.73$, $P < 0.05$), which likely masked potential SN effects.

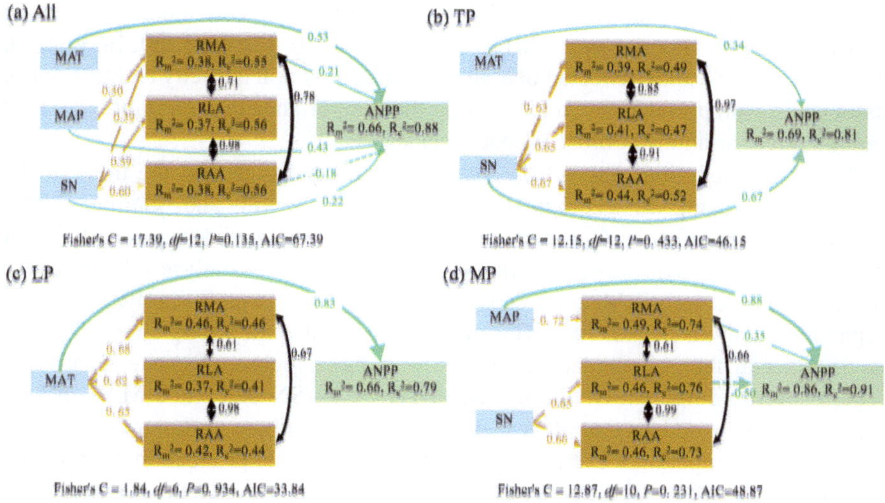

Fig. 8.10 Structural equation model analysis of the direct and indirect climate and soil and root community traits on above net primary productivity (ANPP). Only significant effects ($P < 0.05$) were presented. R^2 values associated with response variables indicate the proportion of variations explained by correlations with other variables. Values associated with solid arrows represent standardized path coefficients. *MAT* mean annual temperature, *AI* aridity index, *SN* soil total nitrogen content, *RMA* total root dry mass per unit land area, *RLA* total root length per unit land area, and *RAA* total root area per unit land area, *TP* Tibetan Plateau, *LP* Loess Plateau, *MP* Mongolian Plateau, *MS* meadow steppe, *TS* typical steppe, *DS* desert steppe

The RCT parameters were consistently positively correlated with ANPP and NPP across different regions. Among these, RMA was particularly influential due to its high explanatory power for productivity variations. Previous studies highlighted that belowground biomass or root biomass fraction reflects the carbon investment in belowground parts and is, therefore, a key trait in determining plant performance and multiple ecosystem functions (Freschet et al. 2015; Kleyer et al. 2018; Freschet et al. 2021b). Kleyer et al. (2018) conducted a trait correlation network analysis and identified belowground biomass as a hub trait because it had the highest betweenness with many other traits and strongly affected whole plant performance. However, ecophysiologists and community ecologists typically neglect the ecological consequences of belowground root biomass on ecosystem functions (Freschet et al. 2015; Kleyer et al. 2018; Freschet et al. 2021b). Theoretically, ecosystem NPP determines the amount of carbon fixation and, thus, the amount of carbon available to build roots; however, aboveground productivity, driven by leaf-level carbon fixation, depends on belowground resource acquisition capacity mediated by roots. Moreover, the framework proposed by Lavorel and Garnier (2002) indicates that NPP is controlled by the relative initial biomass of each species in the community and their functional traits related to carbon assimilation, nutrient uptake, and allocation, which are inherently influenced by resource availability, predominantly determined by root systems. This may explain the positive correlation between ecosystem productivity and root biomass observed at multiple spatiotemporal scales, although biomass allocation varied (Poorter et al. 2015; Yang et al. 2010; Liu et al. 2021a, b, c, d). The RCTs in this study reflected overall nutrient acquisition capacity and their close association with productivity, underscoring their importance in explaining grassland productivity variations at broader spatial scales.

Ecosystem functions depend on environmental conditions and the traits of the species that comprise ecological communities (Lavorel and Garnier 2002; Migliavacca et al. 2021). Our results revealed that when considering the effects of a combination of environmental variables and RCTs in SEM analyses, the influence of RCTs, particularly RMA, on ecosystem productivity was insignificant among separate regions and across sites. This may be attributed to the strong direct effects of environmental factors on plant productivity in these regions. Despite identifying region-specific dominant environmental drivers (e.g., SN in TP and MAT in LP), this study acknowledges that other factors possibly contribute to root trait variation. For example, temperature exerts more direct effects on plant production by influencing the activity of metabolic enzymes, the energy balance of leaves, activation of photosynthesis (Moles et al. 2014; Jing et al. 2015), and modulating plant phenology and the length of the growing season (Lavorel and Garnier 2002; Michaletz et al. 2014; Yan et al. 2022). Previous studies indicated that precipitation rather than temperature primarily drives ecosystem multifunctionality on the Qinghai–Tibet Plateau (Jing et al. 2015). Furthermore, climatic and soil factors may exert stronger indirect effects on ANPP via changes in leaf traits than root traits, which were not considered in the SEM analyses. For instance, community leaf chlorophyll content strongly influence productivity (Zhang et al. 2021a, b, c, d, e). Therefore, future research should simultaneously consider above- and belowground traits to better elucidate their combined effects on ecosystem functions.

Some inherent uncertainties exist in our efforts to develop RCTs and improve the ecosystem productivity of plant communities across wide environmental gradients. First, we only used the ANPP data to determine the effects of environmental gradients on the RCT–productivity correlation owing to limited data. We found that these correlations depend on environmental conditions, limiting the generalizability of correlations between root traits and productivity across diverse ecosystems. Further studies are needed to test the generality and applicability of this method across various biogeographical regions. Notably, root traits play key roles in multiple belowground ecosystem processes and functions, such as soil carbon and nutrient cycling and availability and the abundance and diversity of soil microorganisms (Klumpp and Soussana 2009; Bardgett et al. 2014; Freschet et al. 2021b). Thus, future studies should extend this approach to belowground processes. Despite these limitations, this study provides novel information on ecosystem ecology for scaling from belowground root traits to ecosystem functioning, thereby improving the predictive capacity of ecosystem modeling across ecological scales.

8.6 Conclusions and Perspectives

Incorporating root traits into predicting ecosystem functions is a key challenge and is critical for improving future modeling efforts. The new concept of RCTs was first applied to solve the problem of using root traits to predict ecosystem functioning (using ecosystem productivity as an example) across three distinct plateaus in the Northern Hemisphere. The results revealed that the RCTs of mass, length, and area increased with temperature, moisture, and soil nutrients in different regions, and they had a general correlation with grassland productivity, although environmental conditions modulated these correlations. Our results emphasize the importance of RCTs in linking root traits with ecosystem function in research.

Root community trait parameters should be widely used in studies at the ecosystem level. Future studies should determine the correlations between community-level root traits and the environment at a large scale, investigate the empirical correlations between root traits and ecosystem functioning to promote ecological models of vegetation dynamics, integrate studies on traditional root traits with macroecological studies using new technologies (e.g., ground-penetrating radar and remote sensing), and reveal new perspectives on the correlations between above- and belowground traits. Whether the above- and belowground traits interact or exhibit coordinated responses to a changing environment has been comprehensively debated; however, it remains controversial (Weemstra et al. 2016; Wang et al. 2017a, b, c). Within a community, tree canopies extend fully and capture more light for photosynthesis to meet the demand for stem and root growth; however, belowground roots provide anchorage and resource acquisition. The concept of RCT enables us to reexamine the relationship between roots and shoots from the perspective of above- and belowground plant communities. Furthermore, future studies should determine the belowground dominance of individual species, considering land area using more advanced techniques (e.g., molecular and DNA sequencing-based methods). Thus, community structure assembly and productivity optimization mechanisms can be explored from a new perspective.

In conclusion, the new concept of RCTs may enable the reexamination of the multiple roles of plant roots in community assembly and ecosystem processes while addressing the limitations of traditional scaling methodologies. Researchers may track the dynamics of entire root systems over different spatial and temporal scales in combination with new observation technologies (e.g., ground-penetrating radar, multispectral, and X-ray images). Notably, this new concept of RCTs can enable the incorporation of root traits into the investigations of ecosystem functioning at large scales and improve ecological models, particularly for productivity, nutrient acquisition strategies, and soil carbon cycling.

References

Bai Y, Wu J, Xing Q et al (2008) Primary production and rain use efficiency across a precipitation gradient on the Mongolia plateau. Ecology 89:2140–2153

Bai E, Li S, Xu W et al (2013) A meta-analysis of experimental warming effects on terrestrial nitrogen pools and dynamics. New Phytol 199:441–451

Bardgett RD, Mommer L, De Vries FT (2014) Going underground: root traits as drivers of ecosystem processes. Trends Ecol Evol 29:692–699

Chapin FSIII, Matson PA, Mooney HA (2002) Principles of terrestrial ecosystem ecology. Springer, New York

Croft H, Chen JM, Luo XZ et al (2017) Leaf chlorophyll content as a proxy for leaf photosynthetic capacity. Glob Chang Biol 23:3513–3524

De Long JR, Jackson BG, Wilkinson A et al (2019) Relationships between plant traits, soil properties and carbon fluxes differ between monocultures and mixed communities in temperate grassland. J Ecol 107:1704–1719

Díaz S, Lavorel S, de Bello F et al (2007) Incorporating plant functional diversity effects in ecosystem service assessments. Proc Natl Acad Sci 104:20684–20689

Eissenstat DM (1992) Costs and benefits of constructing roots of small diameter. J Plant Nutr 15: 763–782

Fort F, Freschet GT (2020) Plant ecological indicator values as predictors of fine-root trait variations. J Ecol 108:1565–1577

Freschet GT, Swart EM, Cornelissen JHC (2015) Integrated plant phenotypic responses to contrasting above- and below-ground resources: key roles of specific leaf area and root mass fraction. New Phytol 206:1247–1260

Freschet GT, Pagès L, Iversen CM et al (2021a) A starting guide to root ecology: strengthening ecological concepts and standardising root classification, sampling, processing and trait measurements. New Phytol 232:973–1122

Freschet GT, Roumet C, Comas LH et al (2021b) Root traits as drivers of plant and ecosystem functioning: current understanding, pitfalls and future research needs. New Phytol 232:1123–1158

Garnier E, Cortez J, Billes G et al (2004) Plant functional markers capture ecosystem properties during secondary succession. Ecology 85:2630–2637

He NP, Liu CC, Piao SL et al (2019a) Ecosystem traits linking functional traits to macroecology. Trends Ecol Evol 34:200–210

He L, Chen JM, Liu J et al (2019b) Diverse photosynthetic capacity of global ecosystems mapped by satellite chlorophyll fluorescence measurements. Remote Sens Environ 232:111344

Jing X, Sanders NJ, Shi Y et al (2015) The links between ecosystem multifunctionality and above- and belowground biodiversity are mediated by climate. Nat Commun 6:9159

Kleyer M, Trinogga J, Cebrianpiqueras MA et al (2018) Trait correlation network analysis identifies biomass allocation traits and stem specific length as hub traits in herbaceous perennial plants. J Ecol 107:828–842

Klumpp K, Soussana JF (2009) Using functional traits to predict grassland ecosystem change: a mathematical test of the response-and-effect trait approach. Glob Chang Biol 15:2921–2934

Kong D, Wang J, Wu H et al (2019) Nonlinearity of root trait relationships and the root economics spectrum. Nat Commun 10:2203

Kramer-Walter KR, Bellingham PJ, Millar TR et al (2016) Root traits are multidimensional: specific root length is independent from root tissue density and the plant economic spectrum. J Ecol 104:1299–1310

Laliberté E (2017) Below-ground frontiers in trait-based plant ecology. New Phytol 213:1597–1603

Laughlin DC, Strahan RT, Adler PB et al (2018) Survival rates indicate that correlations between community-weighted mean traits and environments can be unreliable estimates of the adaptive value of traits. Ecol Lett 21:411–421

Lavorel S, Garnier E (2002) Predicting changes in community composition and ecosystem functioning from plant traits: revisiting the holy grail. Funct Ecol 16:545–556

Liu ZP, Shao MA, Wang YQ (2013) Spatial patterns of soil total nitrogen and soil total phosphorus across the entire loess plateau region of China. Geoderma 197:67–78

Liu CC, He NP, Zhang JH et al (2018a) Variation of stomatal traits from cold-temperate to tropical forests and association with water use efficiency. Funct Ecol 32:20–28

Liu J, Milne RI, Cadotte MW et al (2018b) Protect third pole's fragile ecosystem. Science 362: 1368–1368

Liu CC, Li Y, Yan P et al (2021a) How to improve the predictions of plant functional traits on ecosystem functioning? Front Plant Sci 12:622260

Liu XR, Chen HX, Sun TY et al (2021b) Variation in woody leaf anatomical traits along the altitudinal gradient in Taibai Mountain, China. Glob Ecol Conserv 26:e01523

Liu C, Li Y, Xu L et al (2021c) Stomatal arrangement pattern: a new direction to explore plant adaptation and evolution. Front Plant Sci 12:655255

Liu R, Yang X, Gao R et al (2021d) Allometry rather than abiotic drivers explains biomass allocation among leaves, stems and roots of Artemisia across a large environmental gradient in China. J Ecol 109:1026–1040

Lynch JP (2007) Roots of the second green revolution. Aust J Bot 55:493–512

Ma ZQ, Guo DL, Xu XL et al (2018) Evolutionary history resolves global organization of root functional traits. Nature 555:94–97

Ma HZ, Mo LD, Crowther TW et al (2021) The global distribution and environmental drivers of aboveground versus belowground plant biomass. Nat Ecol Evol 5:1110–1037

McCormack ML, Dickie IA, Eissenstat DM et al (2015) Redefining fine roots improves understanding of below-ground contributions to terrestrial biosphere processes. New Phytol 207:505–518

McCormack ML, Guo DL, Iversen CM et al (2017) Building a better foundation: improving root-trait measurements to understand and model plant and ecosystem processes. New Phytol 215: 27–37

Michaletz ST, Cheng DL, Kerkhoff AJ et al (2014) Convergence of terrestrial plant production across global climate gradients. Nature 512:39–43

Migliavacca M, Musavi T, Mahecha MD et al (2021) The three major axes of terrestrial ecosystem function. Nature 598:468–472

Mokany K, Raison RJ, Prokushkin AS (2006) Critical analysis of root: shoot ratios in terrestrial biomes. Glob Chang Biol 12:84–96

Mokany K, Ash J, Roxburgh S (2008) Functional identity is more important than diversity in influencing ecosystem processes in a temperate native grassland. J Ecol 96:884–893

Moles AT, Perkins SE, Laffan SW et al (2014) Which is a better predictor of plant traits: temperature or precipitation? J Veg Sci 25:1167–1180

Muscarella R, Uriarte M (2016) Do community-weighted mean functional traits reflect optimal strategies? Proc R Soc B Biol Sci 283:20152434

Ottaviani G, Keppel G, Gotzenberger L et al (2020a) Linking plant functional ecology to Island biogeography. Trends Plant Sci 25:329–339

Ottaviani G, Molina-Venegas R, Charles-Dominique T et al (2020b) The neglected belowground dimension of plant dominance. Trends Ecol Evol 35:763–766

Poorter H, Niklas KJ, Reich PB et al (2012) Biomass allocation to leaves, stems and roots: meta-analyses of interspecific variation and environmental control. New Phytol 193:30–50

Poorter H, Jagodzinski AM, Ruiz-Peinado R et al (2015) How does biomass distribution change with size and differ among species? An analysis for 1200 plant species from five continents. New Phytol 208:736–749

Pregitzer KS, DeForest JL, Burton AJ et al (2002) Fine root architecture of nine north American trees. Ecol Monogr 72:293–309

Prieto I, Stokes A, Roumet C (2016) Root functional parameters predict fine root decomposability at the community level. J Ecol 104:725–733

Qiu J (2008) The third pole. Nature 454:393–396
Reich PB, Ellsworth DS, Walter MBS et al (1999) Generality of leaf trait relationships: a test across six biomes. Ecology 80:1955–1969
Reich PB, Tjoelker MG, Pregitzer KS et al (2008) Scaling of respiration to nitrogen in leaves, stems and roots of higher land plants. Ecol Lett 11:793–801
Reichstein M, Bahn M, Mahecha MD et al (2014) Linking plant and ecosystem functional biogeography. Proc Natl Acad Sci USA 111:13697–13702
Roumet C, Birouste M, Picon-Cochard C et al (2016) Root structure-function relationships in 74 species: evidence of a root economics spectrum related to carbon economy. New Phytol 210: 815–826
Suding KN, Lavorel S, Chapin FS et al (2008) Scaling environmental change through the community-level: a trait-based response-and-effect framework for plants. Glob Chang Biol 14:1125–1140
Valverde-Barrantes OJ, Freschet GT, Roumet C et al (2017) A worldview of root traits: the influence of ancestry, growth form, climate and mycorrhizal association on the functional trait variation of fine-root tissues in seed plants. New Phytol 215:1562–1573
van der Plas F, Schröder-Georgi T, Weigelt A et al (2020) Plant traits alone are poor predictors of ecosystem properties and long-term ecosystem functioning. Nat Ecol Evol 4:1602–1611
Violle C, Navas ML, Vile D et al (2007) Let the concept of trait be functional! Oikos 116:882–892
Violle C, Reich PB, Pacala SW et al (2014) The emergence and promise of functional biogeography. Proc Natl Acad Sci USA 111:13690–13696
Wang RL, Yu GR, He NP et al (2015) Latitudinal variation of leaf stomatal traits from species to community level in forests: linkage with ecosystem productivity. Sci Rep 5:1–11
Wang RL, Wang QF, Zhao N et al (2017a) Complex trait relationships between leaves and absorptive roots: coordination in tissue N concentration but divergence in morphology. Ecol Evol 7:2697–2705
Wang R, Gibson CD, Berry TD et al (2017b) Photooxidation of pyrogenic organic matter reduces its reactive, labile C pool and the apparent soil oxidative microbial enzyme response. Geoderma 293:10–18
Wang H, Prentice IC, Davis TW et al (2017c) Photosynthetic responses to altitude: an explanation based on optimality principles. New Phytol 213:976–982
Wang J, Gao YZ, Zhang YH et al (2019) Asymmetry in above- and belowground productivity responses to N addition in a semi-arid temperate steppe. Glob Chang Biol 25:2958–2969
Wang R, Chen H, Liu X et al (2020a) Plant phylogeny and growth form as drivers of the altitudinal variation in woody leaf vein traits. Front Plant Sci 10:1735
Wang P, Huang KL, Hu SJ (2020b) Distinct fine-root responses to precipitation changes in herbaceous and woody plants: a meta-analysis. New Phytol 225:1491–1499
Wang RL, Yu GR, He NP (2021) Root community traits: scaling-up and incorporating roots into ecosystem functional analyses. Front Plant Sci 12:690235
Warren JM, Hanson PJ, Iversen CM et al (2015) Root structural and functional dynamics in terrestrial biosphere models – evaluation and recommendations. New Phytol 205:59–78
Weemstra M, Mommer L, Visser EJW et al (2016) Towards a multidimensional root trait framework: a tree root review. New Phytol 211:1159–1169
Weemstra M, Zambrano J, Allen D et al (2021) Tree growth increases through opposing above-ground and below-ground resource strategies. J Ecol 109:3502–3512
Wright IJ, Reich PB, Westoby M et al (2004a) The worldwide leaf economics spectrum. Nature 428:821–827
Wright IJ, Groom PK, Lamont BB et al (2004b) Leaf trait relationships in Australian plant species. Funct Plant Biol 31:551–558
Yan P, Li MX, Yu GR et al (2022) Plant community traits associated with nitrogen can predict spatial variability in productivity. Ecol Indic 140:109001
Yang YH, Fang JY, Ma WH et al (2010) Large-scale pattern of biomass partitioning across China's grasslands. Glob Ecol Biogeogr 19:268–277

Zhang JH, Li MX, Xu L et al (2021a) C:N:P stoichiometry in terrestrial ecosystems in China. Sci Total Environ 795:148849

Zhang JH, Ren TT, Yang JJ et al (2021b) Leaf multi-element network reveals the change of species dominance under nitrogen deposition. Front Plant Sci 12:580340

Zhang Y, He NP, Li MX et al (2021c) Community chlorophyll quantity determines the spatial variation of grassland productivity. Sci Total Environ 801:149567

Zhang ZH, Hou JH, He NP (2021d) Predictability of functional diversity depends on the number of traits. J Resour Ecol 12:332–345

Zhang Y, Migliavacca M, Penuelas J et al (2021e) Advances in hyperspectral remote sensing of vegetation traits and functions. Remote Sens Environ 252:112121

Chapter 9
Ecosystem Traits: Bridging Traditional Functional Traits and Macroecological Research

Functional traits are genetic, relatively stable, and measurable parameters of organisms (plants, animals, and microorganisms), closely related to productivity optimization or environmental adaptation at the species level. With the development of ecology, functional traits have been increasingly used to reveal the adaptation, response, and optimization mechanisms of plants to the environment at different scales, including organs, species, functional types, communities, and ecosystems. Meanwhile, ecosystem ecology (or macroecology) should involve these rapidly developing functional traits and high-level observation technology to expand its application scope and scientifically reveal the response and adaptation of ecosystem structure and function to changing environments. However, linking traditional functional traits with these macro-technologies to resolve ecological problems and adapt to global change is challenging. The mismatch in scale and units between traditional functional traits and macroecological research are the primary problems. Macro-technologies, such as remote sensing, eddy-flux observation, and ecological modeling, are integrated at the community rather than at the species scales. To solve the gaps, researchers have attempted to develop a new concept named ecosystem traits (ESTs). ESTs are any measurable functional traits that can be standardized by unit land area at the community scale, typically in intensity or density, and can reflect the adaptation, reproduction, and productivity optimization of organisms (plants, animals, microbes, and others) to the environment. Theoretically, ESTs in any ecosystem are composed of plant, animal, and microbial community traits, which interact to complete multiple ecosystem functions. ESTs link traditional functional traits and macroecological technologies, particularly leaf traits of plant community traits, and provide a basis for integrating remote sensing and flux observations and model simulations into ecosystems. Furthermore, ESTs are valuable in exploring plant–animal–microbe interactions at the community scale. Notably, they provide a theoretical basis for constructing a new principle of ecosystem ecology using functional traits and developed a new framework of "structure–functional trait–function," solving various challenges in future ecological studies on a large scale.

N. He et al., *Plant Functional Ecology*,
https://doi.org/10.1007/978-981-96-8264-5_9

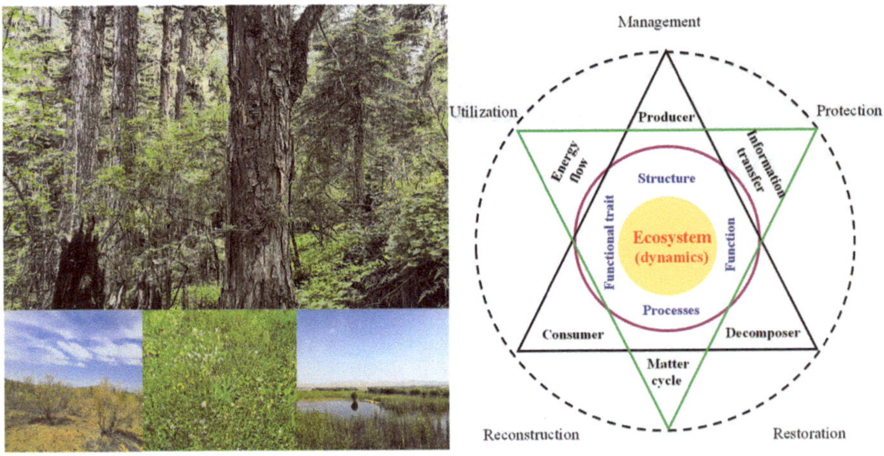

Fig. 9.1 Current research systems and scientific content of ecosystem ecology

An ecosystem is a unified whole composed of organisms and the environment in a particular space, where plants, animals, microorganisms, and the environment interact and are in a relatively stable dynamic equilibrium over a certain period (Tansley 1935). Thereafter, with the dedicated contributions of some predecessors, such as C.S. Elton, G.E. Hutchinson, R.L. Lindeman, E.P. Odum, and E.T Odum, the studies of ecosystem have developed rapidly, as an important discipline of ecology (i.e., ecosystem ecology). In practice, ecosystems are important research objects or scales of ecology, geoscience, environmental science, and other fields and are the most important intermediates linking different (plant, animal, and microbe) communities. Furthermore, it is closely related to the materials, energy sources, and living environments. Consequently, research and scope on various ecosystem-related disciplines have attracted widespread attention from the public and governments (Fig. 9.1). Previous researchers of ecosystem ecology have mainly focused on ecosystem composition (nonbiological material and energy producers, consumers, and decomposers), nutrition structure (food chain and web), and function (material cycle, energy flow, and information transfer) and have made remarkable progress with the help of data collected from organs or species and the rapid development of scaling-up methods (Lindeman 1942; Odum 1957; Chapin and Matson 2011).

In the past three decades, scientists have extensively researched the relationships between ecosystem structure and function and its response and adaptation to external disturbances or global climate change. Biodiversity and ecosystem functioning (BEF), a classic example of ecosystem ecology that examines the relationship between the structure and function of ecosystems, has been referenced in >10,000 scientific papers; however, there is no consensus on the conclusions drawn from these studies (Gravel et al. 2011; Snelgrove et al. 2014; O'Connor et al. 2017). There is a need to determine how ecosystems function and the theoretical basis linking the structure and function of an ecosystem. Previous studies have attempted to use functional traits and their differences to explain various events (Schleuning et al.

2015; Hodapp et al. 2016; Yan et al. 2023b). Although functional traits have been widely recognized in studies on the adaptive mechanisms and productivity optimization at the species level for plants and animals, they have rarely been used in ecosystem studies at the community level, and it is debatable whether they can predict the productivity of natural ecosystems. Previous studies have found that plant functional trait parameters are ineffective in predicting the spatiotemporal variation in productivity in controlled experiments and natural ecosystems (Plas et al. 2020; Zhang et al. 2022a, b). Contrastingly, some previous studies have revealed that plant functional traits can accurately predict the spatiotemporal variation in ecosystem productivity under the scenarios of scientific parameter selection, scaling-up, and scaling-match (Liu et al. 2021a, b, c, d; He et al. 2023; Hugan et al. 2023). Functional traits are primarily studies at the organ or species levels, as traditional definitions have limitations in capturing community-wide traits; therefore, it is essential to investigate the existence of community traits (plants, animals, and microbes) in natural ecosystems and their ecological significance (He et al. 2018a, 2019a, b).

In the classic monograph (i.e., *Principles of Terrestrial Ecosystem Ecology*), ecosystem ecology is defined as the study of the relationships between organisms and their physical environment within the Earth's system (Chapin and Matson 2011). Ecosystem ecology primarily focuses on the interactions between organisms and their environment at the community level rather than at the species level. It is important to understand the components, structure, and function of ecosystems at different spatiotemporal scales and the underlying regulatory mechanisms and influencing factors. Generally, ecosystem ecology, as a branch of ecology, focuses on studying ecosystems at specific spatial scales. It attempts to explore how biological communities and abiotic environments establish mutual connections and interactions through energy flow, material cycles, and information transmission and considers the scientific management of ecosystems and applications in the past three decades (Fig. 9.1). Therefore, the relationship between the biological community and abiotic environment, as well as among different biological communities, is the main component of ecosystem ecology. Practically, studies on organs, species, and functional groups at different spatial scales are typically classified into ecosystem ecology categories. To a certain extent, there are differences among individual, population, community, and ecosystem ecology.

The fundamental concept of functional traits originates from genetics or physiology and is a property with heritable, relatively stable, measurable, and distinct physiological and ecological significance (Cornelissen et al. 2003; Violle et al. 2007; Moretti et al. 2017; He et al. 2019a, b). Although functional traits are widely used at the species level, expanding them to community and ecosystem scales is challenging for ecologists worldwide (Kunstler et al. 2016; Lu et al. 2017). Theoretically, ecosystems have a complex self-organization function to optimize their structure and function to adapt to the changing environment. Understanding how ecosystems adapt and resist changing environments is important. The adaptations at the individual level should be basic; however, adaptation and optimization mechanisms at the community level are also important for a self-organized biological community.

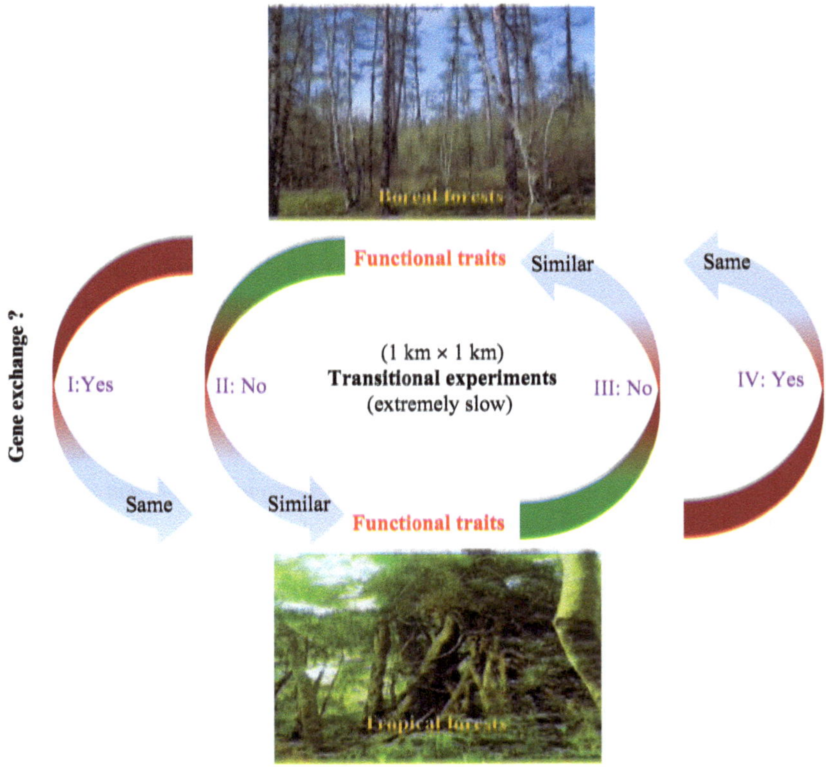

Fig. 9.2 Novel transitional experiments from tropical to boreal forests demonstrating the relative stable correlation between plant functional traits at the community level and climate in the four approaches of transitional experiments (I, II, III, and IV) at extremely slow speed, regardless of internal and external gene exchange

Therefore, it is necessary to explore functional traits and their ecological significance at the community scale, develop basic concepts and theoretical frameworks, improve the framework of ecological studies, and promote the development of related fields (He et al. 2019a, b, 2020a, b, c). For any climate (i.e., specific temperature and precipitation), the functional traits are relatively stable and renewable. Natural plant communities and ecosystem traits should be developed considering the following characteristics: first, any ecosystem can restore its properties in the long-term considering species composition, appearance, and function even after severe destruction; second, novel large-scale transitional experiments from tropical to boreal forests should demonstrate a relatively stable correlation between plant functional traits at the community level and climate (Fig. 9.2). Under extremely slow movement, there was a relatively stable correlation between plant functional traits at the community level and climate in four approaches, regardless of internal and external gene exchange. Therefore, their plant functional traits should be similar to those of the external community to adapt to its specific temperature and precipitation.

9.1 The Urgent Requirement and Theoretical Basis to Develop Ecosystem Traits

Functional traits in natural ecosystems are basic attributes in the adaptation of ecosystems to changing environments. For example, since the birth of humankind, humans have begun to learn and utilize nature and have gradually realized that plant functional traits are the result of long-term adaptation and evolution of plants to the environment. Practically, we have attempted to artificially regulate or screen specific functional traits to increase crop yields and quality. Violle et al. (2007) demonstrated that functional traits have specific adaptive or functional significance across the organ, species, population, community, and ecosystem levels (Violle et al. 2007; He et al. 2023), promoting the concept of functional traits to some extent. Furthermore, their work has been extensively cited in the past 15 years, indicating the recognition of the new concept by scientists worldwide.

Although Violle did not explicitly define the functional traits of ecosystems, it prompts ecologists to cogitate that, as a system of living organisms, an ecosystem may optimize its functions and adapt to a specific environment by continuously adjusting its spatial structure or species composition with the help of the functional traits of plants, animals, and microorganisms subjected to long-term adaptation and evolution (He et al. 2018a, 2019a, b). Moreover, the community traits of plants, animals, and microbes are common and may play important roles in ecosystem function optimization and environmental adaptation through different approaches. This can partly explain the scientific problems mentioned above, such as "structure cannot explain function well as BEF," "how structure explains ecosystem adaptation to the environment, anti-interference, and regeneration ability" (Yan et al. 2023a, b). Specifically, the community traits of plants, animals, and microbes may predict the spatiotemporal variations in ecosystem productivity considering several key community traits in the future (He et al. 2018a, 2019a, b, 2023; Zhang et al. 2021b).

In summary, there is a need for the scientific development of a new concept of functional traits that applies to multiple scales of organs, individuals, populations, communities, and ecosystems. The ecological significance and function of traits at the organ, individual, and population levels are well accepted (Wright et al. 2004a, b; Lu et al. 2017). However, owing to the constraints of traditional definitions of traits, most previous studies on functional traits are limited to the organ, individual, or population levels and few at the community or ecosystem scales (Wang et al. 2016; Zhang et al. 2018a, b; He et al. 2023).

Previous studies have used the simple arithmetic mean method to scale up traits from the organ to community, ecosystem, regional, and global scales (He et al. 2018a; Wieczynski et al. 2019). However, the spatial structure and species composition of natural communities, particularly natural forests, are complex, and significant differences exist in forest structure and composition between different regions. A simple arithmetic mean method may result in significant errors regarding relevant conclusions for natural forests, and the accuracy of such conclusions on a large scale has been widely discussed (Wang et al. 2015; He et al. 2018a; Liu et al. 2018a, b; Zhang et al. 2018a, b, 2021a). Therefore, it is essential to avoid the misunderstanding of "simple arithmetic average at species level be equal to community" to enable studies on functional traits at the community or ecosystem scales to reflect natural occurrences. Therefore, new concepts should be developed for functional traits at the community or ecosystem level to correct the problems associated with traditional concepts.

Furthermore, the development of ecosystem-level functional traits can promote the application of these observation technologies, such as remote sensing observation, flux observation, and model simulation at the macroscale, and link traditional functional traits to macroecology. Scientists can take advantage of the fast development of macro-technologies to promote the rapid development of ecosystem ecology. For example, using remote sensing and LiDAR technologies, parameters such as specific leaf area (SLA), leaf nitrogen content, and leaf spectral characteristics were closely correlated with community traits. These new parameters can provide novel data and insights into ecosystem ecology. Additionally, functional traits at the community level can provide various parameters to validate or optimize parameters for models, which may enhance the comparability and accuracy of remote sensing and flux observations and model simulations by constructing a multidimensional observation system for the sky, space, and the Earth. Most human ecological problems must be solved at the ecosystem, watershed, regional, and global scales. Therefore, it is necessary to address the scientific challenges posed by varying dimensions and scales in traditional functional traits and macroecological research and develop new concepts, methods, and technologies for analyzing functional traits at the ecosystem level to achieve a harmonious balance between functional traits, ecosystem ecology, macroecology, and macro-geoscience in the future (Fig. 9.3).

The rapid development of various technologies at the macroscale is undoubtedly a significant achievement. Ecosystem ecology must contain or accept these new observation technologies, considering their rapid development and novel insights. These space-based remote sensing and flux observations and model simulation technologies are necessary for the development of ecosystem ecology, and ecosystem ecology must adjust itself to efficiently use these technologies. However, most of these sophisticated observation technologies are used in unit space (or unit land area), and the observation time and spatial scale do not simulate ground ecological observations, particularly functional traits. In addition to their high-frequency observations, these macroscale technologies can measure many parameters related to, but not similar to, traditional functional traits (Croft et al. 2017; Xu et al. 2020), which can contribute to new development in ecosystem ecology. In summary, it is crucial to effectively apply these advanced new technologies to meet the contemporary requirements of ecosystem ecology and to develop conceptual frameworks and reshape the classical framework of ecosystem ecology.

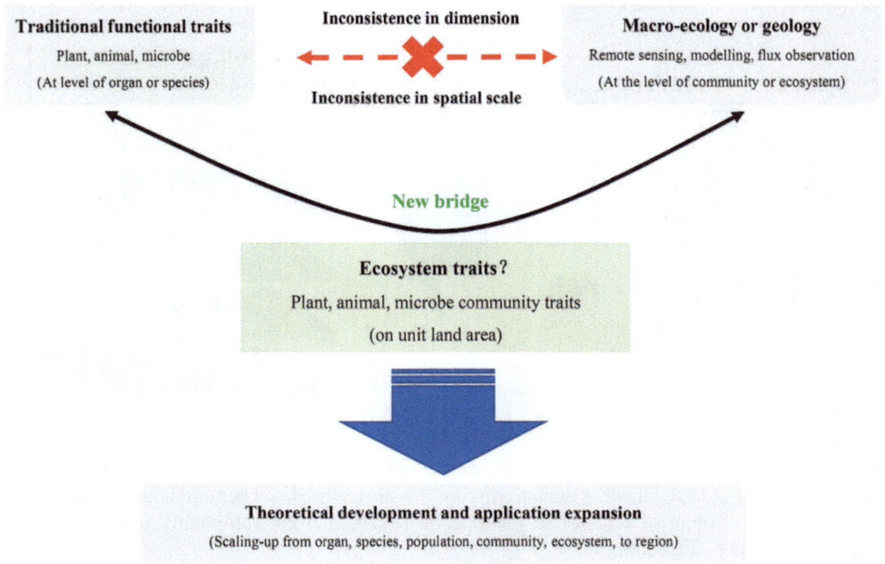

Fig. 9.3 The promising to linking traditional functional traits to macroecological research. Most macroscale observation technologies are currently conducted in unit space (or unit land area), and the observation time and spatial scale do not match the ground ecological observation, particularly as functional traits at the organ or species levels

9.2 What Are Ecosystem Traits?

9.2.1 Definition of Ecosystem Traits

An increasing number of studies on traits and macroecology are focusing on using communities and ecosystems to explore ecological problems on regional or global scales. However, new concepts must be developed to integrate traits measured at the organ level to meet the requirements of the major technologies in macroecology, particularly by matching them at the same spatial scale (Fig. 9.4). Scientists have proposed that the concept of "ecosystem traits" could be used to address these limitations. Ecosystem traits are the characteristics of plants, animals, soil microbes, or other organisms and are calculated as the intensity (or density) normalized per unit land area (He et al. 2019a, b, 2020a, b, c). We hypothesized that these traits provide information on variations in community species composition and structure, including adaptation and distribution of species considering the biotic and abiotic characteristics of the environments and their plasticity, and reflect the optimization of processes that occur during evolution and ecological assembly.

This new definition caters for the dimension or spatial-scale mismatch in the transformation or scaling up of various functional traits to a certain extent. For a specific ecosystem, ecosystem traits comprise a series of biological factors, such as plant, animal, and microbial community traits. Different functional characteristics of

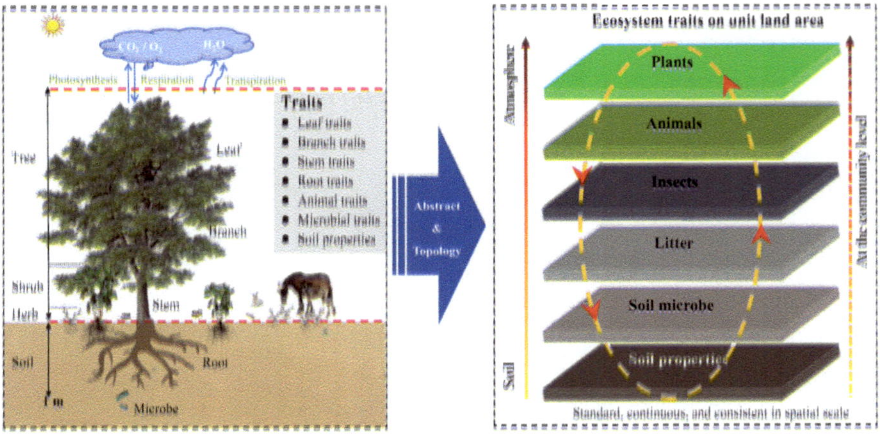

Fig. 9.4 Ecosystem traits representing the characteristics of plants, animals, soil microbes, and other organisms and calculated as the intensity (or density) normalized per unit land area. In nature, ecosystem traits comprise a series of plant, animal, and microbe community traits and others (He et al. 2019a, b, 2020a, b, c)

the community play specific roles or interact to complete multiple ecosystem functions. Plant, animal, and microbial community traits are the core units or operating objects of ecosystem traits (Fig. 9.4). Figure 9.4 illustrates the general relationships among structure, traits, and function at the ecosystem scale. All parameters were measured using different approaches. Functional traits are sampled and measured at the organ level, whereas function is primarily measured or simulated through eddy-flux observations, remote sensing, and ecological models at the ecosystem level, and structure is derived through field investigations or radar (Heinzel and Koch 2011). Researchers can bridge gaps and create connections across different spatial scales by combining organ-level traits and stand structures to develop ecosystem traits.

Ecosystem traits have the following characteristics: (i) they represent community-scale data for plants, animals, and soil microbes; (ii) they are calculated or normalized per unit land area to enable scale matching; (iii) they can be derived from trait measurements at organ or individual levels and typically require community structural variables to enable scaling up; (iv) have specific ecological significance and enable tests of adaptation or optimization of traits at the community or ecosystem levels. For example, scaling leaf area to the community level gives the ecosystem trait leaf area index (LAI), and scaling up of leaf dry mass gives the leaf mass index (LMI). Consequently, LAI and LMI are important ecosystem scaling variables, as ecosystem-scale versions of leaf area-based traits (e.g., leaf stomatal density) per land area are calculated as the products of LAI and SLA-based traits, and ecosystem-scale versions of leaf mass-based traits (e.g., leaf nitrogen and phosphorus concentrations) are calculated as the products of LMI and specific leaf mass-based traits. Thus, LAI and LMI are important ecosystem traits, essential in macroecology, although they have not previously been recognized as ecosystem traits (Gitelson et al. 2003; Heiskanen 2006). The determination and use of ecosystem traits enable scale matching and integration of trait information for comparisons across ecosystems and regions, such that these traits can be used as an interface among ecological modeling, eddy-flux observations, and remote sensing. We observed that many organ-level traits were not sufficiently measured to scale up to the ecosystem level. For example, leaf optical properties (reflectance or hue) can play an important role in energy balance and surface warming, and such data, although currently lacking, could be important if scaled up as ecosystem traits. Furthermore, additional key challenges may be resolved by scaling up important qualitative traits to the ecosystem level (e.g., serrated versus non-serrated leaves, single versus compound leaves, evergreen versus deciduous).

9.2.2 Development of Methods for Expanding Ecosystem Functional Traits

It is challenging for researchers to deduce traits measured at the organ and community levels, and significant progress has been made, particularly in the field of plant community functional traits. Researchers have developed various scaling-up methods depending on the problem: (1) simple arithmetic mean (Eq. 9.1) without considering plant community structure (Wright et al. 2017; Ma et al. 2018). This method is widely used in research on plant functional traits, particularly in large-scale functional trait integration analyses at the national, continental, or global scales. (2) After considering the community structure (Eq. 9.2), the relative richness or importance values of different species were used as the weight coefficient derived from functional traits (Violle et al. 2007). (3) The community weighted average under species and functional trait mismatch scenarios after considering community structure are estimated using Eq. 9.3, which was first used for the latitudinal variation of leaf stomatal and morphological traits at the biome scale (Wang et al. 2015, 2016; Borgy et al. 2017).

$$\mathrm{Trait}_{\mathrm{com-SAM}} = \sum\nolimits_{i=1}^{n} T_i/n \tag{9.1}$$

$$\mathrm{Trait}_{\mathrm{com-CWM}} = \sum\nolimits_{i=1}^{n} p_i T_i \tag{9.2}$$

$$\mathrm{Trait}_{\mathrm{com-CWM}} = \frac{\sum_{i=1}^{n} p_i \sum_{j=1}^{\mathrm{NIV}_i} \left(T_{ij}/NIV_i\right)}{P_{\mathrm{Cover}}} \tag{9.3}$$

where $\mathrm{Trait}_{\mathrm{com}}$ represents the value of community functional traits, which is the mean of different methods; CWM represents the relative biomass of community weighting method (or relative abundance and importance values); p_i is the relative biomass (or relative richness and important value); T_i represents the functional trait value of the i_{th} species in the community; n represents the number of known species

(Eq. 9.1) or the number of species in the community (Eq. 9.2) or the number of species in the community that can match the functional traits in the database (Eq. 9.3); T_{ij} represents the *j*th functional trait value of the *i*th species in the existing database; NIV_i is the repetition number of the functional trait value of the *i*th species in the existing database; and P_{Cover} is the cumulative relative biomass, or relative richness, or importance value of the species in the community which can match the functional traits in the database (Borgy et al. 2017).

After careful analysis, the functional traits in Eqs. 9.1, 9.2, and 9.3 were extended to communities. Because communities have specific corresponding areas, they denote unit land area; however, the conversion of a unit to land area is voluntary; for example, the units of nitrogen and phosphorus contents, SLA, and leaf stomatal density are still g kg^{-1}, mm^2 mg^{-1}, and cm^{-2}, respectively. Generally, the above three methods can be regarded as the average of the species level using different approaches, which can be used to explore the high-level processes and relationships of communities or ecosystems, particularly the interaction, competition, and coexistence between individuals and populations in ecology.

Based on the newly developed plant community functional traits described in Chap. 7, scientists have developed the concept of two-dimensional characteristics of plant community traits. These functional traits characterize the efficiency per unit area (efficiency trait) and the intensity or quantity per unit area (intensity trait) (He et al. 2019a, b, 2023; Zhang et al. 2021b). These plant community traits derived from the above three methods belong to "efficiency traits," and their values are typically positively correlated with specific ecological functions. For example, a plant community with a high leaf nitrogen content or high SLA implies that it will have high productivity per unit of nitrogen content or leaf area rather than ecosystem productivity per year (Yan et al. 2022; Zhang et al. 2022a, b).

Currently, the function or multifunction of the ecosystem scale is almost measured and simulated based on unit area or volume, such as flux observations, model simulations, and remote sensing observations. Therefore, Eqs. 9.1, 9.2, and 9.3 cannot solve the problem of scale mismatch, which is an important reason for the low degree of explanation of plant functional traits (efficiency traits) for spatiotemporal variations in productivity. Therefore, scientifically deducing the functional traits measured at the organ level to the community or ecosystem levels and matching them with natural community or ecosystem functions are huge challenges. Scientifically promoting the acquisition of intensity traits in plant community functional traits, realizing the transformation of unit land area, and then combining the two-dimensional characteristics of plant community traits to reveal the spatiotemporal variations in productivity and its regulatory mechanisms are challenges in ecological studies (Zhang et al. 2021b; He et al. 2022; Yan et al. 2022, 2023a, b).

To resolve these technical problems, researchers have developed a new method for deriving ecosystem functional traits using community structure, SLA for each species, and allometric equation data, which can be standardized into community functional traits per unit land area. Eqs. 9.4 and 9.5 are functional traits for calculating quality and leaf area standardizations, respectively (He et al. 2019a, b; Zhang et al. 2021b; Yan et al. 2022), which consider the complex community structure and

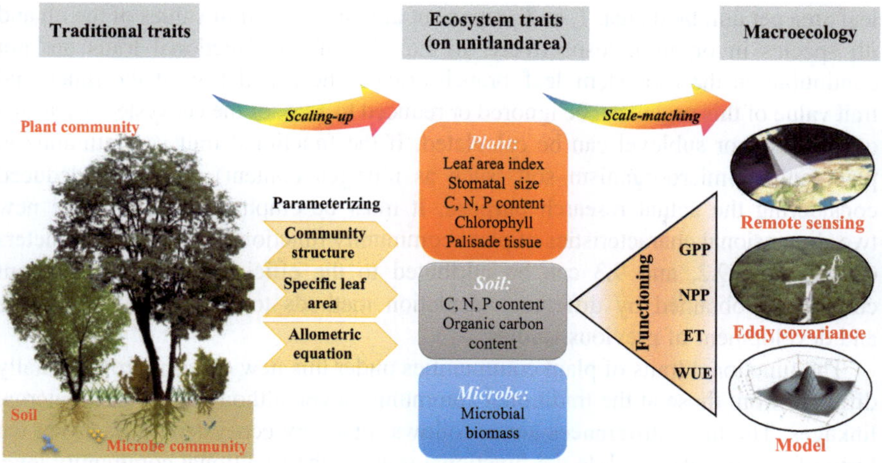

Fig. 9.5 Scaling up and scale-matching of traits measured at the organ level and ecosystem traits per land area. Scientists can first scale up traits measured at the organ level using data on community structure, allometric equation, specific leaf area, and others, standardized on per land area, and then established the quantitative relationship with these parameters using remote sensing, ecological modeling, and eddy-flux observation. These are the important foundations for linking traditional traits to macroecology. *GPP* gross primary productivity, *NPP* net primary productivity, *ET* evapotranspiration, *WUE* water-use efficiency

transform the ecosystem function traits to the unit land area. Therefore, they can be used to explore the correlations among plants, animals, and microorganisms and between biotic and abiotic factors in traditional ecosystem ecology and can be linked with ecosystem function to explore the relationship between functional traits and functions at the ecosystem level and its influencing mechanism (Figs. 9.4 and 9.5).

$$\mathrm{Trait_{com}} = \sum\nolimits_{j=1}^{4} \sum\nolimits_{i=1}^{n} \mathrm{OMI}_{ij} \times T_{ij} \tag{9.4}$$

$$\mathrm{Trait_{com}} = \sum\nolimits_{i=1}^{n} \mathrm{LAI}_{i} \times T_{i} \tag{9.5}$$

These formulae divide functional traits into quality-standardized functional traits (such as nitrogen content per unit leaf mass) and area-standardized functional traits (stomatal density and number per unit leaf area). Equations 9.4 and 9.5 were used to derive quality-normalized functional traits and leaf area-normalized functional traits, respectively, where $\mathrm{Trait_{com}}$ represents the ecosystem-level functional trait value for a specific functional trait (T_i) in an additive form, n represents the number of species in the community, OMI_{ij} represents the quality index of the *j*th organ (leaf, branch, stem, and root) of the *i*th species in the community, and LAI_i represents the leaf area index of the *i*th species in the community. The algorithm was used to deduce the biomass of each organ of the species considering the diameter at breast height, tree height, and univariate or binary growth equation of the species, convert the leaf biomass into leaf area according to the SLA, and standardize the organ biomass or leaf area per unit land area. T_i or T_{ij} represent the functional trait values of the *i*th and *i*th species in organ *j*, respectively. In Eq. 9.4, if the functional traits are not continuous in the root–stem–leaf–branch organs, the calculation of the functional trait value of this organ can be ignored or reduced to zero, or the ecosystem traits at a certain level or sublevel can be calculated. If the functional trait is continuous in plant–animal–microorganism–soil (such as nitrogen content), it can be deduced considering the actual research purpose. It must be emphasized that in the new two-dimensional characteristics of plant community functional traits, the parameters of Eqs. 9.1, 9.2, and 9.3 can be attributed to the efficiency trait of the plant community obtained by different calculation methods to realize the inheritance and development of previous studies.

The functional traits of plant communities under this new concept are essentially different from those at the traditional community level, although they have internal linkages. The main differences are as follows: first, any ecosystem traits are in the form of accumulation, while the functional traits of the traditional community level are in the form of average values; therefore, the scientific questions for the two are different. Second, ecosystem traits require more systematic field-measured data, such as matching data from community structure, species composition, functional traits, and biomass equations, whereas traditional community-level traits can be estimated by species number, richness, or importance value. The internal relationships between them are that if the functional traits at the traditional community level can meet the accurate data requirements, these ecosystem traits and those at the traditional community level can be converted using important parameters such as leaf area (or organ mass). Ecosystem traits are not limited to functional traits per unit land area but can include new functional traits derived from two or more ecosystem functional traits, such as the ratio of nitrogen content to phosphorus content (nitrogen: phosphorus) per unit land area and SLA (the ratio of leaf area to leaf biomass per unit land area). The derived ecosystem traits should be identical to the functional traits at the traditional community level. Furthermore, according to the newly developed two-dimensional characteristics of plant community traits, community functional traits that have not been transformed per unit land area primarily belong to efficiency traits; after subsequent development, the transformation per unit area is more of "intensity traits," and the two are mutually cooperative and complementary resulting in improved plant growth and development (Zhang et al. 2021b; Yan et al. 2022).

9.3 Research on Ecosystem Traits

According to the definition and formula of ecosystem traits, it is necessary to scientifically deduce the functional traits measured at the traditional organ level to the ecosystem level per unit land area. Therefore, field-measured data must adhere to various requirements considering systematics and matching at the community level. Taking plant functional traits as an example, to achieve scaling extrapolation of chlorophyll, nitrogen, and phosphorus contents measured at the organ level, data on community structure, relative biomass, SLA, and chlorophyll, nitrogen, and phosphorus contents for each species are hypothetically required. Different derivative parameters have different systematic requirements for the field-measured data. However, most of the current large-scale plant functional trait databases are compiled using collected public data, with a primary focus on easily measurable individual plant traits (leaf size, thickness, SLA, and carbon and nitrogen contents) (Kattge et al. 2011). There is almost no similar supporting database for conventional leaf morphological, stomatal, anatomical and structure traits, as well as chlorophyll and elemental contents. Although the famous plant trait database (TRY; https://www.try-db.org/) in the world contains a large functional trait data for species, it lacks supporting data for leaves, branches, roots, stems, and community structures. Most analyses based on the TRY database or similar collected data are carried out at the species level because systematic and matching data limitations make it difficult to derive ecosystem traits (He et al. 2018b, 2019a, b). The new functional trait database (such as China_traits) constructed using the "systematicness and compatibility" concept can systematically integrate research on plant community or ecosystem traits (He et al. 2018a, 2020a, b, c).

9.3.1 Application of Ecosystem Traits in Typical Ecosystems

Based on the detailed field-measured data of nine typical forest ecosystems in the north–south transect of eastern China and using Eqs. 9.3, 9.4, and 9.5, researchers have derived some plant functional traits from the organ level to the community level and obtained the corresponding ecosystem traits (He et al. 2019a, b). These parameters included leaf morphological traits, stomatal morphological traits, chlorophyll content, nonstructural carbohydrates, anatomical structure traits, and carbon, nitrogen, and phosphorus contents of different plant organs (leaves, branches, stems, and roots). Previous studies have demonstrated latitudinal variation and the primary influencing factors from the perspective of organs, species, functional groups, communities, and ecosystems (Li et al. 2016, 2018a, b; Wang et al. 2016; He et al. 2018a; Liu et al. 2018a, b; Zhang et al. 2018a, b). These innovative studies on functional traits are systematically deduced from the traditional organ scale to the community and ecosystem scales, which provide replicable cases and a methodological basis for ecosystem traits.

Based on the ecosystem traits derived from measured data, quantitative relationships between several functional traits and productivity have been established in natural forest ecosystems, providing a strong foundation for studying ecosystem traits and functions (Reichstein et al. 2014; Borgy et al. 2017; He et al. 2018b). Specifically, previous studies have found that stomatal density at the plant community level could explain 51% of the spatial variations in water-use efficiency (WUE) (Liu et al. 2018a, b); similarly, the quantitative relationships among SLA, leaf anatomical structure, chlorophyll content, and community gross primary productivity (GPP) were established (He et al. 2018a; Li et al. 2018a, b). Although some plant community traits were significantly correlated with GPP, their individual interpretations were not as expected. Therefore, multiple functional traits of plants should be combined to optimize ecosystem productivity, and the importance of single functional traits should not be blindly exaggerated. Zhang et al. (2022a, b) found that the classical N:P ratio of community leaves cannot directly characterize spatial variation in ecosystem productivity (Zhang et al. 2022a, b). Theoretically, the new development of plant trait networks should be used to put forward the viewpoint that "multiple functional traits cooperate to complete a certain function, or any function is completed by multiple functional traits" (He et al. 2020a, b, c; Li et al. 2022a, b, c, d).

9.3.2 *Spatial Variations in Ecosystem Traits at Regional or National Scales*

Macroecology is typically studied at the watershed or regional scale. It is important to determine whether the ground-measured data of ecosystem traits can be matched with that of watershed or regional scales to expand the application of functional traits. Therefore, scientists have attempted to scale some ecosystem traits at a national scale.

The nitrogen content and its distribution in ecosystems are important for large-scale studies on terrestrial ecosystems. Therefore, researchers have used previously collected aboveground biomass data of >16,000 forests and grasslands and the belowground biomass data of >15,000 forests and grasslands, combined with the supporting data of nitrogen content among different plant organs (leaves, branches, stems, and roots) in the China_traits database, and generated the nitrogen content and density of leaves, branches, stems, roots, 0–20 cm soil layer, and 20–100 cm soil layer per unit area in China (Xu et al. 2020). Furthermore, scientists have conducted several studies to map the spatial variation (content and density per unit land area) of the functional elements of terrestrial ecosystems on the Tibetan Plateau (TP), including carbon, nitrogen, phosphorus, sulfur, potassium, sodium, calcium, and magnesium, at a resolution of 1×1 km (Zhang et al. 2024; Li et al. 2022a, b, c, d; Yu et al. 2022; Zhao et al. 2022a, b, 2023; Jiao et al., 2023; Zhu et al., 2023). With high-resolution spatial data, it is possible to provide new perspectives for improving macroecological research in the future.

9.3.3 *Using Ecosystem Traits to Estimate Productivity Across Scales*

Rapid progress in trait-based ecology has led to opportunities for integrating plant traits into predicting ecosystem primary productivity across various landscape scales. Traditionally, traits have been used to explain physiological productivity at the cell, organ, or plant scales; however, there are challenges in scaling up ecosystem-level carbon fluxes. Scientists have proposed a new framework for "trait-based productivity" (TBP), utilizing data for two-dimensional plant community traits, which are normalized per unit area at the community scale, and external environmental factors (He et al., 2023). Practically, the efficiency of plant community traits ($Trait_{effiency}$), considering Eqs. 9.2 and 9.3, can reflect the mean behavior of species; for example, high leaf nitrogen concentration commonly indicates a community dominated by individuals of fast-growing acquisitive species with a high resource-utilization efficiency (Wright et al., 2010). Contrastingly, the capacity of plant community traits ($Trait_{capacity}$), considering Eqs. 9.4 and 9.5, is multiplied by parameters of stand structural density, that is, the LAI, defined as the total one-sided area of leaf tissue per unit land surface area, equivalent to the number of layers of leaves (Watson, 1947; Asner et al., 2003) or the LMI. Analyses of variation in productivity per unit land area across ecosystems have typically focused on LAI, a key driver of ecosystem biogeochemical cycles (Farquhar et al. 1980). Similarly, mass-based traits can be multiplied by the LMI to predict the productivity per unit land area. Furthermore, ecosystems differ in the efficiency of translating a high $Trait_{capacity}$ value into productivity and thus differ in $Trait_{efficiency}$.

Several previous studies have demonstrated the potential of utilizing plant community traits to predict the spatiotemporal variation in ecosystem productivity from different perspectives. First, plant community traits (leaf area, SLA, and leaf dry matter content), restoration time, species diversity, and soil nutrients could jointly explain 78% of the ANPP variation along forest gradients with different restoration times under selective logging (Li et al. 2021a, b, c). Secondly, Zhang et al. (2021a, b, c, d, e), using leaf chlorophyll content (scaling-up community weighted mean) data in three grassland transects of the TP, Mongolian, and Loess plateaus in China, found that 67% of the spatial variations in grassland productivity can be captured by the framework of "efficiency × quantity × growth length" in all grassland transects, indicating that it was applicable for the three environmentally distinct plateaus (Fig. 7.10). Third, using the leaf nitrogen plant community trait without factoring in climatic factors, a previous study explained 81 and 68% of the spatial variability in annual and monthly GPPs of 73 typical ecosystems in China, respectively (Yan et al. 2022). Fourth, a previous study found that integrating multiple plant community traits with the recently developed TBP theory can effectively predict variations in annual and monthly GPPs across terrestrial ecosystems in China, with R^2 values of 0.87 and 0.73, respectively (Yan et al. 2023b). A previous study has demonstrated that two-dimensional plant community traits are the main factors regulating the spatial variations in terrestrial ecosystem multifunctionality

(EMF) in terrestrial ecosystems in China, which is the maximum productivity, WUE, and carbon-use efficiency of the ecosystem (Yan et al. 2023a). In summary, the new framework presents numerous opportunities for estimating productivity and EMF across scales and for integrating traits into ecological process models to predict the response of terrestrial ecosystems to global change in the future.

9.4 Prospects of Using Ecosystem Traits in Macroecological Research

9.4.1 New Ecosystem Ecology Framework Considering Functional Traits

Ecosystems refer to the composition of plant, animal, and microbial communities, and environmental factors within a specific space. Generally, the structure includes the type, size, horizontal position, spatial position, trophic level, and food web position of biological communities. Ecosystem function or multiple functions should include GPP, NPP, nutrient-use efficiency, WUE, carbon sequestration capacity, and soil and water conservation capacities. Most traditional research on functional traits has been conducted at the organ or individual level owing to the lack of conceptual ecosystem traits. For these ecosystem studies, scientists have typically emphasized parallel structure and function and increasingly formed the classic framework of structure, process, and function (Fig. 9.6). Based on this framework, researchers have conducted many studies on the correlations between ecosystem structure and function and its adaptation and response to exogenous disturbances and have attempted to establish quantitative correlations between ecosystem structure and function to influence production and protection. The new concept of ecosystem traits may help construct a new ecosystem ecology research framework considering functional traits and provide novel insights into solving many problems in ecosystem ecology (Fig. 9.6).

The ecosystem has a complex self-organization ability to optimize its structure and function, effectively adapt to the changing environment, resist external interference, and regenerate (Fig. 9.2). As discussed above, it is difficult for the ecosystem structure to directly realize the corresponding function or reflect adaptation to the environment; however, it can affect the ecosystem function and adaptability by adjusting the species composition owing to each species having specific functional trait and adaptation range and adjusting the functional traits of the community as a whole. Theoretically, plant functional traits play an important role in adaptation and productivity optimization throughout the plant system, from organs to ecosystems (Violle et al. 2007). The development of ecosystem traits, particularly scaling-up methods and available field-measured data, will promote the development of new models at the core of functional traits in ecosystem ecology (Fig. 9.7). In contrast to the classical framework of structure and function, the new framework introduces the

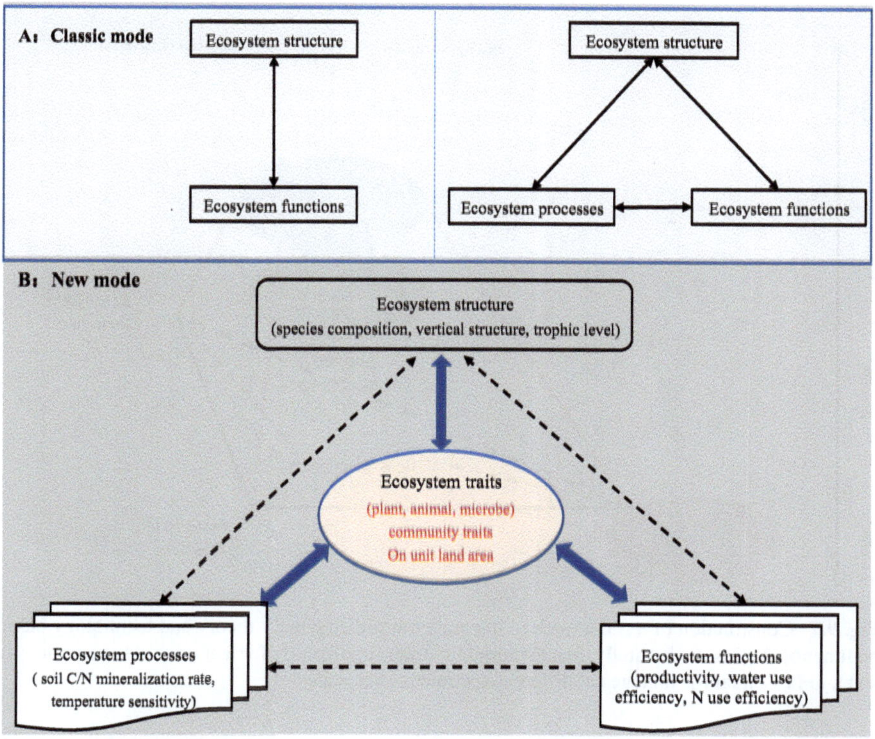

Fig. 9.6 New framework of ecosystem ecology with functional traits as the core. In this study, ecosystem traits comprise a series of plant, animal, and microbial community traits standardized by unit land area at the community scale

concept of functional traits and sets it as a basis to effectively explore the correlations between ecosystem structures, processes, and functions and their formation mechanisms. Furthermore, it reveals the response and adaptation of ecosystem structure, processes, and functions to global change, environmental change, and extreme disturbances.

9.4.2 *Enriching Ecosystem Ecology Research Approaches*

Ecosystem traits comprise several plant, animal, and microbial community traits standardized per land area. Therefore, the concept of ecosystem traits provides a new idea for a comprehensive discussion of the internal relationship, synergy, or divergence of plant, animal, and soil microbes on functional traits at the ecosystem scale and can be used to explore interactions among biological (plant, animal, and microbe) and abiotic factors (soil and climate) and reveal the community assembly and maintenance mechanisms of plant, animal, and soil microbes. Theoretically,

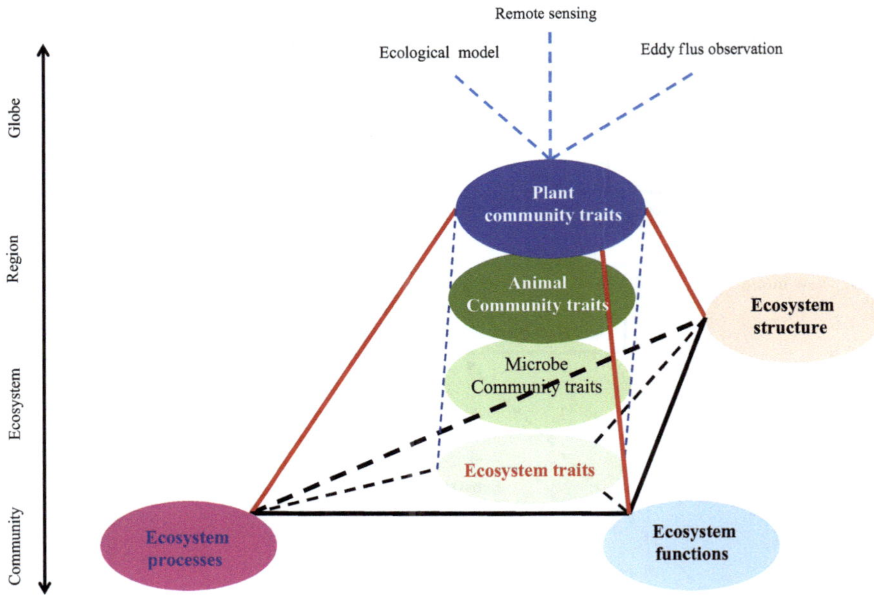

Fig. 9.7 Construction of a framework to integrate multi-disciplinary theory and technology based on functional traits. In this study, plant community traits, particularly for leaf traits at the community level, are essential for linkages at different spatiotemporal scales

scientists can use these multiple datasets at the community scale but not at the organ, species, or population scale to address their concerns regarding ecosystem ecology. Few previous studies have reported the correlations among plant, animal, and soil microbial communities and soil and climate at the ecosystem scale primarily owing to limited available field-measured data.

Most related studies have been limited to specific species (plants, animals, and soil microorganisms) and the impact of other biological or environmental factors at the species level. There has been rapid progress in studies on the functional traits of animals and microbes on a large scale, and researchers have gradually compiled a functional trait database related to reptiles, invertebrates, fish, corals, birds, amphibians, mammals, and fungi (Gallagher et al. 2020). Framing the study of plant, animal, and microorganism interactions on functional traits in substantial ecosystems is a challenge for the future. Notably, plant community traits per land area, particularly leaf traits, can provide a basis for the rapid development of remote sensing and enrich ecosystem ecology research approaches in the future. Overall, the concept of ecosystem traits may yield plant, animal, and microbial community traits on a relatively unified spatial scale and dimension, which may provide a solid theoretical foundation for exploring the correlation between biological-biological and biological-abiotic functional traits at the ecosystem scale (Fig. 9.7).

9.4.3 Promoting the Integrative Development of Macroecology and Geography

Ecosystem traits are a combination of community traits (plant, animal, and microbe) for the standardization of unit land area, which resolves the long-term problem of spatial mismatch between (plant and animal) functional trait data and macroscale observation technologies, such as remote sensing observations, flux observations, model simulations, and big data integration. Furthermore, it can use new data sources, as observed by various macroscale observation technologies, and promote the development of the ecosystem ecology (Fig. 9.8). With the rapid development of these macroscale technologies, more parameters may be produced or used to elucidate the structure, functional traits, and functions of natural ecosystems, such as the LAI, SLA, fluorescence, and community structure. This should be a new focal point in related fields in the future. The direction and development rate depend on the rate of the development of new technologies and how closely they are integrated with research in ecosystem ecology.

Considering technological advancement, the development of macroecology primarily depends on remote sensing observations, flux observations, model simulations, and large data integration. However, regardless of how these macroscale observation techniques are developed, they all require different data measured at the ground level, particularly for functional traits, to verify or test them. Previously, verifying and improving most data from remote sensing, flux observations, and model simulations were difficult because of the spatial mismatch between

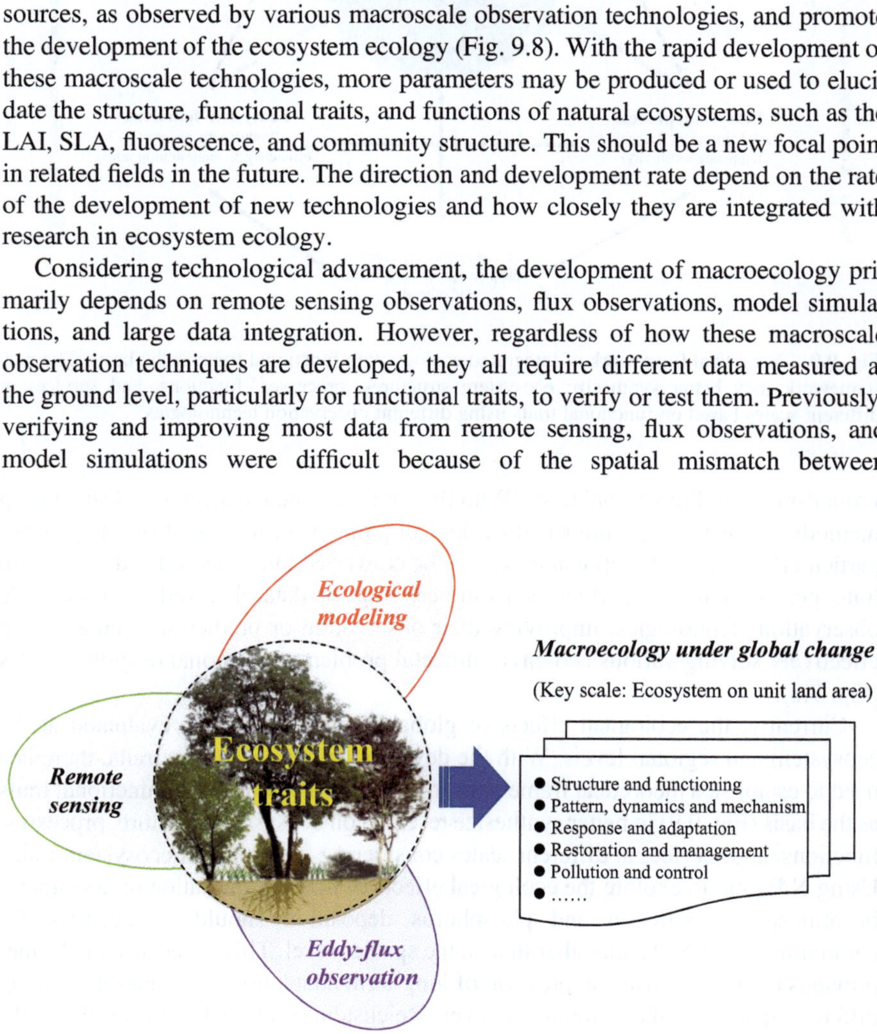

Fig. 9.8 Importance of ecosystem traits for macroecological studies under global change. Ecosystem traits are key to linking the correlation between trait and functioning parameters observed by these advanced technologies (remote sensing, ecological modeling, and eddy-flux observation) or to improve their prediction accuracy

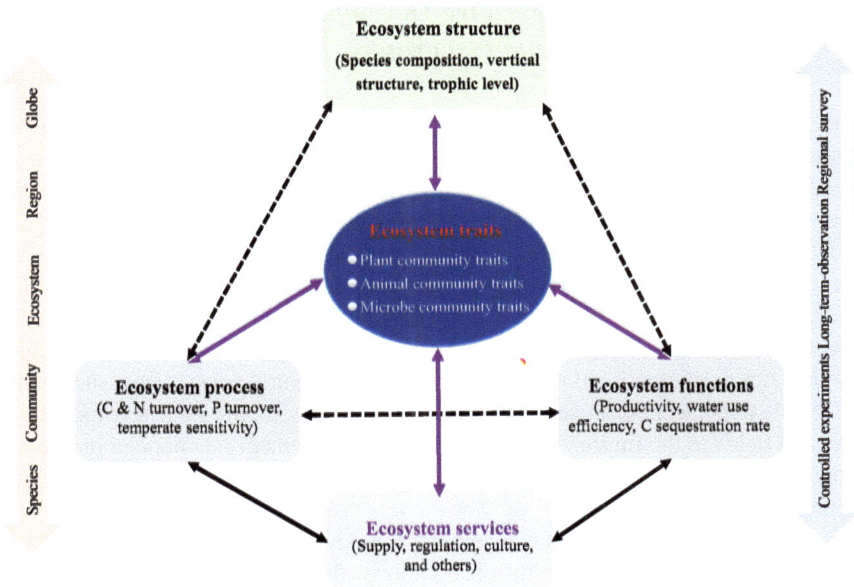

Fig. 9.9 Theoretical framework of integrated ecology with functional traits as the basis. The new framework may better synthesize ecosystem structures, processes, functions, and services at different scales based on functional traits using different observation technologies

ground-measured functional traits. With the support of new concepts and scaling-up methods, an increasing number of ecological parameters measured on the ground, particularly for plant functional traits, can be converted into standardized ecosystem traits per unit land area, which helps in verifying the data observed by macroscale observation technologies, improving their observation or prediction accuracy, and effectively solving various eco-environmental problems at regional or global scales (Fig. 9.8).

Currently, the ecological effects of global change are mostly evaluated at the ecosystem and regional levels. With the development of ecosystem traits, there is a need to establish a theoretical framework for integrated ecology with functional traits as the basis (Fig. 9.9) to better synthesize research on ecosystem structure, processes, functions, and services at different scales considering functional or ecosystem traits. Using N:P ratio to explore the ecological effects of nutrient limitation or asymmetry in atmospheric nitrogen and phosphorus deposition should be accurate for community-level N:P data rather than at the species level. This is because the former provides a comprehensive expression of long-term adaptation and trade-offs among different species at the community level (Reichstein et al. 2014; He et al. 2023). Additionally, more accurate functional traits at the ecosystem level can significantly improve the simulation accuracy of ecological models (Connolly et al. 2013; Lu et al. 2017). Considering the C:N ratio, most ecological models use individual species data or simple arithmetic means of a few species because it is difficult to obtain community-level data, affecting the fitting accuracy of the models to some extent. Future studies should obtain accurate community-level C:N data, reduce the uncertainty of key input parameters, and effectively improve the prediction accuracy of ecological models (Fig. 9.9).

9.5 Major Challenges in the Development of Ecosystem Traits

9.5.1 Standardization of the Terms of Ecosystem Traits Across Multiple Disciplines

Ecosystem ecology considers the ecosystem as its object in a specific spatial range and has various disciplines, such as botany, zoology, animal, microbiology, ecology, meteorology, and pedology. Owing to the differences in the origins and objectives of different disciplines, scientists have developed various terms to describe similar characteristics in an ecosystem (Fig. 9.10). For example, traits originate from genetics and physiology but are primarily used for plant leaves, branches, stems, and roots. However, terms in other disciplines, such as characteristics, properties,

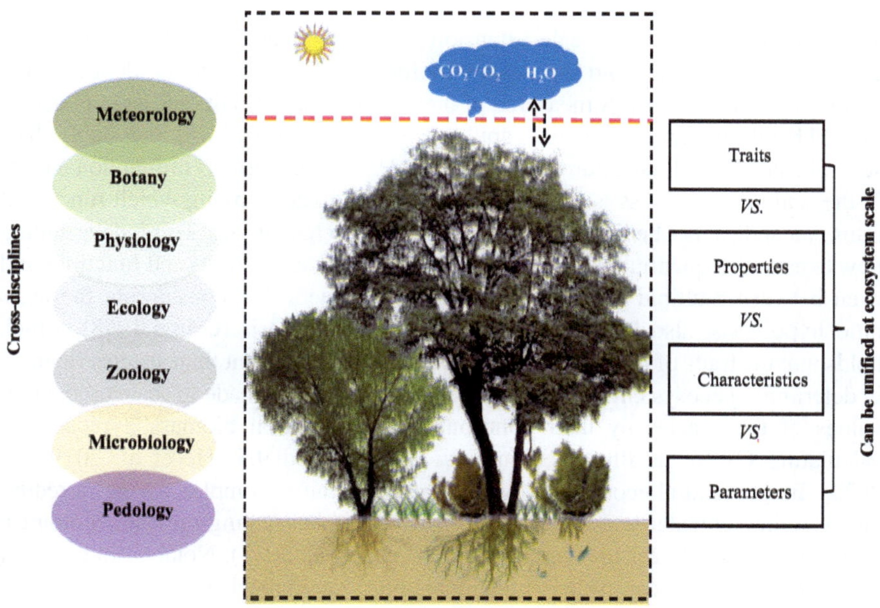

Fig. 9.10 Ecosystem ecology with apparent interdisciplinary features makes it difficult to unify key concepts widely used in different disciplines. As recommended, biological parameters should be uniformly used as functional traits, including plant (leaves, branches, stems, roots, flowers, and fruits), animal, and microbial functional traits. Contrastingly, the abiotic factors in an ecosystem should be described using their original definitions

and parameters, are widely used to describe the different components of an ecosystem. Although these widely used terms are reasonable and traceable in their respective disciplines, the lack of a unified term at the ecosystem scale hinders the development of ecosystem traits. Here, we emphasized the standardization of ecosystem research as a trait and did not negate the value of other terms (characteristics, properties, and parameters) in these relevant disciplines. Therefore, using unified terms in ecosystem ecology is feasible, necessary, and urgently required to promote macroecology development. As the ecosystem is adaptable, resistant, and regenerable, individual-level biological parameters should be uniformly used as functional traits, such as plant (leaves, branches, stems, roots, flowers, fruits, etc.), animal, and microbial functional traits. Contrastingly, abiotic factors in an ecosystem, such as soil, environment, and climate, should be described by their unique definitions, such as characteristics, properties, and parameters. Community traits and their two-dimensional characteristics (efficiency and quantitative traits) should be unified at the community scale, such as plant, animal, and microbial community traits, which combine constitute ecosystem traits in any ecosystem.

9.5.2 Resolving the Nonlinear Effect During Scaling-Up Processes

The nonlinear effect in the scale extension process is a significant challenge for all disciplines. Similarly, from the basic definition of ecosystem traits, deducing the traditional functional traits measured at the organ level per unit land area is necessary. Therefore, developing and applying ecosystem traits will inevitably face several challenges. First, addressing the possible nonlinear effects of functional traits in the scaling-up process is challenging. Secondly, understanding which functional traits can and cannot be linked to ecosystem traits is challenging. Third, determining how to establish quantitative correlations between ecosystem traits and functions and even between multiple functions is a challenge. Practically, considering the mass ratio hypothesis, also known as the dominance hypothesis (Grime 1998), which holds that the traits of the dominant species are more important than species richness in determining ecosystem processes, predictions should be made to scale up the trait values of the species by their contribution to vegetation biomass, typically by calculating CWM, as illustrated in Eqs. 9.3, 9.4, and 9.5 (He et al. 2019a, b, 2023). In any natural ecosystem, every functional trait is complex owing to redundancy, complementarity, or competition, making its scaling-up and ecological effects appear nonlinear to a certain extent (Croft et al. 2017). Notably, in complex natural ecosystems, adaptive and functional optimization mechanisms are completed by coordinating multiple functional traits; therefore, overemphasizing specific functional traits may result in high uncertainty. For example, recent studies have demonstrated that the chlorophyll content of plant communities can only explain 30% of their GPP or NPP (Li et al. 2018a, b), and stomatal morphological traits such as conductance can explain 44% of water utilization efficiency (Liu et al. 2018a, b). Scientists can adopt multiple functional traits to explain variations in ecosystem productivity at different spatial and temporal scales (Li et al. 2021a, b, c; Zhang et al. 2021b; Yan et al. 2022 2023a).

9.5.3 *Rapid Expansion of the Field-Measured Data That Can Be Used to Derive Ecosystem Traits*

As mentioned above, ecosystem traits, particularly plant and animal community traits, require systematic measurement of data in the field. Therefore, the rapid expansion of systematic and matching functional trait data that can be used to derive ecosystem traits is essential. Currently, it is difficult to scale up functional traits to communities on a large scale, even with the TRY database, because of the lack of community structure data and the limited number of species at each site. Therefore, a correlation between plant community traits and species numbers should be established in future research (He et al. 2018b). In some field investigations, scientists have recorded data on systematic community structure and functional traits at specific sites or multiple sites; however, such systematic functional traits or community data are rarely published and successfully gathered owing to limited definitions of traditional functional traits at the species level. Establishing the concept of ecosystem traits will help to preserve such systematic and matching field-investigated data and form an important data source for future studies (He et al. 2019a, b, 2020a, b, c). Furthermore, these long-term experimental plots at ecological stations worldwide and biodiversity plots have all the necessary conditions for conducting "systematic and matching" field investigations of functional traits and corresponding community structures. Therefore, it is possible to obtain a new functional trait database for hundreds or thousands of natural ecosystems, which will significantly promote the development of multidimensional research on functional traits, ecosystem ecology, and macroecology.

In particular, the concept of ecosystem traits provides a new approach for exploring the relationships, synergies, or divergences among plants, animals, and microbes at the community level. However, owing to restrictions in measurement technologies, databases, and analysis methods, most research is limited to plant community functional traits. To date, it is rare to conduct similar research on community traits such as animals and soil microbes. Therefore, extending the concept, method, and idea of plant community traits to animals and microbes is a huge challenge and presents many opportunities in the future.

9.6 Summary

Theoretical frameworks for ecosystem traits have been proposed and developed in several previous studies. Ecosystem traits are defined as any measurable functional trait that can be standardized by unit land area at the community scale, typically in the form of intensity or density, which can partly reflect the adaptation, reproduction, and productivity optimization of organisms (plants, animals, microbes, and others) to the environment. Practically, plant, animal, and microbial community traits are the core of ecosystem traits, and they interact and correlate to complete multiple ecosystem functions. Although ecosystem traits present us with the possibility of multidisciplinary integration and development of functional and ecosystem ecology and macroecological research, there is a need for improvement in methodology and matching field-measured data sources. In particular, the suitability of community traits used for developing plant functional traits for animals and microbes is still in the initial investigation stage. The expansion of systematic and matched functional trait databases in the future will enable researchers to conduct comprehensive studies to address the following questions: (1) What are the spatiotemporal variations, evolutionary patterns, and influencing factors of key community traits (plants, animals, and microorganisms)? (2) How can multiple remote-sensing techniques be incorporated into the study of key community traits? (3) What are the underlying mechanisms that govern ecosystem productivity, material cycling, and energy flow in relation to community traits? (4) How can high-precision spatial data of key community traits be produced and applied to regional ecological problems? In summary, the concept of ecosystem traits has the potential to promote the development of a new framework in ecosystem ecology based on functional traits. It can also promote the interdisciplinary integration of ecosystem ecology, geoscience, and remote sensing science through functional traits with the help of multiple approaches, such as controlled experiments, long-term observation, regional investigation, remote sensing observation, modeling, and data integration.

References

Asner GP, Scurlock JMO, Hicke JA (2003) Global synthesis of leaf area index observations: implications for ecological and remote sensing studies. Glob Ecol Biogeogr 12:191–205

Borgy B, Violle C, Choler P et al (2017) Sensitivity of community-level trait-environment relationships to data representativeness: a test for functional biogeography. Glob Ecol Biogeogr 26: 729–739

Chapin FSIII, Matson PA (2011) Principles of terrestrial ecosystem ecology, 2rd edn. Springer, New York

Connolly J, Bell T, Bolger T et al (2013) An improved model to predict the effects of changing biodiversity levels on ecosystem function. J Ecol 101:344–355

Cornelissen JHC, Lavorel S, Garnier E et al (2003) A handbook of protocols for standardised and easy measurement of plant functional traits worldwide. Aust J Bot 51:335–380

Croft H, Chen JM, Luo XZ et al (2017) Leaf chlorophyll content as a proxy for leaf photosynthetic capacity. Glob Chang Biol 23:3513–3524

Farquhar GD, von Caemmerer S, Berry JA (1980) A biochemical model of photosynthetic CO_2 assimilation in leaves of C_3 species. Planta 149:78–90

Gallagher RV, Falster DS, Maitner BS et al (2020) Open Science principles for accelerating trait-based science across the tree of life. Nat Ecol Evol 4:294–303

Gitelson AA, Vina A, Arkebauer TJ et al (2003) Remote estimation of leaf area index and green leaf biomass in maize canopies. Geophys Res Lett 30:1248

Gravel D, Bell T, Barbera C et al (2011) Experimental niche evolution alters the strength of the diversity-productivity relationship. Nature 469:89–92

Grime JP (1998) Benefits of plant diversity to ecosystems: immediate, filter and founder effects. J Ecol 86:902–910

He NP, Liu CC, Tian M et al (2018a) Variation in leaf anatomical traits from tropical to cold-temperate forests and linkage to ecosystem functions. Funct Ecol 32:10–19

He NP, Liu CC, Zhang JH et al (2018b) Opportunities and challenges for plant trait research: from organs to communities. Acta Ecologic Sinica 38:6787–6796

He NP, Liu CC, Piao SL et al (2019a) Ecosystem traits linking functional traits to macroecology. Trends Ecol Evol 34:200–210

He L, Chen JM, Liu J et al (2019b) Diverse photosynthetic capacity of global ecosystems mapped by satellite chlorophyll fluorescence measurements. Remote Sens Environ 232:111344

He NP, Li Y, Liu CC et al (2020a) Plant trait networks: improved resolution of the dimensionality of adaptation. Trends Ecol Evol 35:908–918

He NP, Liu CC, Xu L et al (2020b) Ecosystem traits open new insights into macroecology: opportunity and challenge. Acta Ecol Sin 40(8):2507–2522

He P, Gleason SM, Wright IJ et al (2020c) Growing-season temperature and precipitation are independent drivers of global variation in xylem hydraulic conductivity. Glob Chang Biol 26: 1833–1841

He NP, Yan P, Liu CC et al (2023) Predicting ecosystem productivity based on plant community traits. Trends Plant Sci 28:43–53

Heinzel J, Koch B (2011) Exploring full-waveform LiDAR parameters for tree species classification. Int J Appl Earth Obs Geoinf 13:152–160

Heiskanen J (2006) Estimating aboveground tree biomass and leaf area index in a mountain birch forest using ASTER satellite data. Int J Remote Sens 27:1135–1158

Hodapp D, Hillebrand H, Blasius B et al (2016) Environmental and trait variability constrain community structure and the biodiversity-productivity relationship. Ecology 97:1463–1474

Hugan JG, Henn JJ, Osterman WHA (2023) Plant traits alone are good predictors of ecosystem properties when used carefully. Nat Ecol Evol 7:332–334

Jiao CL, Zhang JH, Wang XC et al (2023) Plant magnesium on the Qinghai-Tibetan plateau: spatial patterns and influencing factors. Sci Total Environ 862:160743

Kattge J, Diaz S, Lavorel S et al (2011) TRY – a global database of plant traits. Glob Chang Biol 17: 2905–2935

Kunstler G, Falster DS, Coomes DA et al (2016) Plant functional traits have globally consistent effects on competition. Nature 529:204–207

Li NN, He NP, Yu GR et al (2016) Leaf non-structural carbohydrates regulated by plant functional groups and climate: evidences from a tropical to cold-temperate forest transect. Ecol Indic 62: 22–31

Li Y, He NP, Hou JH et al (2018a) Factors influencing leaf chlorophyll content in natural forests at the biome scale. Front Ecol Evol 6:64

Li Y, Liu CC, Zhang JH et al (2018b) Variation in leaf chlorophyll concentration from tropical to cold-temperate forests: association with gross primary productivity. Ecol Indic 85:383–389

Li X, He N, Xu L et al (2021a) Spatial variation in leaf potassium concentrations and its role in plant adaptation strategies. Ecol Indic 130:108063

Li Y, Li Q, Xu L et al (2021b) Plant community traits can explain variation in productivity of selective logging forests after different restoration times. Ecol Indic 131:108181
Li Y, Liu CC, Xu L et al (2021c) Leaf trait networks based on global data: representing variation and adaptation in plants. Front Plant Sci 12:710530
Li X, Li MX, Xu L et al (2022a) Allometry and distribution of nitrogen in natural plant communities of the Tibetan plateau. Front Plant Sci 13:845813
Li Y, Hou J, Xu L et al (2022b) Variation in functional trait diversity from tropical to cold-temperate forests and linkage to productivity. Ecol Indic 138:108864
Li Y, Liu CC, Sack L et al (2022c) Leaf trait network architecture shifts with species-richness and climate across forests at continental scale. Ecol Lett 25:1442–1457
Li J, Chen X, Niklas KJ et al (2022d) A whole-plant economics spectrum including bark functional traits for 59 subtropical woody plant species. J Ecol 110:248–261
Lindeman R (1942) The trophic-dynamic aspect of ecology. Ecology 23:399–417
Liu CC, He NP, Zhang JH et al (2018a) Variation of stomatal traits from cold-temperate to tropical forests and association with water use efficiency. Funct Ecol 32:20–28
Liu J, Milne RI, Cadotte MW et al (2018b) Protect third pole's fragile ecosystem. Science 362: 1368–1368
Liu CC, Li Y, Yan P et al (2021a) How to improve the predictions of plant functional traits on ecosystem functioning? Front Plant Sci 12:622260
Liu XR, Chen HX, Sun TY et al (2021b) Variation in woody leaf anatomical traits along the altitudinal gradient in Taibai Mountain, China. Glob Ecol Conserv 26:e01523
Liu C, Li Y, Xu L et al (2021c) Stomatal arrangement pattern: a new direction to explore plant adaptation and evolution. Front Plant Sci 12:655255
Liu R, Yang X, Gao R et al (2021d) Allometry rather than abiotic drivers explains biomass allocation among leaves, stems and roots of Artemisia across a large environmental gradient in China. J Ecol 109:1026–1040
Lu XJ, Wang YP, Wright IJ et al (2017) Incorporation of plant traits in a land surface model helps explain the global biogeographical distribution of major forest functional types. Glob Ecol Biogeogr 26:304–317
Ma ZQ, Guo DL, Xu XL et al (2018) Evolutionary history resolves global organization of root functional traits. Nature 555:94–97
Moretti M, Dias ATC, de Bello F et al (2017) Handbook of protocols for standardized measurement of terrestrial invertebrate functional traits. Funct Ecol 31:558–567
O'Connor MI, Gonzalez A, Byrnes JEK et al (2017) A general biodiversity-function relationship is mediated by trophic level. Oikos 126:18–31
Odum HT (1957) Trophic structure and productivity of Silver Springs, Florida. Ecol Monogr 27: 55–112
Reichstein M, Bahn M, Mahecha MD et al (2014) Linking plant and ecosystem functional biogeography. Proc Natl Acad Sci USA 111:13697–13702
Schleuning M, Frund J, Garcia D (2015) Predicting ecosystem functions from biodiversity and mutualistic networks: an extension of trait-based concepts to plant-animal interactions. Ecography 38:380–392
Snelgrove PVR, Thrush SF, Wall DH et al (2014) Real world biodiversity-ecosystem functioning: a seafloor perspective. Trends Ecol Evol 29:398–405
Tansley AG (1935) The use and abuse of vegetation concepts and terms. Ecology 16:284–307
van der Plas F, Schröder-Georgi T, Weigelt A et al (2020) Plant traits alone are poor predictors of ecosystem properties and long-term ecosystem functioning. Nat Ecol Evol 4:1602–1611
Violle C, Navas ML, Vile D et al (2007) Let the concept of trait be functional! Oikos 116:882–892
Wang RL, Yu GR, He NP et al (2015) Latitudinal variation of leaf stomatal traits from species to community level in forests: linkage with ecosystem productivity. Sci Rep 5:1–11
Wang RL, Yu GR, He NP et al (2016) Latitudinal variation of leaf morphological traits from species to communities along a forest transect in eastern China. J Geogr Sci 26:15–26

Watson DJ (1947) Comparative physiological studies on the growth of field crops: I. Variation in net assimilation rate and leaf area between species and varieties, and within and between years. Ann Bot 11:41–76

Wieczynski DJ, Boyle B, Buzzard V et al (2019) Climate shapes and shifts functional biodiversity in forests worldwide. Proc Natl Acad Sci 116:587–592

Wright IJ, Reich PB, Westoby M et al (2004a) The worldwide leaf economics spectrum. Nature 428:821–827

Wright IJ, Groom PK, Lamont BB et al (2004b) Leaf trait relationships in Australian plant species. Funct Plant Biol 31:551–558

Wright SJ, Kitajima K, Kraft NJ et al (2010) Functional traits and the growth-mortality trade-off in tropical trees. Ecology 91:3664–3674

Wright IJ, Dong N, Maire V et al (2017) Global climatic drivers of leaf size. Science 357:917–921

Xu L, He NP, Yu GR (2020) Nitrogen storage in China's terrestrial ecosystems. Sci Total Environ 709:136201

Yan P, Li MX, Yu GR et al (2022) Plant community traits associated with nitrogen can predict spatial variability in productivity. Ecol Indic 140:109001

Yan P, Fernández-Martínez M, Van Meerbeek K et al (2023a) The essential role of biodiversity in the key axes of ecosystem function. Glob Chang Biol 29:4569–4585

Yan P, Zhang JH, He NP et al (2023b) Functional diversity and soil nutrients regulate the interannual variability in gross primary productivity. J Ecol 111:1094–1106

Yu C, Xu L, Li MX, He NP (2022) Phosphorus storage and allocation in vegetation on the Tibetan plateau. Ecol Indic 145:109636

Zhang JH, Zhao N, Liu CC et al (2018a) C:N:P stoichiometry in China's forests: From organs to ecosystems. Funct Ecol 32:50–60

Zhang JH, He NP, Liu CC et al (2018b) Allocation strategies for nitrogen and phosphorus in forest plants. Oikos 127:1506–1514

Zhang JH, Li MX, Xu L et al (2021a) C:N:P stoichiometry in terrestrial ecosystems in China. Sci Total Environ 795:148849

Zhang JH, Ren TT, Yang JJ et al (2021b) Leaf multi-element network reveals the change of species dominance under nitrogen deposition. Front Plant Sci 12:580340

Zhang Y, He NP, Li MX et al (2021c) Community chlorophyll quantity determines the spatial variation of grassland productivity. Sci Total Environ 801:149567

Zhang ZH, Hou JH, He NP (2021d) Predictability of functional diversity depends on the number of traits. J Resour Ecol 12:332–345

Zhang Y, Migliavacca M, Penuelas J et al (2021e) Advances in hyperspectral remote sensing of vegetation traits and functions. Remote Sens Environ 252:112121

Zhang JH, Hedin LO, Li MX et al (2022a) Leaf N:P ratio does not predict productivity trends across natural terrestrial ecosystems. Ecology 103:e3789

Zhang M, Gao HR, Mo WY et al (2022b) Variation in leaf vein traits along environmental gradient in loess plateau grasslands. Acta Ecol Sin 42(19):8082–8093

Zhang JH, Wang XM, Hou JH et al (2024) High-resolution community-level sodium variation on the Tibetan plateau: content, density, and storage. Sci Total Environ 944:173766

Zhao WZ, Xiao CW, Li MX et al (2022a) Variation and adaptation in leaf sulfur content across China. J Plant Ecol 15:743–755

Zhao WZ, Xiao CW, Li MX et al (2022b) Spatial variation and allocation of sulfur among major plant organs in China. Sci Total Environ 844:157155

Zhao WZ, Xiao CW, Li MX et al (2023) Biogeographic patterns of sulfur in the vegetation of the Tibetan plateau. Journal of geophysical research. Biogeosciences 128:e2022JG007051

Zhu XY, Hou JH, Li MX et al (2023) High-resolution spatial distribution of vegetation biomass and its environmental response on Qinghai-Tibet plateau: intensive grid-field survey. Ecol Indic 149:110167

MIX
Papier aus verantwortungsvollen Quellen
Paper from responsible sources
FSC® C105338

If you have any concerns about our products,
you can contact us on
ProductSafety@springernature.com

In case Publisher is established outside the EU,
the EU authorized representative is:
Springer Nature Customer Service Center GmbH
Europaplatz 3, 69115 Heidelberg, Germany

Printed by Libri Plureos GmbH
in Hamburg, Germany